METHODS IN MOLECULAR BIOLOGY™

Series Editor
John M. Walker
School of Life Sciences
University of Hertfordshire
Hatfield, Hertfordshire, AL10 9AB, UK

For other titles published in this series, go to
www.springer.com/series/7651

Cell-Free Protein Production

Methods and Protocols

Edited by

Yaeta Endo

Cell-Free Science and Technology Research Center,
Ehime University, Ehime, Japan

Kazuyuki Takai

Cell-Free Science and Technology Research Center,
Ehime University, Ehime, Japan

and

Takuya Ueda

The Department of Medical Genome Sciences, Graduate School of Frontier Sciences,
The University of Tokyo, Kashiwa, Chiba, Japan

Editors
Yaeta Endo
Cell-Free Science and Technology Research Center
Ehime University
Ehime
Japan
yendo@eng.ehime-u.ac.jp

Kazuyuki Takai
Cell-Free Science and Technology Research Center
Ehime University
Ehime
Japan
takai@eng.ehime-u.ac.jp

Takuya Ueda
The Department of Medical Genome Sciences
Graduate School of Frontier Sciences
The University of Tokyo
Kashiwa, Chiba
Japan
ueda@k.u-tokyo.ac.jp

ISSN 1064-3745 e-ISSN 1940-6029
ISBN 978-1-60327-330-5 e-ISBN 978-1-60327-331-2
DOI 10.1007/978-1-60327-331-2
Springer New York Dordrecht Heidelberg London

Library of Congress Control Number: 2009933999

Printed on acid-free paper

Humana Press is part of Springer Science+Business Media (www.springer.com)

Preface

Understanding how living organisms handle and transfer their genetic information at the molecular level is the major purpose of molecular biology. This, of course, includes the understanding of proteins that constitute the major part of the activity of living cells. However, the analysis of protein molecules has been running behind that of nucleic acids. This is clear because proteins are much more difficult to handle than nucleic acids. We have experienced an unprecedented decade during which the data on gene sequences and expression patterns has accumulated rapidly. This, on the other hand, clearly revealed the dazzling complexity of the living cells created by the various functions of proteins. It is obvious that the analyses of proteins are essential for a "molecular" understanding of life.

Test-tube reactions in the absence of living cells have played very important roles in the history of biochemistry and molecular biology. The discovery of cell-free fermentation by E. Buchner in 1897 is one of the most important historical moments in the early period of biochemistry. P. C. Zamecnik and other researchers have developed a method for measuring cell-free incorporation of radiolabelled amino acids into protein fraction by the year of 1954. This catalyzed the deciphering of the genetic code in the 1960s. The invention in the middle of the 1980s of polymerase chain reaction, which is another cell-free reaction, dramatically improved methods for the preparation of DNA molecules and accelerated the analyses and engineering of DNA sequences and genomes in the 1990s.

The cell-free protein synthesis technology experienced a revolution in the 1990s and has been further improved significantly. It has become a powerful tool ("RIBOENGINE", driven by the ribosomes) for the production of proteins for functional and structural analyses of proteins. We are sure that the technology will contribute much to the most basic research in biology: the compilation of the "dictionaries" for understanding the meanings of the "words" of gene sequences in the "books" of the genomes. These dictionaries would be useful for understanding the stories of the lives of cells and individuals in detail and as a whole, and would be helpful for reading the scenario of the history of evolution of life on Earth. Of course, the technology is providing a flexible platform for many other aspects of gene expression, some of which may lead to new technologies in future.

This volume has been edited for readers who wish to use cell-free protein expression systems for their purposes in biochemistry, molecular biology, and biotechnology. The first several chapters are devoted to basic methods for different systems. The following chapters describe examples of applications of basic methods to the functional and structural analyses of proteins. Then, some challenges involving the preparation of proteins that are generally difficult to prepare in their functional forms are described. Specific applications of cell-free technologies to protein engineering and some methods that are expected to constitute a part of future technologies appear in the last part of the volume.

We will see in this volume, that these cell-free techniques are beginning to be used for the analyses of genome products, development of new vaccines, development of new functional proteins, reconstitution of cellular molecular processes, and synthetic biology. We believe that these research fields will be accelerated by cell-free technologies and will expand dramatically during the first half of this century. We refer to these fields as the "cell-free science and technology" fields.

In particular, reconstitution of cellular molecular systems and synthetic biology approaches, through which we could benefit from the accumulating genome and cDNA sequence information, will be quite important to our understanding of life. These constitutive approaches are required now because various platforms for analytical approaches are now available as a result of the expanding post-genomic research around the world, and because the analytical and constitutive approaches are the "two wheels of a cart" to the understanding of life in terms of chemistry and physics. For example, membrane proteins may be produced in the cell-free extract on some lipid bilayer, and the analyses of the products could promote the understanding of the functions and mechanisms related to the proteins. The knowledge obtained from the analytical experiments might be useful for partial reconstitution of translocation of the newly synthesized polypeptides, which in turn might contribute to production of other membrane proteins in their active forms. The analyses of the produced proteins could then be useful for a more complete reconstitution of biological membranes. Then, it might facilitate production of proteins with controlled post-translational modifications. The methods for cell-free production of many different proteins in parallel and totally reconstituted protein synthesis systems, such as those presented in this volume, will be essential for this kind of step-by-step reconstitution work in the future. We hope many of the readers will join us in enjoying cell-free sciences and technologies.

Matsuyama, Ehime, Japan — *Yaeta Endo*
Matsuyama, Ehime, Japan — *Kazuyuki Takai*
Kashiwa, Chiba, Japan — *Takuya Ueda*

Contents

Contributors

EIJI ANDO • *Clinical & Biotechnology Business Unit, Analytical & Measuring Instruments Division, Life Science Business Department, Shimadzu Corporation, Kyoto, Japan*

FRANK BERNHARD • *Centre for Biomolecular Magnetic Resonance, Institute for Biophysical Chemistry, University of Frankfurt/Main, Frankfurt/Main, Germany*

VOLKER DÖTSCH • *Centre for Biomolecular Magnetic Resonance, Institute for Biophysical Chemistry, University of Frankfurt/Main, Frankfurt/Main, Germany*

FLORIAN DURST • *Centre for Biomolecular Magnetic Resonance, Institute for Biophysical Chemistry, University of Frankfurt/Main, Frankfurt/Main, Germany*

YAETA ENDO • *Cell-Free Science and Technology Research Center and Venture Business Laboratory, Ehime University, Ehime, Japan*
RIKEN Genomic Sciences Center, Yokohama, Japan

TORU EZURE • *Clinical & Biotechnology Business Unit, Analytical & Measuring Instruments Division, Life Science Business Department, Shimadzu Corporation, Kyoto, Japan*

BRIAN G. FOX • *Department of Biochemistry, Center for Eukaryotic Structural Genomics, College of Agricultural and Life Sciences, University of Wisconsin, Madison, WI, USA*

MICHAEL A. GOREN • *Department of Biochemistry, Center for Eukaryotic Structural Genomics, College of Agricultural and Life Sciences, University of Wisconsin, Madison, WI, USA*

HIROYUKI HORI • *Department of Materials Science and Biotechnology, Graduate School of Sciences and Engineering, Ehime University, Ehime, Japan*

HIROAKI IMATAKA • *Department of Materials Science and Chemistry, Graduate School of Engineering, University of Hyogo, Hyogo, Japan*
RIKEN Systems and Structural Biology Center, Yokohama, Japan

MASAAKI ITO • *Clinical & Biotechnology Business Unit, Analytical & Measuring Instruments Division, Life Science Business Department, Shimadzu Corporation, Kyoto, Japan*

TAKASHI KANAMORI • *The Department of Medical Genome Sciences, Graduate School of Frontier Sciences, The University of Tokyo, Kashiwa, Chiba, Japan*

TAKUYA KANNO • *Cell-Free Science and Technology Research Center and Venture Business Laboratory, Ehime University, Ehime, Japan*

TAKANORI KIGAWA • *RIKEN Systems and Structural Biology Center, Yokohama, Japan*

ICHIZO KOBAYASHI • *Department of Medical Genome Sciences, Graduate School of Frontier Science, Graduate Program in Biophysics and Biochemistry, Graduate School of Science, and Institute of Medical Science, University of Tokyo, Tokyo, Japan*

TOMINARI KOBAYASHI • *RIKEN Systems and Structural Biology Center, Yokohama, Japan*

TOSHIYUKI KOHNO • *Mitsubishi Kagaku Institute of Life Sciences, Machida, Tokyo, Japan*
YUTETSU KURUMA • *The Department of Medical Genome Sciences, Graduate School of Frontier Sciences, The University of Tokyo, Kashiwa, Chiba, Japan*
FRANK LÖHR • *Centre for Biomolecular Magnetic Resonance, Institute for Biophysical Chemistry, University of Frankfurt/Main, Frankfurt/Main, Germany*
SHIN-ICHI MAKINO • *Department of Biochemistry, Center for Eukaryotic Structural Genomics, College of Agricultural and Life Sciences, University of Wisconsin, Madison, WI, USA*
JOHN L. MARKLEY • *Department of Biochemistry, Center for Eukaryotic Structural Genomics, College of Agricultural and Life Sciences, University of Wisconsin, Madison, WI, USA*
TOMOAKI MATSUURA • *Department of Bioinformatic Engineering, Graduate School of Information Science and Technology, Osaka University, Osaka, Japan*
SATOSHI MIKAMI • *RIKEN Systems and Structural Biology Center, Yokohama, Japan*
KEN-ICHI MIYAZONO • *Department of Applied Biological Chemistry, Graduate School of Agricultural and Life Sciences, University of Tokyo, Tokyo, Japan*
HIDEAKI NANAMIYA • *Cell-Free Science and Technology Research Center, Ehime University, Ehime, Japan*
KAZUYA NISHIKAWA • *Department of Biomolecular Science, Faculty of Engineering, Gifu University, Gifu, Japan*
AKIRA NOZAWA • *Cell-Free Science and Technology Research Center, Ehime University, Ehime, Japan*
HIROYUKI OHASHI • *The Department of Medical Genome Sciences, Graduate School of Frontier Sciences, The University of Tokyo, Kashiwa, Chiba, Japan*
SATOSHI OHNO • *Department of Biomolecular Science, Faculty of Engineering, Gifu University, Gifu, Japan*
SINA RECKEL • *Centre for Biomolecular Magnetic Resonance, Institute for Biophysical Chemistry, University of Frankfurt/Main, Frankfurt/Main, Germany*
TATSUYA SAWASAKI • *Cell-Free Science and Technology Research Center and Venture Business Laboratory, Ehime University, Ehime, Japan*
RIKEN Genomic Sciences Center, Yokohama, Japan
MASAMITSU SHIKATA • *Clinical & Biotechnology Business Unit, Analytical & Measuring Instruments Division, Life Science Business Department, Shimadzu Corporation, Kyoto, Japan*
YOSHIHIRO SHIMIZU • *The Department of Medical Genome Sciences, Graduate School of Frontier Sciences, The University of Tokyo, Kashiwa, Chiba, Japan*
VLADIMIR A. SHIROKOV • *Institute of Protein Research, Russian Academy of Sciences, Pushchino, Moscow Region, Russia*
SOLMAZ SOBHANIFAR • *Centre for Biomolecular Magnetic Resonance, Institute for Biophysical Chemistry, University of Frankfurt/Main, Frankfurt/Main, Germany*
TAKESHI SUNAMI • *Department of Bioinformatic Engineering, Graduate School of Information Science and Technology, Osaka University, Osaka, Japan*
HIROAKI SUZUKI • *Department of Bioinformatic Engineering, Graduate School of Information Science and Technology, Osaka University, Osaka, Japan*

TAKASHI SUZUKI • *Clinical & Biotechnology Business Unit, Analytical & Measuring Instruments Division, Life Science Business Department, Shimadzu Corporation, Kyoto, Japan*

TOSHIHARU SUZUKI • *ATP Synthesis Regulation Project, International Cooperative Research Project (ICORP), Japan Science and Technology Corporation (JST), Tokyo, Japan*

KAZUYUKI TAKAI • *Cell-Free Science and Technology Research Center and Venture Business Laboratory, Ehime University, Ehime, Japan*

SATORU TAKEO • *Cell-Free Science and Technology Research Center, Ehime University, Ehime, Japan*

MASARU TANOKURA • *Department of Applied Biological Chemistry, Graduate School of Agricultural and Life Sciences, University of Tokyo, Tokyo, Japan*

TAKAFUMI TSUBOI • *Cell-Free Science and Technology Research Center and Venture Business Laboratory, Ehime University, Ehime, Japan*

MOTOMI TORII • *Department of Molecular Parasitology, Ehime University Graduate School of Medicine, Ehime, Japan*

YUZURU TOZAWA • *Cell-Free Science and Technology Research Center and Venture Business Laboratory, Ehime University, Ehime, Japan*

TAKUYA UEDA • *The Department of Medical Genome Sciences, Graduate School of Frontier Sciences, The University of Tokyo, Kashiwa, Chiba, Japan*

MIKI WATANABE • *Department of Medical Genome Sciences, Graduate School of Frontier Science, University of Tokyo, Tokyo, Japan*

TAKASHI YOKOGAWA • *Department of Biomolecular Science, Faculty of Engineering, Gifu University, Gifu, Japan*

TETSUYA YOMO • *Department of Bioinformatic Engineering, Graduate School of Information Science and Technology, Osaka University, Osaka, Japan*

Chapter 1

Cell-Free Protein Preparation Through Prokaryotic Transcription–Translation Methods

Takanori Kigawa

Abstract

We have been developing and using an *Escherichia coli* cell extract-based coupled transcription–translation cell-free system. The development includes many different issues such as cell extract preparation, template construction, reaction condition, reaction format, and automation. These developments improved the efficiency, productivity, and throughput of our prokaryotic cell-free system, enabling us to use the system as one of the standard expression methods in our group. Our system certainly has the largest successful applications especially to the protein production for the structure determination, among the existing cell-free protein synthesis systems.

Key words: Prokaryotic coupled transcription–translation, *Escherichia coli* cell extract

1. Introduction

Cell-free protein synthesis has become widely considered as one of the useful protein expression methods. Its low productivity problem has been solved by the modification of reaction conditions (1–5) and the development of the continuous methods (6–10). In addition to the conventional *E. coli*, rabbit, and wheat germ systems, new cell-free systems based on eukaryotic cell extracts have been developed (11–14).

In this chapter, we present an *E. coli* cell extract-based coupled transcription–translation system, which we have developed to be efficient and productive, and routinely used for many kinds of applications. One of the major advantages of our cell-free system is that PCR-amplified linear DNA fragment can be directly used as a template even for protein production in milligram quantity for the structure determination (15).

Y. Endo et al. (eds.), *Cell-Free Protein Production: Methods and Protocols,* Methods in Molecular Biology, vol. 607
DOI 10.1007/978-1-60327-331-2_1, © Humana Press, a part of Springer Science+Business Media, LLC 2010

2. Materials

2.1. Cell Preparation

1. *E. coli* BL21 codon-plus RIL strain (Stratagene).
2. 2× YT medium: 1.6 % (w/v) BactoTryptone, 1.0 % (w/v) Yeast extract, 0.5 % (w/v) NaCl. Autoclave at 121°C for 20 min before use.
3. S30 buffer (A): 10 mM Tris–acetate buffer (pH 8.2) containing 14 mM $Mg(OAc)_2$, 60 mM KOAc, 1 mM DTT, and 0.5 mL/L 2-mercaptoethanol. Prepare just before use.
4. Chloramphenicol.
5. Absolve (DuPont/NEN).
6. 50 mL NALGENE centrifuge tube.
7. 500 mL culture flask.
8. 2 L baffled culture flask.
9. Polytron cell homogenizer (KINEMATICA.AG, PT-MR3100).

2.2. Cell Disruption

1. S30 buffer (B): This is the same as the S30 buffer (A) except for the absence of 2-mercaptoethanol. Prepare just before use.
2. Preincubation buffer: 293 mM Tris–OAc buffer (pH 8.2) containing 9.2 mM $Mg(OAc)_2$, 13.2 mM ATP (pH 7.0), 84 mM Phospho*enol*pyruvate (pH 7.0), 4.4 mM DTT, 40 μM each of 20 amino acids, and 6.7 U/mL Pyruvate Kinase (Sigma P7768). Prepare just before use.
3. 50 mL tube (IWAKI 2341-050, etc.).
4. Glass beads (B. Brawn, $\phi = 0.17$–0.18 mm).
5. Spatula.
6. Dialysis tubing (Spectra/Por 2.1, MWCO = 15,000, $\phi = 14$ mm).
7. Multi-beads shocker (YASUI KIKAI, http://www.yasuikikai.co.jp/).

2.3. Extract Validation

1. LMCPY: 155 mM HEPES-KOH buffer (pH 7.5) containing 0.45 mg/mL *E. coli* total tRNA (Roche), 10.8 % (w/v) PEG 8000, 0.48 mg/mL tyrosine, 0.54 M potassium glutamate, 4.7 mM DTT, 3.28 mM ATP (pH 7.0), 2.33 mM GTP (pH 7.0), 2.33 mM CTP (pH 7.0), 2.33 mM UTP (pH 7.0), 9.3 μg/mL Folinic acid•Ca, 1.72 mM cAMP•Na, 74 mM NH_4OAc, and 217 mM creatine phosphate. Store at −20°C.
2. AA(−Y +DTT): 20 mM each of glutamine, asparagine, arginine, tryptophan, lysine, histidine, phenylalanine, isoleucine, glutamic acid, proline, asparatic acid, glycine, valine, serine,

alanine, threonine, cysteine, methionine, leucine, and 10 mM DTT. Store at –20°C.

3. 1.6 mM Mg(OAc)2; Store at –20°C.
4. T7 RNA polymerase (*see* Note 1); Store at –20°C.
5. 3.75 mg/mL Creatine Kinase; Store at –80°C.
6. Color reagent: 0.4 mg/mL 5, 5′-Dithiobis-2-nitrobenzoic acid, 0.1 mM Acetyl coenzyme A, and 0.1 mM chloramphenicol. Prepare just before use.
7. (1×) CAT dilution buffer: 1.0 M Tris–HCl buffer containing 1% (w/v) PEG 8000, 10 % (v/v) glycerol, 0.5 mM DTT. Store at –20°C.
8. Template DNA.

3. Methods

We usually use *E. coli* BL21 codon-plus RIL strain (BL21 CP) as the source of the cell extract. Some part of this protocol is optimized to the BL21 CP strain, and it should be modified if the other strains are used. To validate the activity of the prepared S30 extract, the optimum magnesium concentration of chloramphenicol acetyl transferase (CAT) protein synthesis, using the batch mode of cell-free protein synthesis, is usually investigated.

3.1. Preparation of Cells

1. Prepare a 100 mL 2× YT medium containing 34 μg/mL chloramphenicol (Cm) in 500 mL culture flask.
2. Inoculate 200 μL of the glycerol stock of the BL21 CP into the medium and incubate at 37°C in an incubator in linear shaking mode at 130 rpm (r=3.5 cm) overnight.
3. Prepare six 500 mL 2× YT medium containing 34 μg/mL Cm in 2 L baffled flask.
4. Inoculate 5 mL of overnight culture into each of six 500 mL medium.
5. Incubate at 30°C or lower with circular shaking at 160 rpm (r=2.5 cm).
6. During incubation, prepare 4 L of S30 buffer (A) and chill it in ice-water. Dispense 20 mL of S30 buffer (A) and store at –20°C for the next day.
7. When OD_{600} reaches at 3 (usually in 3–4 h), chill immediately the culture flasks in ice-water.
8. Harvest the cells by centrifugation at 7,500×g for 10 min at 4°C.

9. Suspend the cells with S30 buffer (A) (*see* Note 2), and then harvest the cells by centrifugation at 7,500 × *g* for 10 min at 4°C.
10. Repeat this wash step twice more.
11. Discard supernatant, and measure the wet weight of the cells (*see* Note 3).
12. Freeze cells immediately and completely by immersing a tube in liquid nitrogen at least for 2 min, and then store it at –80°C for 1–3 days (*see* Note 4).
13. Clean 50 mL NALGENE centrifuge tube by immersing in 2% (v/v) Absolve solution overnight. On the day tubes are used, rinse thoroughly with the Milli-Q water to remove AbSolve completely (*see* Note 5).

3.2. Cell Disruption

1. Prepare 3 L of S30 buffer (B).
2. Thaw the frozen cells and a 50 mL aliquot of S30 buffer (A) in a water bath at room temperature. The temperature of the cells must be maintained below 4°C while thawing. Thus, the cells should be transferred to ice before they are completely thawed, in order to keep the temperature below 4°C.
3. Add approximately 20 mL of chilled S30 buffer (A) into the thawed cells, and suspend them gently with a sterilized spatula. Pour the cells into a chilled centrifuge tube when they are completely suspended.
4. Centrifuge at 16,000 × *g* for 30 min at 4°C. During this step, add DTT to the ice-cold S30 buffer (B).
5. Remove the supernatant completely with a pipette, and measure the weight of the cell pellet (*see* Note 6).
6. If necessary, adjust the weight of cells to 7 g per tube.
7. Add 8.9 mL of the ice-cold S30 buffer (B), and suspend the cells with a spatula.
8. Add 22.7 g of glass beads for each tube.
9. Disrupt the cells with a cell disrupting instrument (Multi-beads shocker) with the program: 30 s on, 30 s off, 30 s on, 30 s off, 30 s on (*see* Note 7).
10. Immediately centrifuge the disrupted cells at 30,000 × *g* for 30 min at 4°C. During the centrifugation, thaw the reagents for Preincubation buffer.
11. Transfer the supernatant to another chilled centrifuge tube (*see* Note 8).
12. Centrifuge the supernatant again at 30,000 × *g* for 30 min at 4°C. Prepare Preincubation buffer during the centrifugation (*see* Note 9).

13. Transfer the supernatant (crude extract) to a 50 mL tube (*see* Notes 8 and 10).
14. Measure the volume of the crude extract, add 0.3 volume of Preincubation buffer, and mix well by gentle pipetting to avoid foaming.
15. Incubate with gentle swirling in a water bath at 37°C for 80 min. Meanwhile, cut dialysis tubing into 15 cm lengths, and wash with the Milli-Q water.
16. Transfer the extract into the dialysis membrane, and dialyze it four times against S30 buffer (B) for 45 min each at 4°C. The volume of each dialysis buffer is 750 mL per 7 g of tightly packed cells.
17. Centrifuge the extract at 4,000 × *g* for 10 min at 4°C.
18. Transfer the supernatant (S30 extract) to a chilled tube, and mix well by gentle pipetting to avoid foaming.
19. Dispense the S30 extract into 0.2–2.0 mL aliquots, and immediately and completely freeze them by immersing them in liquid nitrogen for at least 2 min.
20. Store the S30 extract in liquid nitrogen or below –80°C (*see* Note 11).

3.3. Validating the Prepared Extract Using the Cell-Free Protein Synthesis Reaction

1. Thaw all of the reagents required for the cell-free protein synthesis, according to Table 1.1.
2. Dispense $Mg(OAc)_2$ into 0.6 mL tubes, according to Table 1.1.
3. Mix template DNA (pUC19-CAT), LMCPY, and AA(–Y +DTT) (Premix A).
4. Mix S30 extract and T7 RNA polymerase (Premix B).
5. Add Creatine kinase (CK) into the Premix A.
6. Dispense the Premix A into the solutions.
7. Dispense the Premix B into the solutions. Mix gently and thoroughly by pipetting, to avoid foaming.
8. Incubate at 37°C for 60 min. During the incubation period, prepare the color reagent. At the end of the incubation period, place the reaction tubes on ice.
9. Add 270 μL of the (1×) CAT dilution buffer into each reaction mixture.
.10. Add 10 μL of each diluted reaction mixture from the last step into 990 μL of the (1×) CAT dilution buffer (*see* Note 12).
11. Add 3 μL of the solution from the last step into 400 μL of color reagent (*see* Note 13).
12. Incubate the color reagent mixed at the last step at 37°C for 30 min.

Table 1.1
Composition of the batch mode cell-free protein synthesis reaction mixture (Premix A and B)

Final conc.	Reagent	Stock conc.	#1	#2	#3	#4	#5	#6
Final magnesium ion concentration (mM)			5.3	8.0	10.7	13.3	16.0	10.7
5.3–16.0 mM	$Mg(OAc)_2$	0.4 M	0.4	0.6	0.8	1.0	1.2	0.8
	H_2O		2.0	1.8	1.6	1.4	1.2	1.6
6.7 μg/mL	pUC19-CAT	0.2 mg/mL	0.6	0.6	0.6	0.6	0.6	0
	H_2O		3.4	3.4	3.4	3.4	3.4	4.0
	LMCPY		11.2	11.2	11.2	11.2	11.2	11.2
1.0 mM each	AA(−Y +DTT)	20 mM each	1.5	1.5	1.5	1.5	1.5	1.5
0.25 mg/mL	CK	3.75 mg/mL	2.0	2.0	2.0	2.0	2.0	2.0
0.24 vol.	S30 extract		7.2	7.2	7.2	7.2	7.2	7.2
66.7 μg/mL	T7 RNA polymerase	10 mg/mL	0.2	0.2	0.2	0.2	0.2	0.2
	Total (μL)		30.0	30.0	30.0	30.0	30.0	30.0

13. Measure A_{412} and calculate the CAT protein content in the cell-free reaction mixture, using the following formulas (16). As the S30 extract prepared from the BL21 CP strain contains endogenous CAT protein, the CAT protein productivity of the cell-free reaction at different magnesium concentrations (samples 1–5 of Table 1.1) can be obtained by subtracting the background CAT content in the reaction mixture without template DNA (sample 6 of Table 1.1). An example result is shown in Figs. 1.1 and 1.2 (*see* Note 14).

$$\text{CAT [units/mL]} = (A_{412} - A_{412}\text{blk}) \times 1000 \times 1000 / T / e_{412}$$
$$\text{CAT [}\mu\text{ g/mL (assay mixture)]} = \text{CAT [units/mL]} / \text{A}$$
$$\text{CAT [}\mu\text{ g/mL (Cell - free reaction mixture)]} = \text{CAT [}\backslash\mu\text{ g/mL (assay mixture)]} / (\text{V/S}) \times \text{D}$$

where e_{412} [/M/cm] is the molar extinction coefficient for reduced DTNB at 412 nm (13,600/M/cm), *A* [units/μg] is specific activity for CAT (150 units/mg), *T* [min] is color development time (30 min), *V* [μL] is sample volume used for assay (3 μL), *S* [μL] is assay scale (400 μL), *D* is sample dilution factor (1,000).

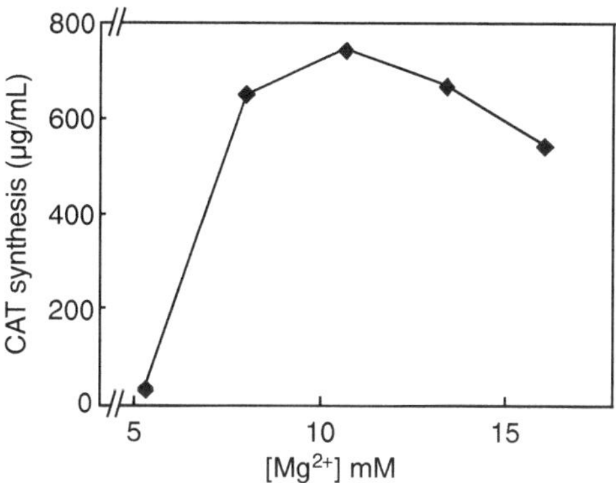

Fig. 1.1. Typical magnesium concentration dependency profile of CAT production by the batch mode of cell-free protein synthesis.

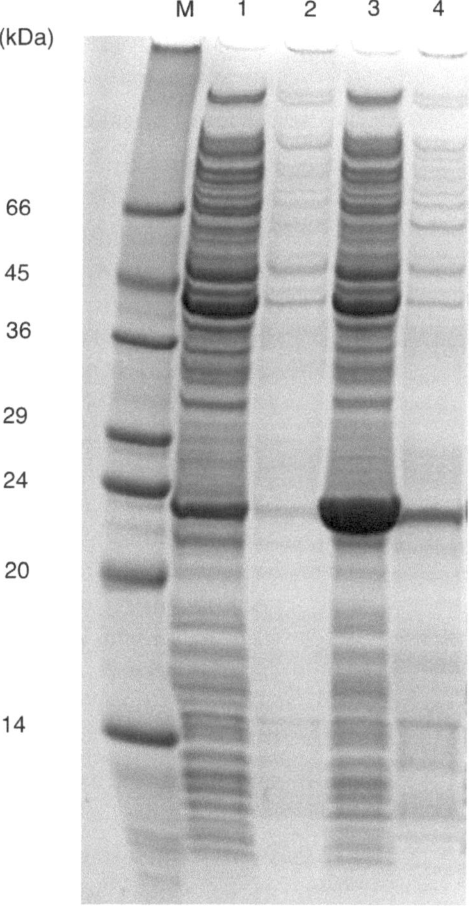

Fig. 1.2. SDS-PAGE analysis of the CAT protein expressed using the plasmid template by the batch mode (*lanes* 1 and 2) and the dialysis mode (*lanes* 3 and 4). The expression reaction was divided into supernatant (*lanes* 1 and 3) and precipitate (*lanes* 2 and 4) fractions. On each lane, the sample equivalent to 2.5 μL of the reaction mixture was loaded. The gel was stained with Coomassie Brilliant Blue.

4. Notes

1. We usually prepare the T7 RNA polymerase used for the cell-free synthesis according to reported procedures (17–19), mainly because the enzymes from commercial companies are not sufficiently concentrated. However, based on our tests, highly concentrated lots from the following companies have almost the same activity as ours for the cell-free synthesis: Ambion, Takara, Toyobo (not suitable for the linear template in our assays), and Promega.
2. We use Polytron cell homogenizer to easily homogenate cell pellets.
3. In the typical case, about 16 g of loosely packed cells are obtained from 3 L culture.
4. From our experience, longer storage more than 3 days results in lower activity of the extract.
5. It should be taken care that AbSolve is strong basic solution.
6. Approximately 7 g of tightly packed cells are usually obtained from 10 g of loosely packed cells.
7. Pratt's protocol (20), on which our protocol is based, uses a French press (Aminco) to disrupt the *E. coli* cells. However, in our experience, it is difficult to obtain reproducible cell disruption results by using a French press in a laboratory-scale preparation, mainly because it is quite difficult to disrupt *E. coli* cells uniformly. We now employ the disruption with small glass beads, which is usually used to disrupt cells with rigid cell walls, such as yeast and small algae, and thus always obtain highly active cell extracts. We had previously used the MSK cell homogenizer (B. Braun) for this purpose, and then switched to the Multi-beads shocker because of better sample cooling during disruption and higher capacity. In our tests, sonication, which is another popular cell disruption method, especially for *E. coli* cells, is not suitable for extract preparation, because of sample heating and difficulty of management.
8. Be careful not to pour any pellet material, and avoid making air bubbles.
9. Pyruvate kinase should be added just before use.
10. About 4.9 mL of crude extract are usually obtained from 7 g of tightly packed cells.
11. The S30 extract remains active at least for 1 year if stored in liquid nitrogen.
12. The final dilution of the reaction mixture is 1,000-fold.

13. The color reagent mixed with the (1×) CAT dilution buffer should be used for reference.
14. In our experience, the optimum magnesium concentration of the cell-free system that uses the S30 extract prepared according to this protocol is in the range of 10–14 mM, and the CAT productivity at the optimum magnesium concentration is in the range of 600–800 μg/mL reaction mixture. It is important to monitor the protein products by SDS-PAGE analysis (Fig. 1.2). We usually examine (a) the optimum magnesium concentration by the batch system using the plasmid template and (b) the protein productivity of the dialysis systems as well as the batch system, using both the plasmid and linear templates.

Acknowledgments

The author would like to thank Takashi Yabuki, Natsuko Matsuda, Eiko Seki, and Takayoshi Matsuda for development of our cell-free system.

References

1. Kim, D.-M., Kigawa, T., Choi, C.-Y., and Yokoyama, S. (1996) A highly efficient cell-free protein synthesis system from *Escherichia coli*. *Eur. J. Biochem.* 239, 881–886.
2. Yabuki, T., Kigawa, T., Dohmae, N., Takio, K., Terada, T., Ito, Y., Laue, E. D., Cooper, J. A., Kainosho, M., and Yokoyama, S. (1998) Dual amino acid-selective and site-directed stable-isotope labeling of the human c-Ha-Ras protein by cell-free synthesis. *J. Biomol. NMR* 11, 295–306.
3. Kim, D. M., and Swartz, J. R. (1999) Prolonging cell-free protein synthesis with a novel ATP regeneration system. *Biotechnol. Bioeng.* 66, 180–188.
4. Madin, K., Sawasaki, T., Ogasawara, T., and Endo, Y. (2000) A highly efficient and robust cell-free protein synthesis system prepared from wheat embryos: plants apparently contain a suicide system directed at ribosomes. *Proc. Natl. Acad. Sci. USA.* 97, 559–564.
5. Kim, T. W., Kim, D. M., and Choi, C. Y. (2006) Rapid production of milligram quantities of proteins in a batch cell-free protein synthesis system. *J. Biotechnol.* 124, 373–380.
6. Spirin, A. S., Baranov, V. I., Ryabova, L. A., Ovodov, S. Y., and Alakhov, Y. B. (1988) A Continuous Cell-Free Translation System Capable of Producing Polypeptides in High Yield. *Science* 242, 1162–1164.
7. Davis, J., Thompson, D., and Beckler, G. S. (1996) *Promega Notes Magazine* 56, 14–18.
8. Kim, D. M., and Choi, C. Y. (1996) A semi-continuous prokaryotic coupled transcription/translation system using a dialysis membrane. *Biotechnol. Prog.* 12, 645–649.
9. Kigawa, T., Yabuki, T., Yoshida, Y., Tsutsui, M., Ito, Y., Shibata, T., and Yokoyama, S. (1999) Cell-free production and stable-isotope labeling of milligram quantities of proteins. *FEBS Lett.* 442, 15–19.
10. Sawasaki, T., Hasegawa, Y., Tsuchimochi, M., Kamura, N., Ogasawara, T., Kuroita, T., and Endo, Y. (2002) A bilayer cell-free protein synthesis system for high-throughput screening of gene products. *FEBS Lett.* 514, 102–105.
11. Tarui, H., Imanishi, S., and Hara, T. (2000) A novel cell-free translation/glycosylation system prepared from insect cells. *J. Biosci. Bioeng.* 90, 508–514.
12. Katzen, F., and Kudlicki, W. (2006) Efficient generation of insect-based cell-free translation extracts active in glycosylation and signal sequence processing. *J. Biotechnol.* 125, 194–197.

13. Mikami, S., Masutani, M., Sonenberg, N., Yokoyama, S., and Imataka, H. (2006) An efficient mammalian cell-free translation system supplemented with translation factors. *Protein Expr. Purif.* 46, 348–357.
14. Wakiyama, M., Kaitsu, Y., and Yokoyama, S. (2006) Cell-free translation system from Drosophila S2 cells that recapitulates RNAi. *Biochem. Biophys. Res. Commun.* 343, 1067–1071.
15. Seki, E., Matsuda, N., Yokoyama, S., and Kigawa, T. (2008) Cell-free protein synthesis system from Escherichia coli cells cultured at decreased temperatures improves productivity by decreasing DNA template degradation. *Anal. Biochem.* 377, 156–161.
16. Shaw, W. V. (1975) Chloramphenicol acetyltransferase from chloramphenicol-resistant bacteria. *Methods Enzymol.* 43, 737–755.
17. Davanloo, P., Rosenberg, A. H., Dunn, J. J., and Studier, F. W. (1984) Cloning and expression of the gene for bacteriophage T7 RNA polymerase. *Proc. Natl. Acad. Sci. USA* 81, 2035–2039.
18. Grodberg, J., and Dunn, J. J. (1988) ompT Encodes the Escherichia coli Outer Menbrane Protease That Cleaves T7 RNA Polymerase during Purification. *J. Bacteriol.* 170, 1245–1253.
19. Zawadzki, V., and Gross, H. J. (1991) Rapid and simple purification of T7 RNA polymerase. *Nucleic Acids Res.* 19, 1948.
20. Pratt, J. M. (1984) Coupled Transcription-Translation in Prokaryotic Cell-Free System, in *Transcription and translation* (Hames, B. D., and Higgins, S. J., Eds.), pp. 179–209, IRL Press, Oxford, Washington DC.

Chapter 2

PURE Technology

Yoshihiro Shimizu and Takuya Ueda

Abstract

The *Escherichia coli*-based reconstituted cell-free protein synthesis system, which we named the PURE (Protein synthesis Using Recombinant Elements) system, provides several advantages compared with the conventional cell-extract-based system. Stability of RNA or protein is highly improved because of the lack of harmful degradation enzymes. The system can be easily engineered according to purposes or the proteins to be synthesized, by manipulating the components in the system. In this chapter, we describe the construction and exploitation of the PURE system. Methods for preparing and assembling the components composing the PURE system for the protein synthesis reaction are shown.

Key words: Cell-free protein synthesis system, In vitro translation system, PURE system, Ribosome, tRNA, Translation factor

1. Introduction

A cell-free protein synthesis system is a useful tool for the protein production or protein engineering (1). Since proteins can be synthesized only by the addition of the template DNA to the reaction mixtures, it provides an easy way for the rapid expression of various proteins. Furthermore, several applied technologies utilizing this system enable the synthesis of active membrane proteins (2, 3), incorporation of unnatural amino acids into the proteins (4), or screening of proteins with desired function from the large library pool (5). These features ensure its availability in the field of protein sciences or biotechnology.

Generally, the system is constructed with cell extract, so-called S30 fraction, which contains essential components for the protein synthesis. Through the recent development of the S30-directed cell-free protein synthesis system, milligram quantity of the protein production has been achieved (6, 7). However, the use of such crude extract for the basis of the system encounters several problems.

Y. Endo et al. (eds.), *Cell-Free Protein Production: Methods and Protocols,* Methods in Molecular Biology, vol. 607
DOI 10.1007/978-1-60327-331-2_2, © Humana Press, a part of Springer Science+Business Media, LLC 2010

Rapid shortage of energy sources or degradation of nucleic acids and proteins results in poor reproducibility of the system. Lack of information on the ingredients of the system causes low reliability upon the extension to the applied technologies.

To develop reliable system, we developed the cell-free protein synthesis system reconstituted with purified factors and enzymes required for the translation (8). This system, designated as the PURE (Protein synthesis Using Recombinant Elements) system, is based on the *E. coli* translation apparatus. Purified ribosomes, tRNAs, translation factors, aminoacyl-tRNA synthetases, and several other enzymes for the accomplishment of the protein synthesis composes this system. In contrast to the S30 system, all of the components of the system are identified and, therefore, it is more easily controlled. Reduced nucleases are advantageous for the screening system utilizing the cell-free protein synthesis system (9). Expanded or reconfigured genetic code can be achieved by regulating the codon of mRNA and tRNA or a specific release factor (4, 8, 10). In this report, we describe how to construct the PURE system. Methods for preparation of PURE system components and protein synthesis using this system are described.

2. Materials

2.1. Preparation of His-Tagged PURE System Components

1. Isopropyl β-D-1-thiogalactopyranoside (IPTG, Wako). Dissolve at 100 mM and store at –20°C (see Note 1).
2. Buffer A: 50 mM Hepes-KOH, pH 7.6, 1 M ammonium chloride, 10 mM magnesium chloride, 7 mM 2-mercaptoethanol. Store at 4°C (see Note 2).
3. Buffer B: 50 mM Hepes-KOH, pH 7.6, 100 mM potassium chloride, 10 mM magnesium chloride, 500 mM imidazole, pH 7.6, 7 mM 2-mercaptoethanol. Store at 4°C.
4. HT buffer: 50 mM Hepes-KOH, pH 7.6, 100 mM potassium chloride, 10 mM magnesium chloride, 7 mM 2-mercaptoethanol. Store at 4°C.
5. Stock buffer: 50 mM Hepes-KOH, pH 7.6, 100 mM potassium chloride, 10 mM magnesium chloride, 30% glycerol, 7 mM 2-mercaptoethanol. Store at 4°C.
6. 100 mM nickel sulfate.
7. HiTrap chelating column (2×5 mL, GE Healthcare) (see Note 3).
8. Protein assay kit based on Bradford method (Bio-Rad).
9. 2 mg/ml bovine serum albumin (BSA) standard for the protein assay (Pierce).
10. Digital Sonifier (Branson).

2.2. Preparation of Ribosome

1. Suspension buffer: 10 mM Hepes-KOH, pH 7.6, 50 mM potassium chloride, 10 mM magnesium acetate, 7 mM 2-mercaptoethanol. Store at 4°C.
2. Buffer C: 20 mM Hepes-KOH, pH 7.6, 1.5 M ammonium sulfate, 10 mM magnesium acetate, 7 mM 2-mercaptoethanol. Store at 4°C.
3. Buffer D: 20 mM Hepes-KOH, pH 7.6, 10 mM magnesium acetate, 7 mM 2-mercaptoethanol. Store at 4°C.
4. Cushion buffer: 20 mM Hepes-KOH, pH 7.6, 30 mM ammonium chloride, 10 mM magnesium acetate, 30% sucrose, 7 mM 2-mercaptoethanol. Store at 4°C.
5. Ribosome buffer: 20 mM Hepes-KOH, pH 7.6, 30 mM potassium chloride, 6 mM magnesium acetate, 7 mM 2-mercaptoethanol. Store at 4°C.
6. GD/X syringe filter (0.45 mm, PVDF, Whatman).
7. HiTrap Butyl FF column (2 × 5 mL, GE Healthcare).
8. High-pressure homogenizer (EmulsiFlex-C5, Avestin).

2.3. Preparation of Template DNA for the Protein Synthesis

1. Template-specific forward primer: 5′-AAGGAGATATACCA *ATG*NNNNNNNNNNNNNNNNNNNN-3′. The underlined is a nucleotide sequence identical to a 5′-terminus of the target gene. The Italicized represents an initiation codon.
2. Template-specific reverse primer: 5′-TATTCA*TTA* NNNNNNNNNNNNNNNNNNNN-3′. The underlined is a nucleotide sequence complementary to a 3′-terminus of the target gene. The Italicized represents a termination codon.
3. Universal primer: 5′-GAAATTAATACGACTCACTATA GGGAGACCACAACGGTTTCCCTCTAGAAA TAATTTTGTTTAACTTTAAGAAGGAGATA TACCA-3′. The underlined represents a T7 promoter sequence.
4. TaKaRa Ex Taq (Takara-Bio).
5. Wizard SV Gel and PCR Clean-Up System (Promega).

2.4. Translation Reaction

1. 1 mM each 20 amino acids. Dissolve 20 amino acids in water and store at −20°C.
2. 700 OD_{260}/ml *E. coli* tRNA mixtures. Dissolve *E. coli* tRNA mixtures (Roche) in water and store at −20°C (see Note 4).
3. 100 mM ATP, GTP, UTP, CTP. NTPs are dissolved in water and the pH is adjusted to 7.0 by the addition of sodium hydroxide. Store at −20°C.
4. 500 mM creatine phosphate. Creatine phosphate (sodium salt) is dissolved in water and stored at −20°C.

5. 1 mg/ml formyl donor. 25 mg Folinic acid (calcium salt) is dissolved in 2 mL of 50 mM 2-mercaptoethanol. After the addition of 220 μL hydrochloric acid, the solution is incubated for 3 h at room temperature. The solution is diluted to 1 mg/ml by water and store at –20°C.
6. 1 M Hepes-KOH, pH 7.6. Store at 4°C.
7. 1 M magnesium acetate. Store at room temperature.
8. 2 M potassium glutamate. Store at –20°C in aliquots.
9. 500 mM spermidine. Store at –20°C in aliquots.
10. 500 mM DTT. Store at –20°C in aliquots.
11. Prepared his-tagged PURE system components, ribosome, and the template DNA for the protein synthesis.
12. Microcon YM-100 (Millipore).
13. SYPRO Orange protein gel stain (Bio-Rad).
14. 7.5% acetic acid.
15. Typhoon 9410 (GE Healthcare).

3. Methods

3.1. Preparation of His-Tagged PURE System Components

1. Construction of overexpression plasmids of the His-tagged PURE system components is shown in Table 2.1. Amplify each gene by PCR from corresponding genome and clone into an appropriate vector described in Table 2.1. Transform the plasmids into an *Escherichia coli* strain BL21/DE3 (pET vector) or BL21/pREP4 (pQE vector).
2. Grow cells to an $A_{600nm} = 0.7$ in 2 L of LB broth at 37°C. Add IPTG to a final concentration of 0.1 mM and additionally grow for 3 h. Harvest cells by centrifugation (see Note 5).
3. Resuspend the cells in buffer A and disrupt them by sonication. Remove cell debris by centrifugation at 20,000 × *g*.
4. Apply the supernatant to a 10 mL Ni^{2+} column (see Note 3). Wash the column with 100 mL of buffer A–buffer B (95:5 (vol/vol); 25 mM imidazole) with a 2 mL/min flow rate. Elute His-tagged protein with a linear gradient from buffer A–buffer B (95:5 (vol/vol); 25 mM imidazole) to buffer A–buffer B (20:80 (vol/vol); 400 mM imidazole) with a 1 mL/min flow rate for 80 min.
5. Subject the eluted fractions to SDS–PAGE and recover the fractions containing His-tagged protein (Fig. 2.1).
6. Dialyze the recovered fractions against HT buffer (2 × 3 h) and stock buffer (1 × 3 h). Determine the concentration of purified His-tagged protein by Bradford method using BSA as a standard.

Table 2.1
Construction of overexpression plasmids of the His-tagged PURE system components

Components	Source	Expression vector	Terminus for His-tag attachment
Alanyl-tRNA synthetase	*E. coli*	pQE30 (SphI/HindIIII)	N
Arginyl-tRNA synthetase	*E. coli*	pET16b (NdeI/BamHI)	N
Asparaginyl-tRNA synthetase	*E. coli*	pET16b (NdeI/BamHI)	N
Aspartate-tRNA synthetase	*E. coli*	pET21a (NdeI/XhoI)	C
Cysteinyl-tRNA synthetase	*E. coli*	pET21a (NdeI/XhoI)	C
Glutaminyl-tRNA synthetase	*E. coli*	pET21a (NdeI/XhoI)	C
Glutamyl-tRNA synthetase	*E. coli*	pET21a (NdeI/XhoI)	C
Glycyl-tRNA synthetase	*E. coli*	pET21a (NdeI/XhoI)	C
Histidyl-tRNA synthetase	*E. coli*	pET21a (NdeI/XhoI)	C
Isoleucyl-tRNA synthetase	*E. coli*	pET21a (NdeI/HindIII)	N
Leucyl-tRNA synthetase	*E. coli*	pET21a (XbaI/XhoI)	C
Lysyl-tRNA synthetase	*E. coli*	pET21a (NdeI/XhoI)	C
Methionyl-tRNA synthetase	*E. coli*	pET21a (XbaI/XhoI)	C
Phenylalanyl-tRNA synthetase	*E. coli*	pQE30 (SphI/HindIII)	N
Prolyl-tRNA synthetase	*E. coli*	pET21a (NdeI/XhoI)	C
Seryl-tRNA synthetase	*E. coli*	pET21a (XbaI/XhoI)	C
Threonyl-tRNA synthetase	*E. coli*	pQE30 (BamHI/HindIII)	N
Tryptophanyl-tRNA synthetase	*E. coli*	pET21a (NdeI/XhoI)	C
Tyrosyl-tRNA synthetase	*E. coli*	pET21a (NdeI/XhoI)	C
Valyl-tRNA synthetase	*E. coli*	pET21a (NdeI/XhoI)	C
Methionyl-tRNA formyltransferase	*E. coli*	pET21a (NdeI/XhoI)	C
Initiation factor 1	*E. coli*	pQE30 (BamHI/HindIII)	N
Initiation factor 2	*E. coli*	pQE30 (BamHI/HindIII)	N
Initiation factor 3	*E. coli*	pQE30 (BamHI/HindIII)	N
Elongation factor G	*E. coli*	pQE60 (MunI/BglII)	C
Elongation factor Tu	*E. coli*	pQE60 (EcoRI/BglII)	C
Elongation factor Ts	*E. coli*	pQE60 (NcoI/BglII)	C
Release factor 1	*E. coli*	pQE60 (BamHI/HindIII)	
Release factor 3	*E. coli*	pQE30 (EcoRI/BamHI)	

(continued)

Table 2.1 (continued)

Components	Source	Expression vector	Terminus for His-tag attachment
Ribosome recycling factor	*E. coli*	pQE60 (EcoRI/BamHI)	C
Creatine kinase	Rabbit	pET15b (NdeI/BamHI)	N
Myokinase	Yeast	pET15b (NdeI/BamHI)	N
Nucleotide diphosphate kinase	*E. coli*	pQE30 (BamHI/HindIII)	N
Pyrophosphatase	*E. coli*	pET15b (NdeI/BamHI)	N
T7 RNA polymerase	T7 phage	pQE30 (BamHI/HindIII)	N

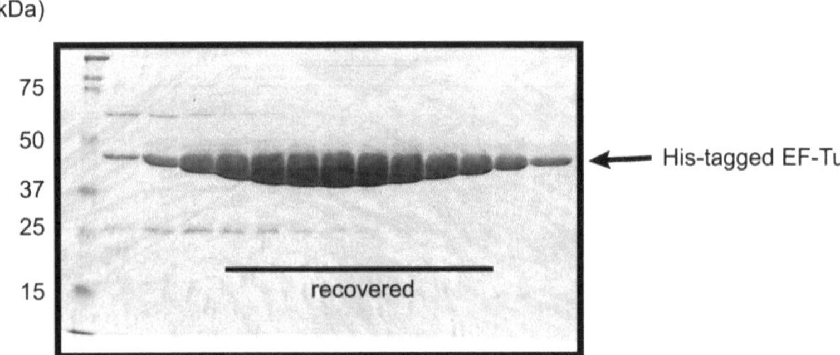

Fig. 2.1. SDS–PAGE analysis of the eluted His-tagged protein. Eluted fractions containing His-tagged protein by the Ni^{2+} column chromatography is analyzed by SDS–PAGE. Samples with high homogeneity and high concentration are recovered as a purified PURE system component. The figure shows the result of His-tagged EF-Tu purification.

3.2. Preparation of Ribosome

1. Grow *Escherichia coli* A19 strain to mid-log phase in LB broth at 37°C. Harvest the cells by centrifugation (see Note 5).
2. Suspend the cells in an equal volume of suspension buffer and disrupt them by high-pressure homogenizer at 7,000–10,000 psi. Remove the cell debris by centrifugation at 20,000 × *g* for 30 min.
3. Add ammonium sulfate to the supernatant to a final concentration of 1.5 M (see Note 6) and dissolve it with stirring at 4°C for 30 min. Remove the precipitated fraction by centrifugation at 20,000 × *g*. Filter the supernatant through a 0.45 μm membrane to completely eliminate the precipitated fraction.
4. Use butyl column for the subsequent hydrophobic chromatography. Equilibrate 10 mL butyl column with buffer C. Load 1,000 U at OD_{260} of the recovered supernatant onto the column and wash with 100 mL of buffer C–buffer D (80:20

(vol/vol); 1.2 M ammonium sulfate) with a 2 mL/min flow rate. Elute the fraction containing ribosome with 35 mL of buffer C–buffer D (50:50 (vol/vol); 0.75 M ammonium sulfate) with a 2 mL/min flow rate and recover the eluate (see Note 7).

5. Overlay the recovered fraction onto 35 mL of cushion buffer in a polycarbonate tube (void volume: 70 mL) for the Beckman type 45 Ti rotor. Pellet down the ribosome by the ultracentrifugation at 36,000 rpm (100,000 × g) for 16 h.
6. After ultracentrifugation, dissolve the pelletted ribosome in 200–300 mL of ribosome buffer. Store the recovered ribosome at −80°C in aliquots until use (see Note 8).

3.3. Preparation of Template DNA for the Protein Synthesis

1. The outline of the method to prepare template DNA for the protein synthesis is illustrated in Fig. 2.2 (see Note 9). Using template-specific forward primer and reverse primer, amplify the gene encoding target protein by PCR.
2. Further amplify the gene by PCR using universal primer and template-specific reverse primer.
3. Purify the amplified gene and dilute it to 0.2 pmol/μL.

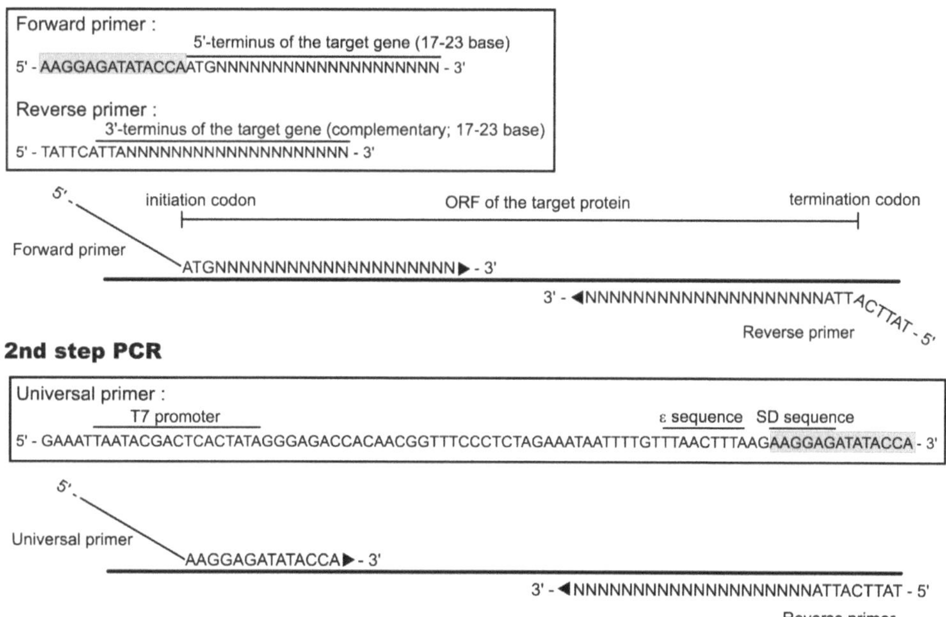

Fig. 2.2. Preparation of template DNA for the protein synthesis. A DNA template is prepared by two-step PCR. An ORF of the target protein is amplified by first step PCR and T7 promoter, epsilon sequence originated from bacteriophage T7 (12), and SD (Shine–Dalgarno) sequence are attached by second step PCR.

3.4. Translation Reaction and Detection of the Synthesized Protein

1. Prepare solution A: Solution A contains 0.2 mM each 20 amino acids, 108 OD_{260}/ml *E. coli* tRNA mixtures, 4 mM ATP, 4 mM GTP, 2 mM CTP, 2 mM UTP, 40 mM creatine phosphate, 20 μg/ml formyl donor, 100 mM Hepes-KOH, pH 7.6, 200 mM potassium glutamate, 26 mM magnesium acetate, 4 mM spermidine, 2 mM DTT.
2. Prepare solution B: Solution B contains 690 μg/mL alanyl-tRNA synthetase, 20 μg/mL arginyl-tRNA synthetase, 220 μg/mL asparaginyl-tRNA synthetase, 80 μg/mL aspartate-tRNA synthetase, 12 μg/mL cysteinyl-tRNA synthetase, 38 μg/mL glutaminyl-tRNA synthetase, 126 μg/mL glutamyl-tRNA synthetase, 96 μg/mL glycyl-tRNA synthetase, 8 μg/mL histidyl-tRNA synthetase, 400 μg/mL isoleucyl-tRNA synthetase, 40 μg/mL leucyl-tRNA synthetase, 64 lysyl-tRNA synthetase, 21 μg/ml methionyl-tRNA synthetase, 170 μg/ml phenylalanyl-tRNA synthetase, 100 μg/ml prolyl-tRNA synthetase, 19 μg/ml seryl-tRNA synthetase, 63 μg/ml threonyl-tRNA synthetase, 11 μg/ml tryptophanyl-tRNA synthetase, 6 μg/ml tyrosyl-tRNA synthetase, 18 μg/ml valyl-tRNA synthetase, 200 μg/ml methionyl-tRNA formyltransferase, 100 μg/ml initiation factor 1 (IF1), 400 μg/ml initiation factor 2 (IF2), 100 μg/ml initiation factor 3 (IF3), 500 μg/ml elongation factor G (EF-G), 1,000 μg/ml elongation factor Tu (EF-Tu), 500 μg/ml elongation factor Ts (EF-Ts), 100 μg/ml release factor 1 (RF1), 100 μg/ml release factor 3 (RF3), 100 μg/ml ribosome recycling factor (RRF),

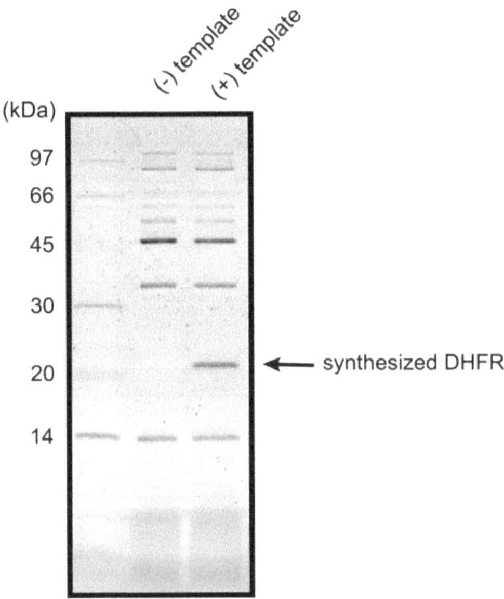

Fig. 2.3. Detection of the synthesized protein. Synthesized protein is analyzed by SDS–PAGE. The figure shows the result of dihydrofolate reductase (DHFR) synthesis.

40 μg/ml creatine kinase, 30 μg/ml myokinase, 11 μg/ml nucleotide diphosphate kinase, 10 μg/ml pyrophosphatase, T7 RNA polymerase, and 12 μM ribosome.

3. Synthesize the protein. Mix 25 μL solution A, 19 μL water, 5 μL solution B, and 1 μL 0.2 pmol/μL in a reaction tube and incubate it at 37°C for 1–2 h (see Note 10).
4. Apply all the reaction mixtures to Microcon YM-100 (see Note 11).
5. Centrifuge at 1,500 × *g* at 4°C for 30 min (see Note 12).
6. Subject the permeate to the SDS–PAGE (see Note 13). Stain the gel with SYPRO Orange protein gel stain, wash with 7.5% acetic acid, and visualize with Typhoon 9410 (Fig. 2.3) (see Note 14).

4. Notes

1. Unless otherwise specified, all solutions should be prepared in water that has a resistivity of 18.2 MΩ-cm.
2. 2-Mercaptoethanol should be added to the prechilled buffer before use.
3. Before column chromatography, apply 15 mL of 100 mM nickel sulfate solution to the column with a syringe and wash away the unbound solution with water. After the column chromatography, the column can be restored by washing with the buffer containing 20 mM sodium phosphate, pH 7.2, 500 mM sodium chloride, and 50 mM ethylenediamine-*N,N,N′,N′*-tetraacetic acid (EDTA). When you reuse the column, it is important to wash away EDTA with water before immobilization of the nickel ion.
4. Put in the water directly into the vial and dissolve RNA in it. Do not use spatula since the RNA is sensitive to the nucleases.
5. Unless harvested cells are immediately processed for purification, freeze the cells with liquid nitrogen and store at –80°C.
6. Total amount of the ammonium sulfate (g) is the value that the volume of supernatant (mL) is multiplied by 0.222. If the volume of supernatant is 100 mL, add 22.2 g ammonium sulfate into the supernatant.
7. Although the total volume of the eluate is limited by the void volume of a polycarbonate tube used in the following ultracentrifugation, all the ribosome adsorbed on the column can be eluted with this volume.

8. Recovery rate is about 20%, which means 200 U at OD_{260} of the ribosome (24 nmol) is recovered from 1,000 U at OD_{260} of the supernatant after the ammonium sulfate precipitation.
9. A plasmid DNA can also be used for the template DNA. The DNA should encode T7 promoter, SD sequence, the ORF of the protein, and T7 terminator. If the plasmid does not encode T7 terminator, a digested plasmid by specific restriction enzyme at a position downstream of the ORF of the protein can be used.
10. If the detection of the synthesized protein by SDS–PAGE is not necessary for your purpose, following protocol can be abbreviated.
11. This protocol is to remove the ribosome proteins from the reaction mixtures that may hinder the detection of the synthesized protein on the SDS–PAGE. Other methods such as the use of radioisotope-labeled amino acids and BODIPY-Lys-$tRNA_{Lys}$ (Promega) are also available for the detection of the synthesized protein (8, 11).
12. Insoluble or extremely large proteins cannot pass through the membrane. If the synthesized proteins possess such characteristics, other methods described in Note 11 are recommended to try.
13. It is recommended to prepare a negative control sample that does not contain the template DNA or mRNA.
14. Any other equipment that can detect the fluorescence can be used alternatively.

References

1. Shimizu, Y., Kuruma, Y., Ying, B. W., Umekage, S., and Ueda, T. (2006) Cell-free translation systems for protein engineering. *FEBS J.* **273**, 4133–4140.
2. Ozaki, Y., Suzuki, T., Kuruma, Y., Ueda, T., and Yoshida, M. (2008) UncI protein can mediate ring-assembly of c-subunits of FoF1-ATP synthase in vitro. *Biochem. Biophys. Res. Commun.* **367**, 663–666.
3. Katzen, F., Fletcher, J. E., Yang, J. P., Kang, D., Peterson, T. C., Cappuccio, J. A., et al. (2008) Insertion of membrane proteins into discoidal membranes using a cell-free protein expression approach. *J. Proteome Res.* **7**, 3535–3542.
4. Hartman, M. C., Josephson, K., Lin, C. W., and Szostak, J. W. (2007) An expanded set of amino acid analogs for the ribosomal translation of unnatural peptides. *PLoS ONE* **2**, e972.
5. Leemhuis, H., Stein, V., Griffiths, A. D., Hollfelder, F. (2005) New genotype-phenotype linkages for directed evolution of functional proteins. *Curr. Opin. Struct. Biol.* **15**, 472–478.
6. Kigawa, T., Yabuki, T., Yoshida, Y., Tsutsui, M., Ito, Y., Shibata, T. et al. (1999) Cell-free production and stable-isotope labeling of milligram quantities of proteins. *FEBS Lett.* **442**, 15–19.
7. Madin, K., Sawasaki, T., Ogasawara, T., and Endo, Y. (2000) A highly efficient and robust cell-free protein synthesis system prepared from wheat embryos: plants apparently contain a suicide system directed at ribosomes. *Proc. Natl. Acad. Sci. U.S.A.* **97**, 559–564.
8. Shimizu, Y., Inoue, A., Tomari, Y., Suzuki, T., Yokogawa, T., Nishikawa, K. et al. (2001)

Cell-free translation reconstituted with purified components. *Nat. Biotechnol.* **19**, 751–755.

9. Ohashi, H., Shimizu, Y., Ying, B. W., and Ueda, T. (2007) Efficient protein selection based on ribosome display system with purified components. *Biochem. Biophys. Res. Commun.* **352**, 270–276.
10. Murakami, H., Ohta, A., Ashigai, H., and Suga, H. (2006) A highly flexible tRNA acylation method for non-natural polypeptide synthesis. *Nat. Methods* **3**, 357–359.
11. Shimizu, Y., Kanamori, T., and Ueda, T. (2005) Protein synthesis by pure translation systems. *Methods* **36**, 299–304.
12. Olins, P. O., and Rangwala, S. H. (1989) A novel sequence element derived from bacteriophage T7 mRNA acts as an enhancer of translation of the *lacZ* gene in *Escherichia coli. J. Biol. Chem.* **264**, 16973–16976.

Chapter 3

The Cell-Free Protein Synthesis System from Wheat Germ

Kazuyuki Takai and Yaeta Endo

Abstract

The wheat-germ cell-free protein synthesis system had been one of the most efficient eukaryotic cell-free systems since it was first developed in 1964. However, radio-labeled amino acids had long been essential for detection of the products. Since the discovery of a method for prevention of the contamination by a protein synthesis inhibitor originated from endosperm, the wheat cell-free system has found a wide variety of applications in postgenomic high-throughput screening, structural biology, medicine, and so on. In this chapter, we describe a method for preparation of the cell-free extract and a standard protein synthesis method, as the methods for the applications are found in later chapters.

Key words: Wheat, Embryos, Batch-mode, Radioisotope labeling

1. Introduction

The first wheat-germ cell-free translation system was developed by Marcus and Feeley in 1964 (1). They could measure the incorporation of radioactive amino acids into the polypeptide fraction catalyzed by the isolated ribosomes. A crude extract from commercially available wheat germ was used later for protein synthesis dependent on the exogenous addition of messages (2). With some modifications (3), this had been the prototype of the conventional wheat-germ cell-free translation systems for a long time, and had been one of the most efficient eukaryotic cell-free translation systems.

Many plant seeds contain ribosome inactivating proteins (RIPs) that inactivate the ribosomes by depurinating the ribosomal RNA at the sarcin–ricin loop. The RIP from wheat was found very early and is called "tritin" (4, 5). Earlier papers reported that tritin inactivates mammalian ribosomes but not the

Y. Endo et al. (eds.), *Cell-Free Protein Production: Methods and Protocols,* Methods in Molecular Biology, vol. 607
DOI 10.1007/978-1-60327-331-2_3, © Humana Press, a part of Springer Science+Business Media, LLC 2010

wheat-germ ribosomes (5), while it had been known to be localized mainly in endosperm and to be present in the wheat-germ extracts. This had explained the reason why the wheat-germ cell-free translation system was efficient. However, the fact was that radioactive amino acids were essential for detection of the translation products.

In 1988, A. S. Spirin's group reported that the *E. coli* and wheat crude extracts can sustain their protein synthesis activities for as long as 20 h if the reactions were performed by the "continuous flow" method, by which the stoichiometric inhibition caused by exhaustion of substrate and accumulation of byproducts can be avoided (6). Many researchers were shocked by the report and some of them tested the method by their own hands. It was found that the results were surely reproducible, though some catalytic inhibition that is different from the stoichiometric inhibition occurs in the case of the wheat system (7, 8). Since then, Endo's group had worked on the assumptions that tritin had been damaging the wheat-germ ribosomes during prolonged incubation in the extract, and that the ribosomes are quite robust in nature. While many different methods were tested for excluding or inactivating the endosperm inhibitors, it was finally found that extensive selection and purification of the embryo particles before homogenization was the most promising method for reproducible, efficient cell-free translation (9).

Now, the wheat cell-free protein synthesis system is widely used for the preparation of proteins as you can see in later chapters in this volume. With this method, we could synthesize many different proteins in parallel encoded by mRNAs prepared by transcription of PCR-generated DNA templates, which is very useful for screening of proteins from a large library by function (10–13). We could also prepare large amounts of proteins suitable for structural analyses and antigen production by adding mRNAs transcribed from purified plasmid DNAs (10, 14, 15). It has been demonstrated that the protein synthesis reaction could last for more than 2 weeks, if mRNA and substrates are supplied properly (10).

It has been observed that many, much more than in the case of the *E. coli* cell-free system, of the produced eukaryotic proteins are soluble in the wheat extract. This is probably because eukaryotes have much more multidomain proteins than prokaryotes, and because the eukaryotic protein synthesis apparatus should have adapted so that it prevents misfolding due to interdomain misinteraction during elongation of newly synthesized polypeptide chains (16). This better folding in eukaryotes may at least in part be due to the slower elongation rate (17, 18). This eukaryotic nature of the wheat system, we believe, is benefiting high-throughput production of eukaryotic proteins in functional conformations, as seen in the later chapters. Recently, more than

13,000 human proteins were produced with the wheat cell-free system, and their biological activities were assessed (19). This clearly showed the advantage of the wheat cell-free system in producing human proteins in active conformations.

In this chapter, we first describe the method for the preparation of the wheat extract. The method for preparation of the extract is a slight modification of the method by Erickson and Blobel (3) except that the embryo particles are extensively washed before homogenization. We then describe a standard method for protein synthesis with a radio-labeled amino acid. As you have already found in the contents of this volume, a lot of modifications of this standard protein synthesis protocol are described in this volume. You will find the modifications are very small: it is one of the most prominent features of the wheat-germ cell-free gene expression method that protocols for many different purposes could be devised by small modifications of the core methods.

2. Materials

2.1. Purification of Wheat Embryo Particles

1. Wheat seeds (any strain may be useful).
2. Cyclohexane-carbon tetrachloride mixture (1:2.5, v/v).
3. 0.5% Nonidet P-40 (NP-40).

2.2. Preparation of the Cell-Free Extract

1. Liquid nitrogen.
2. A pestle and mortar, dry-heat-sterilized and precooled in –80°C.
3. Extraction buffer (EB): 40 mM HEPES-KOH (pH 7.6), 100 mM KOAc, 5 mM $Mg(OAc)_2$, 2 mM $CaCl_2$, 4 mM DL-dithiothreitol (DTT), and 0.3 mM each of the standard 20 amino acids.
4. Dialysis buffer (DB): 30 mM HEPES-KOH (pH 7.6), 100 mM KOAc, 2.7 mM $Mg(OAc)_2$, 0.4 mM spermidine, 2.5 mM DTT, 0.3 mM amino acids, 1.2 mM ATP, 0.25 mM GTP, and 16 mM creatine phosphate.
5. Sephadex G-25 columns (e.g., GE Healthcare Microspin G-25 columns).
6. Amicon Ultra-15 Centrifugal Filter Units (Millipore Corp.).

2.3. mRNA Preparation

1. Template DNA: A 1 μg/μL solution of pEU plasmid *(10)* harboring the target gene, which may be prepared with VIOGENE Ultrapure Plasmid DNA Extraction Maxiprep (or Midiprep) System without RNase (see Acknowledgments), or a crude PCR product with the plasmid sequence ranging from the SP6

promoter to the sequence 500-bp downstream of the stop codon amplified with ExTaq DNA polymerase (Takara Biochemicals, Japan).

2. 5× TB: 80 mM HEPES-KOH (pH 7.8), 16 mM $Mg(OAc)_2$, 2 mM spermidine, and 10 mM DTT.
3. 25 mM NTPs: aqueous solution containing ATP, CTP, GTP, and UTP, pH 7–8, which should not be confused with dNTPs.
4. SP6 RNA polymerase (10–20 units/μL, Promega).
5. RNase inhibitor from human placenta (40 units/μL, Promega).
6. Formamide Loading Dye: 0.1% Bromophenol Blue and 0.1% Xylene Cyanol FF in deionized formamide.

2.4. Radio-Labeled Protein Synthesis

1. 4× DB: fourfold concentrated solution of DB.
2. 1 mg/mL creatine kinase.
3. mRNA solution (1–5 mg/mL in water).
4. [^{14}C(U)]Leucine (>10 GBq/mmol).
5. 0.2 M NaOH: dispense 5-μL aliquots in small tubes.
6. Filter paper (Whatman 3MM or Advantec 51A): cut into 1- to 1.4-cm square pieces and number them with pencil; penetrate a 28-mm steel straight pin through each piece at the center and stand the pins on a polystylene foam board covered by aluminum foil by sticking the pin at the tip on the board.
7. Ice-cold 5% trichloroacetic acid (TCA) solution in water (see Note 1).
8. Liquid scintillation cooktail: 4 g/l PPO (2,5-diphenyloxazol) and 0.05 g/L POPOP (2,2′-phenylene-bis(5-phenyloxazol)) dissolved in toluene by an overnight stirring.
9. Counting vials (Wheaton #986542 or Perkin-Elmer #6000477) containing 10 mL of the liquid scintillation cooktail (see Note 2).

3. Methods

3.1. Purification of Wheat Embryo Particles

1. Grind the wheat seeds (typically 6 kg for 6 mL extract) by a Fritsch Rotor Speed Mill Model P-14.
2. Sieve the sample by a sieve shaker (Fritsch A-3 PRO shaker with 710-, 850-, and 1,000-μm mesh sieves), and collect the particles that passed through the 1,000-μm mesh but not the 710-μm mesh.
3. Remove flittering flakes of bran by dropping the sample from the height of a few feet into another container repeatedly.

4. Put the sample into a mixture of cyclohexane and carbon tetrachloride (1:2.5, v/v), and let denser particles to sink down.
5. Collect the floating particles by a mesh skimmer, and put the particles on a sheet of Kim Wipe in a fume hood to dry them overnight.
6. Select yellow particles by eye using toothpicks: remove carefully those particles with brownish color and those with much white matter.
7. Wrap the embryo particles (typically 2 g) in gauze.
8. Dip it into Milli-Q water (typically 100 mL) and knead it gently but thoroughly to remove the white matter from the embryos.
9. Change the water and repeat washing twice more.
10. Sonicate the sample in 0.5% NP-40 with the Bronson Model 2210 sonicator (degas mode, 2–3 min).
11. Sonicate it twice more in Milli-Q water.
12. Put the embryo particles into a Buchner funnel and wash them with Milli-Q (500 mL) thoroughly.

3.2. Preparation of the Cell-Free Extract

1. Grind the embryo particles (2 g) into fine powder in the chilled mortar with liquid nitrogen poured into the mortar occasionally to cool the sample.
2. Add 2 mL of chilled EB to suspend the sample.
3. Centrifuge the suspension at a low speed to remove lipid (2,000 × *g*, 1 h).
4. Centrifuge the sample at 30,000 × *g* for 30 min at 4°C.
5. Recover the supernatant.
6. Pass the sample through a Sephadex G-25 column preequilibrated with EB.
7. Collect the flow through fraction and repeat gel filtration with DB.
8. Centrifuge again at 30,000 × g for 10 min and recover the supernatant.
9. Adjust the UV absorption at 260 nm to 240 nm with DB. If needed, concentrate the S30 fraction with an Amicon Ultra-15 filter unit before adding DB.
10. Store at −80°C until use in aliquots.

3.3. mRNA Preparation

1. Mix 10 µL of 5× TB, 4 µL of 25 mM NTPs, 0.5 µL of RNase inhibitor, 3 µL of SP6 RNA polymerase, 5 µL of the template DNA, and Milli-Q water in a 50-µL solution.
2. Incubate the solution at 42°C for 2.5 h.

3. Pick up a 1-μL aliquot for analysis.
4. Store the remaining solution at –80°C if you do not use it immediately.
5. Mix the aliquot with 5 μL of Formamide Loading Dye, incubate it at 65°C for 15 min, and analyze the sample on a standard agarose gel (see Note 3).

3.4. Cell-Free Translation in the Batch Mode with (^{14}C) Leucine

1. Before starting the reaction, measure the background scintillation of the counting vials for 1 min with a liquid scintillation counter. Prepare the tubes containing 5 μL of 0.2 M NaOH for stopping the reaction (see below).
2. For a 50-μL reaction, mix 9.5 μL of 4× DB, 20.5 μL of water, 2 μL of 1 mg/mL creatine kinase, 1 μL of (^{14}C)Leu, and 12 μL of the extract (see Note 4).
3. Add 5 μL of the mRNA solution and mix thoroughly on ice.
4. Pick 5 μL of the reaction mixture as the 0-min sample and transfer it to one of the 0.2 M NaOH tubes.
5. Incubate the reaction tube at 26°C. Pick a 5-μL aliquot and mix it with the NaOH solution at the predetermined time. Typically reaction will stop within 3 h.
6. Incubate the stopped samples at 26°C for at least 15 min. This will hydrolyze aminoacyl-tRNAs.
7. Apply the 10-μL samples onto the filter paper pieces.
8. Rinse the paper pieces by soaking them in chilled 5% TCA for 15 min.
9. Separate the TCA solution from the paper pieces and repeat rinsing twice more.
10. Soak the paper pieces in ethanol for 10 min at room temperature. Repeat once.
11. Dry the paper pieces under a drying lamp.
12. Put the paper pieces at the bottom of each counting vials containing the scintillation cooktail, and count the radioactivity with a liquid scintillation counter (e.g., Aloka LSC-5100).

4. Notes

1. To prepare 5% TCA, dilute a solution of 100% TCA, which can be prepared by adding 227 mL of water into a bottle of 500 g TCA powder and by shaking the bottle overnight at room temperature.

2. Static electricity may sometimes give very high counts in the scintillation counting. This could in most cases be avoided by using the vials whose catalog numbers are shown.
3. Typical transcription of a pEU plasmid stops partially near the replication origin. Therefore, several bands will be observed on the gel. A standard DNA loading dye containing glycerol could be used, but the formamide-based dye solution gives sharper bands on the gel, probably due to spermidine in the sample.
4. DB contains nonradioactive Leu. The concentration is usually around a hundred times higher than the radioactive Leu added to the reaction.

Acknowledgments

This work is supported in part by the Special Coordination Funds for Promoting Science and Technology (Y. E.) and in part by a Grant-in-Aid for Scientific Research on the Priority Areas (No. 20034040 to K. T.) by the Ministry of Education, Culture, Sports, Science and Technology, Japan.

Recently, the reagents in VIOGENE Ultrapure Plasmid DNA Extraction Kit in Subsection 2.3 (item 1) were changed. With the new reagents, RNA is not removed well from the plasmid DNA preparation without using RNase. Therefore, it is recommended to purify plasmid DNAs with a standard plasmid preparation kit using RNase and to purify them further before use by phenol/chloroform extraction followed by ethanol precipitation.

References

1. Marcus, A., and Feeley, J. (1964) Activation of protein synthesis in the imbibitions phase of seed germination. *Proc. Natl. Acad. Sci. U.S.A.* **51**, 1075–1079.
2. Roberts, B. E., and Paterson, B. M. (1973) Efficient translation of tobacco mosaic virus RNA and rabbit globin 9 S RNA in a cell-free system from commercial wheat germ. *Proc. Natl. Acad. Sci. U.S.A.* **70**, 2330–2334.
3. Erickson, A. H., and Blobel, G. (1983) Cell-free translation of messenger RNA in a wheat germ system. *Methods Enzymol.* **96**, 38–50.
4. Roberts, W. K., and Stewart, T. S. (1979) Purification and properties of a translation inhibitor from wheat germ. *Biochemistry* **18**, 2615–2621.
5. Coleman, W. H., and Roberts, W. K. (1981) Factor requirements for the tritin inactivation of animal cell ribosomes. *Biochim. Biophys. Acta* **654**, 57–66.
6. Spirin, A. S., Baranov, V. I., Ryabova, L. A., Ovodov, S. Y., and Alakhov, Y. B. (1988) A continuous cell-free translation system capable of producing polypeptides in high yield. *Science* **242**, 1162–1164.
7. Kigawa, T., and Yokoyama, S. (1991) A continuous cell-free protein synthesis system for coupled transcription-translation. *J. Biochem.* **110**, 166–168.
8. Endo, Y., Otsuzuki, S., Ito, K., and Miura, K. (1992) Production of an enzymatic active protein using a continuous flow cell-free translation system. *J. Biotechnol.* **25**, 221–230.

9. Madin, K., Sawasaki, T., Ogasawara, T., and Endo, Y. (2000) A highly efficient and robust cell-free protein synthesis system prepared from wheat embryos: plants apparently contain a suicide system directed at ribosomes. *Proc. Natl. Acad. Sci. U.S.A.* **97**, 559–564.
10. Sawasaki, T., Ogasawara, T., Morishita, R., and Endo, Y. (2002) A cell-free protein synthesis system for high-throughput proteomics. *Proc. Natl. Acad. Sci. U.S.A.* **99**, 14652–14657.
11. Sawasaki, T., Hasegawa, Y., Tsuchimochi, M., Kamura, N., Ogasawara, T., Kuroita, T., and Endo, Y. (2002) A bilayer cell-free protein synthesis system for high-throughput screening of gene products. *FEBS Lett.* **514**, 102–105.
12. Masaoka, T., Nishi, M., Ryo, A., Endo, Y., and Sawasaki, T. (2008) The wheat germ cell-free based screening of protein substrates of calcium/calmodulin-dependent protein kinase II delta. *FEBS Lett.* **582**, 1795–1801.
13. Kobayashi, T., Kodani, Y., Nozawa, A., Endo, Y., and Sawasaki, T. (2008) DNA-binding profiling of human hormone nuclear receptors via fluorescence correlation spectroscopy in a cell-free system. *FEBS Lett.* **582**, 2737–2744.
14. Miyazono, K., Watanabe, M., Kosinski, J., Ishikawa, K., Kamo, M., Sawasaki, T., et al. (2007) Novel protein fold discovered in the PabI family of restriction enzymes. *Nucleic Acids Res.* **35**, 1908–1918.
15. Tsuboi, T., Takeo, S., Iriko, H., Jin, L., Tsuchimochi, M., Matsuda, S., et al. (2008) Wheat germ cell-free system-based production of malaria proteins for discovery of novel vaccine candidates. *Infect. Immun.* **76**, 1702–1708.
16. Hirano, N., Sawasaki, T., Tozawa, Y., Endo, Y., and Takai, K. (2006) Tolerance for random recombination of domains in prokaryotic and eukaryotic translation systems: Limited interdomain misfolding in a eukaryotic translation system. *Proteins* **64**, 343–354.
17. Netzer, W. J. and Hartl, F. U. (1997). Recombination of protein domains facilitated by co-translational folding in eukaryotes. *Nature* **388**, 343–349.
18. Hartl, F. U., and Hayer-Hartl, M. (2002). Molecular chaperones in the cytosol: from nascent chain to folded protein. *Science* **295**, 1852–1858.
19. Goshima, N., Kawamura, Y., Fukumoto, A., Miura, A., Honma, R., Satoh, R., Wakamatsu, A., Yamamoto, J., Kimura, K., Nishikawa, T., Andoh, T., Iida, Y., Ishikawa, K., Ito, E., Kagawa, N., Kaminaga, C., Kanehori, K., Kawakami, B., Kenmochi, K., Kimura, R., Kobayashi, M., Kuroita, T., Kuwayama, H., Maruyama, Y., Matsuo, K., Minami, K., Mitsubori, M., Mori, M., Morishita, R., Murase, A., Nishikawa, A., Nishikawa, S., Okamoto, T., Sakagami, N., Sakamoto, Y., Sasaki, Y., Seki, T., Sono, S., Sugiyama, A., Sumiya, T., Takayama, T., Takayama, Y., Takeda, H., Togashi, T., Yahata, K., Yamada, H., Yanagisawa, Y., Endo, Y., Imamoto, F., Kisu, Y., Tanaka, S., Isogai, T., Imai, J., Watanabe, S., and Nomura, N. (2008) Human protein factory for converting the transcriptome into an in vitro-expressed proteome. *Nat. Methods* **5**, 1011–1017.

Chapter 4

A Cell-Free Protein Synthesis System from Insect Cells

Toru Ezure, Takashi Suzuki, Masamitsu Shikata, Masaaki Ito, and Eiji Ando

Abstract

The Transdirect *insect cell* is a newly developed in vitro translation system for mRNA templates, which utilizes an extract from cultured *Spodoptera frugiperda* 21 (Sf21) insect cells. An expression vector, pTD1, which includes a 5′-untranslated region (UTR) sequence from a baculovirus polyhedrin gene as a translational enhancer, was also developed to obtain maximum performance from the insect cell-free protein synthesis system. This combination of insect cell extract and expression vector results in protein productivity of about 50 μg/mL of the translation reaction mixture. This is the highest protein productivity yet noted among commercialized cell-free protein synthesis systems based on animal extracts.

Keywords: Translation, *Spodoptera frugiperda* 21, Cell-free protein synthesis system, pTD1 vector, Insect cell extract

1. Introduction

Most of the recombinant proteins produced in a baculovirus expression system have been shown to be functionally similar to authentic proteins, because the insect cells can carry out many types of posttranslational modifications. Therefore, we developed a cell-free protein synthesis system from *Spodoptera frugiperda* 21 (Sf21) insect cells, which are widely used as the host for baculovirus expression systems, and commercialized it as the Transdirect *insect cell.*

We have demonstrated that this insect cell-free protein synthesis system is one of the most effective protein synthesis systems among those based on animal extracts (1). Furthermore, it has the potential to perform eukaryote-specific protein modifications such as protein *N*-myristoylation and prenylation (2, 3). Thus, we

Y. Endo et al. (eds.), *Cell-Free Protein Production: Methods and Protocols,* Methods in Molecular Biology, vol. 607
DOI 10.1007/978-1-60327-331-2_4, © Humana Press, a part of Springer Science+Business Media, LLC 2010

expect that the insect cell-free protein synthesis system will be a useful method for postgenomic studies. In this chapter, we describe standard protocols to synthesize proteins of interest using the insect cell-free protein synthesis system.

2. Materials

2.1. Construction of an Expression Clone

1. Primers for amplification of the target cDNA (see Subheading 3.1): Store at −20°C.
2. KOD-plus DNA polymerase (TOYOBO, Osaka, Japan): Store at −20°C (see Note 1).
3. 50× TAE buffer: Mix 242 g of Tris base, 57.1 mL of glacial acetic acid, and 100 mL of 0.5 M EDTA (pH 8.0), and adjust to 1,000 mL with water. Store at room temperature.
4. 1 Kb DNA Ladder (BIONEER, Korea): Store at −20°C.
5. Phenol/Chloroform/Isoamyl Alcohol 25:24:1 Saturated with 10 mM Tris–HCl, pH 8.0, 1 mM EDTA (Sigma, St. Louis, MO): Store at 4°C.
6. 3 M sodium acetate, pH 5.2: Store at room temperature.
7. T4 polynucleotide kinase (TOYOBO): Store at −20°C.
8. Restriction endonucleases (*Eco*RV, *Eco*RI, *Sac*I, *Kpn*I, *Bam*HI, and *Xba*I). Store at −20°C.
9. MinElute® PCR Purification Kit (QIAGEN, Maryland, USA): Store at room temperature.
10. pTD1 vector, a component of the Transdirect *insect cell* kit (Shimadzu, Kyoto, Japan). Store at −20°C or below. The map of the pTD1 vector is shown in Fig. 4.1.
11. Quick Ligation™ Kit (NEW ENGLAND BioLabs, Ipswich, MA): Store at −20°C.
12. Chemically competent cells *Escherichia coli* DH5α (TAKARA Bio, Shiga, Japan): Store at −80°C.
13. Ampicillin sodium salt (SIGMA) is dissolved in distilled water at 100 mg/mL and stored at −20°C.
14. Luria-Bertani (LB, 1.0% polypeptone, 0.5% yeast extract, 1.0% NaCl, pH 7.0) medium containing 100 μg/mL ampicillin (LB-amp) and LB-amp agar medium (LB containing 1.0% agar): Store at 4°C.
15. GenElute™ plasmid miniprep kit (SIGMA): Store at room temperature. Resuspension solution should be stored at 4°C after the addition of RNase A.

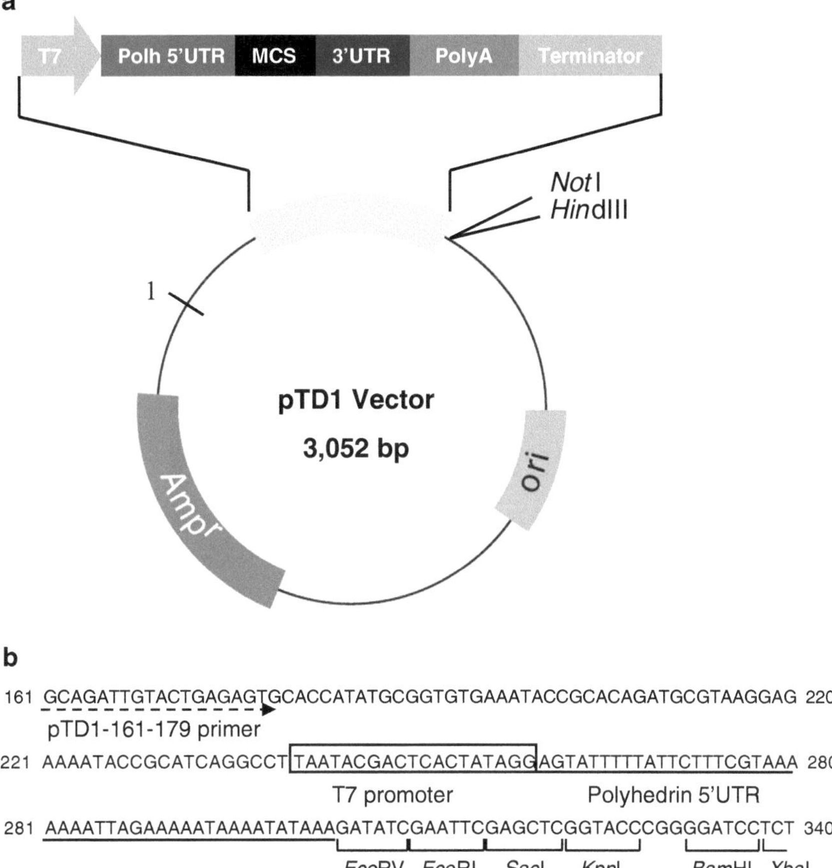

Fig. 4.1. The expression vector pTD1: (**a**) pTD1 vector map; (**b**) DNA sequence of the pTD1 vector around the multiple cloning sites.

16. BigDye® Terminator v3.1 Cycle Sequencing Kit (Applied Biosystems, Warrington, UK): Store at –20°C.
17. Primers for DNA sequencing (see Subheading 3.1): Store at –20°C.

2.2. Preparation of mRNA

1. Restriction endonucleases (*Hin*dIII and *Not*I) are stored at –20°C.
2. T7 RiboMAX™ Express Large Scale RNA Production System (Promega, Madison, WI) (see Note 2): Store at –20°C.
3. NICK™ Columns (GE Healthcare, Buckinghamshire, UK): Store at room temperature.

4. 3 M potassium acetate, pH 5.5 (Ambion, Austin, TX): Store at room temperature.
5. TE buffer (10 mM Tris–HCl, pH 8.0, 1 mM EDTA): Store at room temperature.
6. MOPS buffer (20×): 400 mM MOPS, 100 mM NaOAc, 20 mM EDTA (adjust to pH 7.0 by NaOH). Store at 4°C in the shade.

2.3. In Vitro Translation and Detection of Synthesized Proteins

1. Transdirect *insect cell* (Shimadzu): Store at -80°C.
2. Purified mRNA (see Subheading 3.2). Store at -80°C.
3. FluoroTect™ Green$_{Lys}$ in vitro Translation Labeling System (Promega): Store at –80°C.
4. SDS–PAGE running buffer (10×): 250 mM Tris, 1.92 M glycine, 1% SDS: Store at room temperature.
5. SDS–PAGE loading buffer (4×): 200 mM Tris–HCl (pH 6.8), 8% SDS, 32% Glycerol, 0.008% bromophenol blue 8% 2-melcapto ethanol. Store at room temperature.
6. Precast gel: c-PAGEL® (ATTO, Tokyo, Japan): Store at 4°C.
7. Prestained molecular weight markers: Full range RAINBOW™ (GE Healthcare). Store at –20°C.

3. Methods

The Transdirect *insect cell* kit is an in vitro translation system for mRNA templates. We developed and optimized a method to prepare the insect cell extract, the concentrations of the reaction components, and an expression vector pTD1 (1, 4). The pTD1 vector contains all factors involved in mRNA and protein synthesis, including the T7 promoter sequence required for mRNA synthesis, the polyhedrin 5′-untranslated region (UTR), which enhances the translation reaction and multiple cloning sites (MCS) (see Fig. 4.1). The complete DNA sequence of the pTD1 vector is registered in the following DNA databank: DDBJ/GenBank®/EMBL Accession Number AB194742.

To obtain maximal protein productivity, it is necessary to construct an expression clone in which a protein coding region (open reading frame, mature region, domain, etc.) obtained from a cDNA of interest is inserted into the MCS of the pTD1 vector. Typically, expression of the target protein at about 35–50 μg/mL of the translation reaction mixture can be obtained by using mRNA transcribed from the expression clone and the Transdirect *insect cell* kit. Furthermore, the expression clone can be effectively combined with other eukaryotic cell-free protein synthesis systems, such as rabbit reticulocyte lysate and wheat germ systems (see Note 3).

3.1. Construction of the Expression Clone for the Insect Cell-Free Protein Synthesis System

1. Procedures for construction of the expression clone follow classical molecular cloning methods. The overall cloning strategy is shown in Fig. 4.2.
2. Design and synthesize two primers, an N-terminal primer and a C-terminal primer, for amplification of protein coding region of the target cDNAs. The N-Terminal primer should have the initiation codon at its 5′-terminus. In the case of the C-terminal primer, the restriction endonuclease recognition sequence should be introduced upstream of the stop codon, and an additional sequence (at least two bases) should be added to the 5′-end (see Note 4).

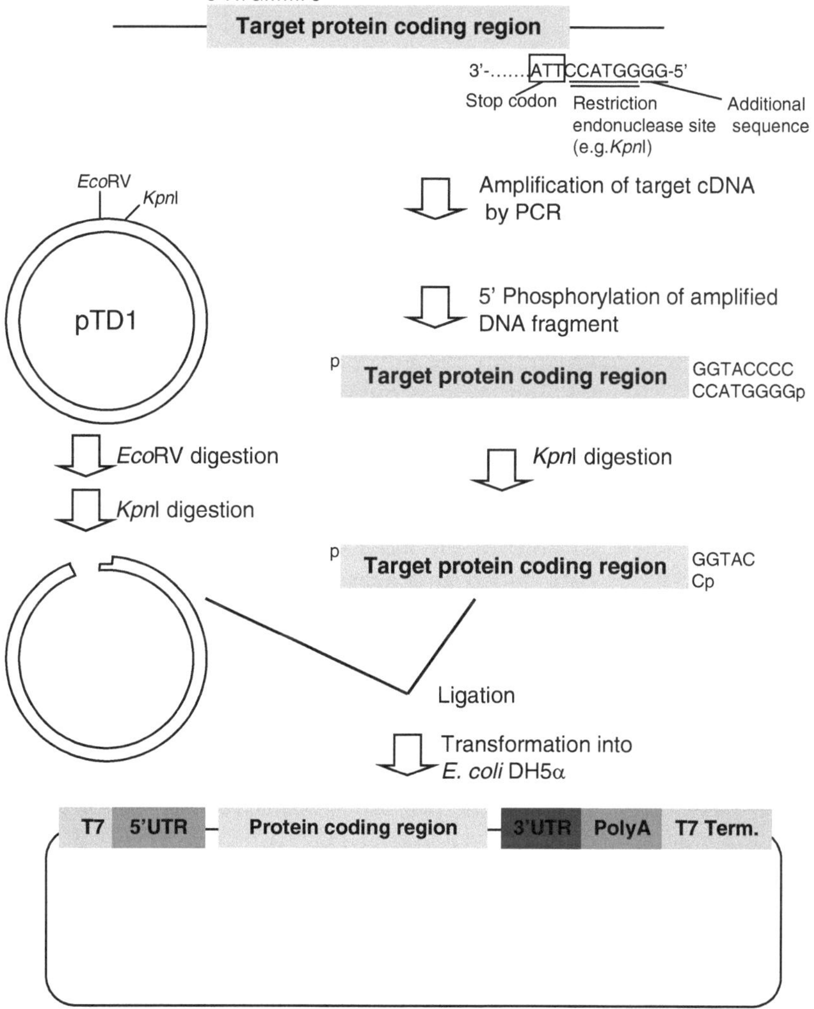

Fig. 4.2. Strategies to construct an expression clone. This figure shows procedures to construct an expression clone when a *Kpn*I site is introduced into a C-terminal primer.

3. Perform PCR using these primers, KOD-plus, and cDNA of the target gene as the template and check the size of amplified DNA by gel electrophoresis.
4. Purify the amplified DNA fragment by phenol/chloroform extraction and ethanol precipitation.
5. Treat the purified DNA fragment by T4 polynucleotide kinase at 37°C for 1 h, then purify the DNA fragment by ethanol precipitation.
6. Suspend the precipitate in distilled water, and digest the DNA fragment using the restriction enzyme that recognizes the appropriate sequence in the C-terminal primer at 37°C for 2 h.
7. Purify the DNA fragment using the MinElute® PCR Purification Kit.
8. Quantitate the DNA fragment by absorbance at 260 nm using a spectrophotometer, and use it as an insert.
9. Digest the pTD1 vector with *Eco*RV at 37°C for 2 h (see Note 5).
10. After ethanol precipitation, digest the pTD1 vector with another restriction endonuclease to produce the same cohesive end as that introduced in the C-terminal primer.
11. Purify the digested pTD1 vector using a MinElute® PCR Purification Kit.
12. Quantitate the DNA concentration using a spectrophotometer, and use it as the vector.
13. Mix the vector and the insert at a ratio of about 1:10 (mol/mol), and incubate them with T4 DNA ligase at 25°C for 5 min.
14. Transform the ligation sample into *E. coli* DH5α and incubate on LB-amp agar plates at 37°C overnight (see Note 6).
15. Cultivate single colonies of the transformants in LB-amp medium at 37°C overnight.
16. Extract the plasmids using a GenElute™ plasmid miniprep kit, linearize them with an appropriate restriction enzyme, and check their size by agarose gel electrophoresis.
17. Confirm the plasmid DNA sequence using a pTD1-161-179 primer (5′-GCAGATTGTACTGAGAGTG-3′) for N-terminal sequencing and a pTD1-412-394 primer (5′-ACAACGCACA GAATCTAGC-3′) for C-terminal sequencing. The annealing temperature of these primers is 50°C (see Notes 7 and 8).

3.2. Preparation of mRNA

1. Linearize the expression clone using an appropriate restriction endonuclease downstream from the T7 terminator sequence (see Notes 9 and 10).

2. Purify the digested expression clone by phenol/chloroform extraction and ethanol precipitation (see Note 11).
3. Dissolve the pellet in sterilized distilled water and quantitate the DNA concentration by spectrophotometer, then use as the template for mRNA synthesis (see Note 12).
4. Perform the in vitro transcription reaction using the T7 RiboMAX™ Express Large Scale RNA Production System (see Note 13) at 37°C for 30 min (see Note 14). Use 5 μg of DNA template for 100 μL of transcription reaction. Typically, about 500 μg of purified mRNA is obtained from 100 μL of transcription reaction. This yield corresponds to about 1.5 mL of translation reaction.
5. After the incubation, the synthesized mRNA should be purified immediately using NICK™ Columns, which are gel filtration columns. This purification step is necessary to remove salts and unincorporated NTPs. We recommend performing this treatment to achieve stable and highly reproducible translation reactions.
6. Procedures to set up the NICK™ Columns are as follows. First, remove the column cap and pour off the excess liquid. Rinse the column with 3 mL of sterilized distilled water. Remove the bottom cap and place it in a column stand. Equilibrate the gel with 3 mL of sterilized distilled water and flush completely. These procedures should be carried out during the in vitro transcription reaction.
7. Apply 100 μL of the transcriptional reaction mixture on top of the gel, and flush completely. If the reaction scale of the in vitro transcription is less than 100 μL, fill up the reaction mixture to 100 μL with sterilized distilled water before applying to the column.
8. Add 400 μL of sterilized distilled water, and then flush completely. Discard the eluate.
9. Before elution of the mRNA fraction, place a new 1.5 mL tube under the column.
10. Add 400 μL of sterilized distilled water, and collect the eluate.
11. Add 40 μL of 3 M potassium acetate (see Note 15) and 950 μL of ethanol to the eluate. Mix thoroughly and centrifuge for 20 min at 21,500 g, 4°C.
12. Discard the supernatant and then rinse the pellet with 70% ethanol. Do not dry the pellet completely, so it will dissolve mRNA in water easily. Dissolve the pellet in 100 μL of sterilized distilled water. If the reaction scale of the in vitro transcription is less than 100 μL, dissolve the pellet with sterilized distilled water in an equal volume of the in vitro transcription reaction.

13. After the purification, measure the absorbance at 260 nm of the purified mRNA solution using a spectrophotometer. Dilute 2 μL of the purified mRNA solution into 500 μL of TE buffer, and then measure the absorbance. Use TE as the blank. The mRNA concentration is determined by the following equation: mRNA concentration (mg/mL) = A_{260} value × 0.04 × 250.
14. About 3–6 mg/mL mRNA is usually obtained by the above method.
15. Confirm the purity and size of synthesized mRNA by gel electrophoresis as described in the following protocols.
16. Prepare a 1.0% agarose gel in 1× TAE (see Note 16).
17. Mix 10 μL of 20× MOPS buffer, 30 μL of 37% formaldehyde, and 80 μL of deionized formamide, and use this as an RNA sample buffer.
18. Mix 8 μg of the mRNA sample and 11 μL of the RNA sample buffer, and adjust to 20 μL with sterilized distilled water.
19. Treat the mRNA sample at 65°C for 15 min, and immediately place it on ice.
20. Perform electrophoresis, and visualize the mRNA sample by ethidium bromide staining. If the RNA band is smeared or not visible, possible causes may be degradation of mRNA by RNase contamination.

3.3. In Vitro Translation

1. Kit components of the Transdirect *insect cell* kit are the Insect Cell Extract (yellow cap) (see Notes 17–19), Reaction Buffer (blue cap), 4 mM Methionine (red cap), pTD1 vector (green cap), and the Control DNA (white cap).
2. Procedures to set up a translation reaction mixture should be carried out on ice.
3. Thaw the Reaction Buffer, 4 mM Methionine, and Insect Cell Extract. The reaction Buffer and 4 mM Methionine can be thawed at room temperature.
4. Assemble the reaction components (see Table 4.1; Note 20). Gently mix by pipetting up and down. If necessary, centrifuge briefly to return the sample to the bottom of the tube.
5. Incubate the translation reaction mixture at 25°C for 5 h.
6. The protein productivity of the Transdirect *insect cell* kit is about 50 μg/mL of the translation reaction mixture (see Note 21).

3.4. Detection of a Synthesized Protein

1. Generally, it is difficult to detect synthesized proteins by CBB staining. To confirm the expression of the target protein, we usually perform fluorescent labeling of the in vitro translation products using the FluoroTect™ Green$_{Lys}$ in vitro Translation Labeling System (FluoroTect) (see Note 22).

Table 4.1
Reaction components

mRNA	16 μg
4 mM methionine	1 μL
Reaction buffer	15 μL
Insect cell extract	25 μL
Sterilized distilled water	Adjust to 50 μL

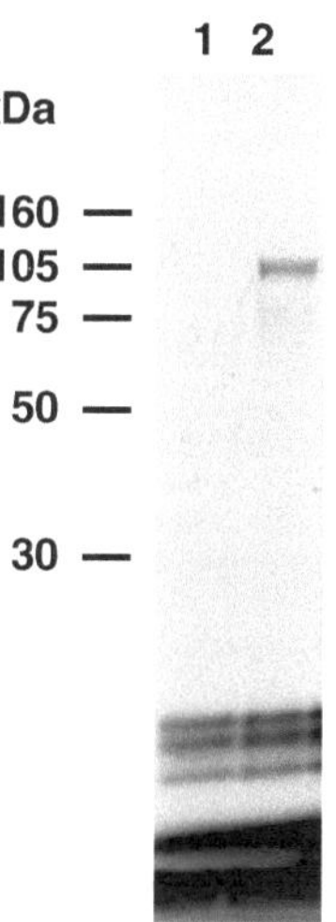

Fig. 4.3. Detection of a synthesized protein by fluorescent labeling. Cell-free protein synthesis was carried out with or without the use of mRNA transcribed from a linearized expression clone containing the β-galactosidase gene and the synthesized protein was labeled by FluoroTect. The translational reaction mixtures were resolved by 12.5% SDS–PAGE. Detection of labeled protein was performed using a laser-based fluorescent scanner (FX pro, Bio-Rad, Hercules, CA). *Lanes* 1 and 2 represent negative control (absence of mRNA) and β-galactosidase, respectively.

2. Add 1 μL of the FluoroTect solution to 50 μL of the translation reaction mixture described in Table 4.1, and incubate at 25°C for 5 h.
3. After the translation reaction, add 2 μL of SDS–PAGE loading buffer (4×) to 6 μL of the reaction mixture. Incubate at 70°C for 3 min.
4. Resolve the sample by SDS–PAGE.
5. Detect the fluorescent-labeled protein using a laser-based fluorescent scanner. An experimental example is shown in Fig. 4.3 (see Note 23).

4. Notes

1. Other DNA polymerases, which are high fidelity enzymes that do not have terminal transferase activity, also can be used.

2. It has been confirmed that protein synthesis can be performed effectively using mRNA prepared using the following kits: AmpliScribe™ T7-*Flash*™ Transcription Kit (Epicentre), AmpliScribe™ T7 Transcription Kit (Epicentre), CUGA®7 in vitro Transcription Kit (Nippon Genetech), MEGA script® T7 High Yield Transcription Kit (Ambion), RiboMAX™ Large Scale RNA Production System-T7 (Promega), RNAMaxx™ High Yield Transcription Kit (Stratagene), and the ScriptMAX™ Thermo T7 Transcription Kit (Toyobo).
3. The pTD1 vector should not be used as the expression vector for an *E. coli* cell-free protein synthesis system because this vector does not contain Shine–Dalgarno sequence.
4. We recommend introducing the *Kpn*I recognition sequence into the C-terminal primer if the target cDNA does not have a *Kpn*I site, because this strategy has been shown to have the highest cloning efficiency.
5. Generally, translation efficiency gradually decreases depending on the length between the initiating codon of the target cDNA and the polyhedrin 5′-untranslated region, which contains a translational enhancer sequence. To obtain the highest translation efficiency, the initiating codon of the target cDNA should be inserted into the *Eco*RV site of the pTD1 vector.
6. This system does not utilize blue-white selection.
7. Deletion mutation of the initiating codon (especially "A") has sometimes been observed for this cloning strategy.
8. To clarify whether mutations occur during PCR, the overall nucleotide sequence of the insert DNA should be confirmed.
9. For linearization of expression clones, we recommend using *Hin*dIII or *Not*I. Restriction enzymes, *Cfr*10I, *Eco*52I, *Eco* T14I, *Nde*I, *Pvu*II, *Sca*I, and *Stu*I may be used.
10. PCR-generated DNA templates can be used in transcription reactions. In such cases, the following primers are recommended: A pTD1-161-179 primer (5′-GCAGATTGTACTGAGAGTG-3′) and a pTD1-845-827 primer (5′-GGAAACAGCTATGACCATG-3′). Their annealing temperature is 50°C.
11. This step is very important to avoid RNase contamination.
12. At least 125 μg/mL of the linearized DNA template is required for in vitro transcription reactions.
13. Before use, RiboMAX™ Express T7 2× Buffer must be dissolved completely by warming the buffer at 37°C and mixing well.
14. The suggested incubation times should be adhered to. In the case of a long template (more than about 2 kb), an excessive reaction time may cause precipitation. In this case, mRNA will not be collected. To avoid this problem: I: shorten the reaction

time to 20 min, II: decrease the quantity of DNA template to 70–80%, III: use a PCR-generated DNA template.

15. Do not use sodium acetate for precipitation of synthesized mRNA. Sodium ion inhibits the translation reaction.
16. We usually use a nondenaturing agarose gel.
17. We confirmed that the Insect Cell Extract is stable against freeze-thawing up to eight times. After use, the extract should be immediately stored at –80°C.
18. Insect Cell Extract is sensitive to CO_2. After opening the package, avoid prolonged exposure to CO_2 (e.g., dry ice).
19. We confirmed that the Insect Cell Extract has the ability to perform eukaryote-specific protein modifications, such as *N*-myristoylation (2) and prenylation (3). To obtain such modified proteins effectively, specific substrates for each protein modification should be added to the translation reaction mixture.
20. The addition of RNase inhibitor (50 units) to the translation reaction mixture (50 μL) may improve translational efficiency (Recommended products: Promega Code No. N2611).
21. Generally, it is difficult to synthesize membrane proteins having multiple transmembrane domains.
22. Radio-isotope labeling or western blotting may also be used to detect synthesized proteins.
23. For detection of proteins having molecular masses less than 20 kDa, it is necessary to treat the translation reaction mixture with RNase A, because unincorporated FluoroTect tRNA migrates at about 20 kDa.

Acknowledgments

We thank Dr. Osamu Ohara and Dr. Takahiro Nagase, Kazusa DNA Research Institute, for valuable discussions. We also thank Dr. Sumiharu Nagaoka, Kyoto Institute of Technology, Dr. Toshihiko Utsumi, University of Yamaguchi, and Dr. Susumu Tsunasawa, Shimadzu Corporation, for helpful discussions.

References

1. Ezure, T., Suzuki, T., Higashide, S., Shintani, E., Endo, K., Kobayashi, S., Shikata, M., Ito, M., Tanimizu, K., and Nishimura, O. (2006) Cell-free protein synthesis system prepared from insect cells by freeze-thawing. *Biotechnol. Prog.* **22**, 1570–1577.
2. Suzuki, T., Ito, M., Ezure, T., Shikata, M., Ando, E., Utsumi, T., Tsunasawa, S., and Nishimura, O. (2006) N-Terminal protein modifications in an insect cell-free protein synthesis system and their identification by mass spectrometry. *Proteomics* **6**, 4486–4495.

3. Suzuki, T., Ito, M., Ezure, T., Shikata, M., Ando, E., Utsumi, T., Tsunasawa, S., and Nishimura, O. (2007) Protein prenylation in an insect cell-free protein synthesis system and identification of products by mass spectrometry. *Proteomics* 7, 1942–1950.

4. Suzuki, T., Ito, M., Ezure, T., Kobayashi, S., Shikata, M., Tanimizu, K., and Nishimura, O. (2006) Performance of expression vector, pTD1, in insect cell-free translation system. *J. Biosci. Bioeng.* **102**, 69–71.

Chapter 5

Cell-Free Protein Synthesis Systems with Extracts from Cultured Human Cells

Satoshi Mikami, Tominari Kobayashi, and Hiroaki Imataka

Abstract

Cell-free protein synthesis systems have been established with extracts from cultured human cells, HeLa, and hybridoma cells. The former cell line is used to prepare extracts for robust translation, whereas the extract from the latter is primarily employed for expression of glycoproteins. Productivity of both systems can be enhanced by addition of K3L and GADD34, factors that diminish phosphorylation of eIF2α. The coupled transcription/translation system is also available as a convenient tool, particularly for the production of large recombinant proteins.

Key words: Human cells, Translation initiation, Glycosylation, Hybridoma, HeLa cells, GADD34, K3L, Coupled transcription/translation

1. Introduction

Among mammalian cell-based in vitro protein synthesis systems, the rabbit reticulocyte lysate (RRL) may be the most popular system. However, commercially available RRLs are expensive with varied activities depending on supplied lots, and preparation of RRL by a researcher's own hands is not an easy task. Furthermore, RRL is not able to glycosylate proteins unless combined with microsomes from dog pancreas, which are not easily and reliably obtained. Recent progress in the cell-free protein synthesis systems from mammalian (human) cultured cells has been solving these problems (1–3). Also, human cells-derived in vitro transcription-coupled translation system is now available with such a high performance as to synthesize very large (~250 kDa) proteins in a reasonable quantity (50–180 μg/mL) (4) (Fig. 5.1). In this chapter, we describe the procedure of

Y. Endo et al. (eds.), *Cell-Free Protein Production: Methods and Protocols,* Methods in Molecular Biology, vol. 607
DOI 10.1007/978-1-60327-331-2_5, © Humana Press, a part of Springer Science+Business Media, LLC 2010

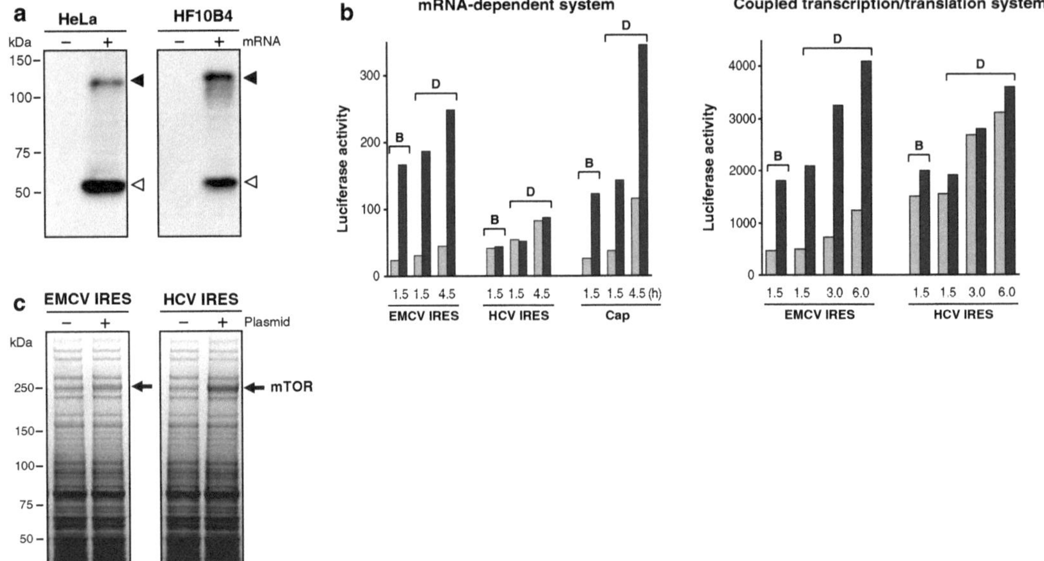

Fig. 5.1. Synthesis of proteins by mammalian cell-derived cell-free systems. (**a**) Glycosylation. EMCV-IRES-HIVgp120-HA-polyA mRNA was incubated in the HeLa or HF10B4 (hybridoma) cell-derived cell-free mRNA-dependent translation system (batch method) supplemented with GADD34/K3L. Following SDS–PAGE, translated products were detected by Western blot with anti-HA antibody. *Open* and *closed arrowheads* indicate unglycosylated and glycosylated forms, respectively. (**b**) Renilla luciferase was synthesized by the HeLa cell-derived cell-free mRNA-dependent translation system (EMCV-IRES, HCV-IRES, or cap dependent system, *left panel*) or by the cell-free coupled transcription/translation system (EMCV-IRES or HCV-IRES dependent system, *right panel*) not supplemented (*gray bar*) or supplemented (*black bar*) with GADD34/K3L (B: batch method; D: dialysis method). Following incubation, an aliquot was used for luciferase assay. (**c**) Robustness. The mTOR protein was synthesized by the HeLa cell-derived cell-free coupled transcription/translation system (EMCV-IRES or HCV-IRES dependent system using the dialysis method) supplemented with GADD34/K3L. Following incubation for 12 h, an aliquot was separated by SDS–PAGE followed by CBB-staining.

culturing human cells (HeLa and hybridoma cells) to the cell-free protein synthesis systems (Figs. 5.2 and 5.3).

2. Materials

2.1. Cell Culture

1. HeLa S3 cells and HF10B4 cells (human hybridoma) are purchased from RIKEN BRC, Japan (see Note 1).
2. Minimum essential medium for suspension culture S-MEM (Invitrogen), SMEM (Sigma-Aldrich) or JMEM (Sigma-Aldrich) is used for culturing HeLa S3, while E-RDF medium (Kyokuto, Japan) is for HF10B4. Store at 4°C.
3. Fetal calf serum and calf serum are obtained from Sigma-Aldrich, Invitrogen, or other manufactures (see Note 2). Store in 50 mL aliquots at –20°C.

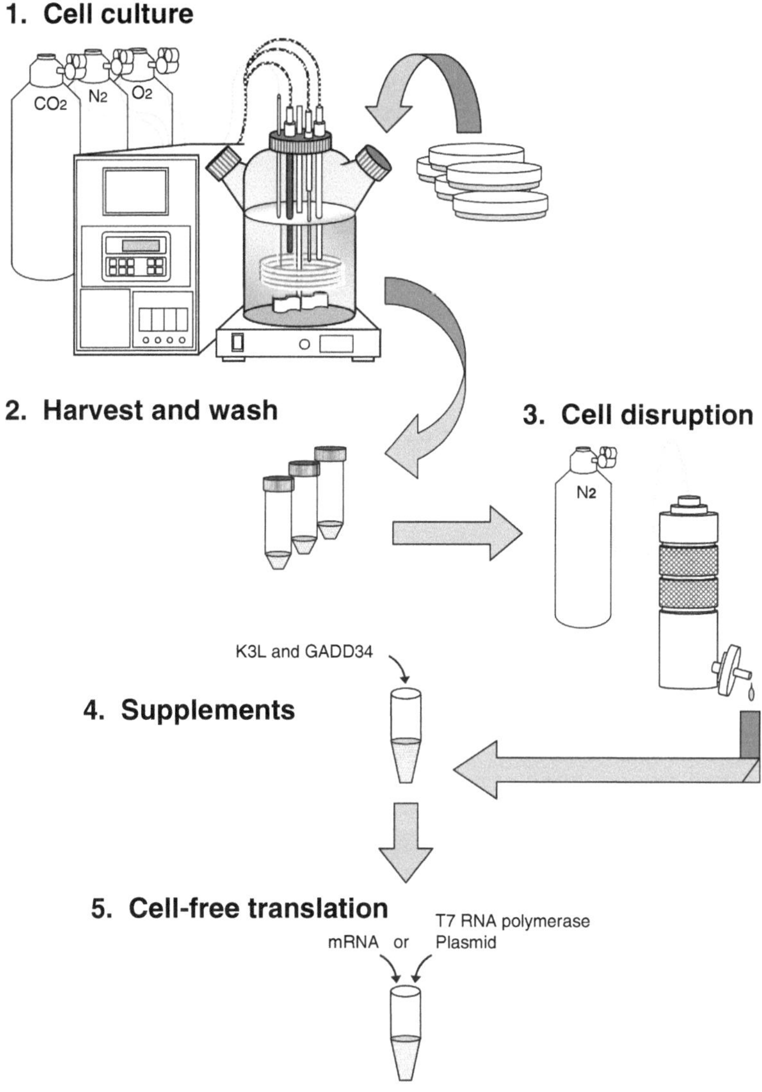

Fig. 5.2. Cartoon depicting cell culture to cell-free translation.

4. GlutaMAX (Invitrogen). Store at 4°C.
5. Cellmaster Model 1700 (Wakenyaku, Japan).

2.2. Cell Extracts

1. Washing buffer: 35 mM HEPES-KOH, pH 7.5, 140 mM NaCl, and 11 mM glucose. Store at 4°C.
2. Extraction buffer: 20 mM HEPES-KOH, pH 7.5, 45 mM potassium acetate, 45 mM KCl, 1.8 mM magnesium acetate, and 1 mM DTT (dithiothreitol). Store at 4°C without DTT. Add DTT just before use.

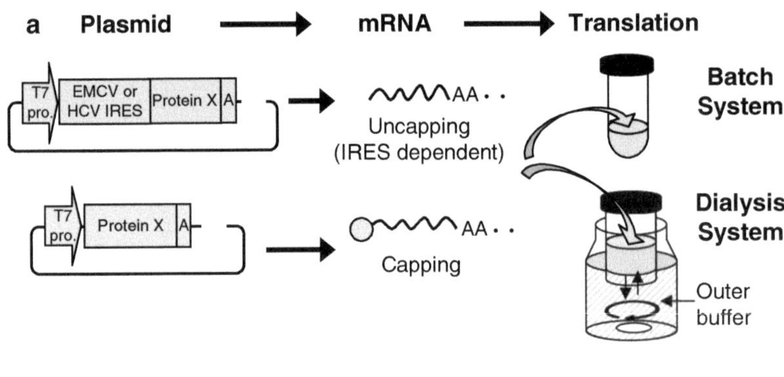

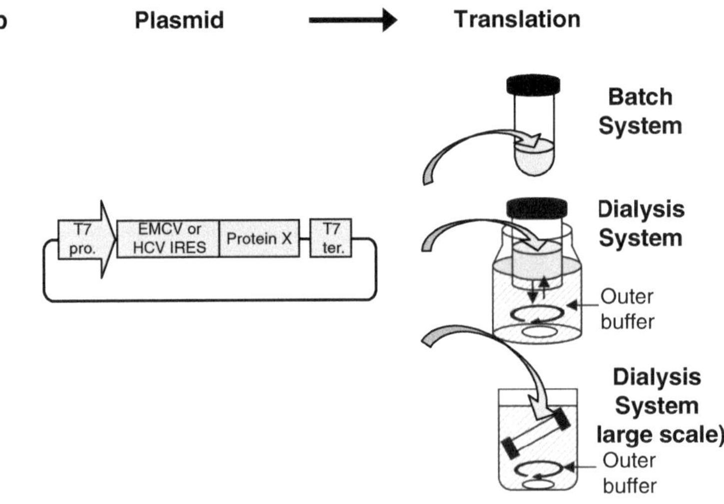

Fig. 5.3. Cartoon depicting different incubation systems. (**a**) mRNA-dependent system. mRNA is synthesized in vitro and purified. The purified mRNA is incubated either by the batch system or the dialysis system. (**b**) Coupled transcription/translation system. An appropriate plasmid is directly incubated either by the batch system or the dialysis system.

3. High potassium buffer: 20 mM HEPES-KOH, pH 7.5, 945 mM potassium acetate, 945 mM KCl, 1.8 mM magnesium acetate, and 1 mM DTT. Store at 4°C without DTT. Add DTT just before use.
4. Mini-Bomb cell disruption chamber (KONTES, USA).

2.3. mRNA and Plasmids

1. mRNA for mRNA-dependent translation: T7, T3, or SP6 RNA polymerase; plasmids harboring T7, T3, or SP6 RNA polymerase promoter (Fig. 5.3) (see Note 3).
2. Plasmids for transcription-coupled translation system: pUC-T7-EMCV-IRES and pUC-T7-HCV-IRES (4) (Fig. 5.3) (see Note 4).
3. mRNA synthesis kits: RiboMAX large scale RNA production system (Promega) or mMESSAGEmMACHINE (Ambion).

4. RNA purification column: Chroma spin+TE-30 column (Clontech).
5. Reagents for purifying plasmids: TE (10 mM Tris–HCl, pH 7.5, 1 mM EDTA), Proteinase K (TAKARA).

2.4. Cell-Free Translation

1. ATP, GTP, UTP, CTP, and creatine phosphate (Sigma-Aldrich): Each nucleotide and creatine phosphate are dissolved in 240 mM HEPES-KOH, pH 7.5 to the final concentrations of 200 mM each and 1.0 M, respectively. Store at −80°C.
2. Twenty amino acids: The amino acid mixture consisting of 20 amino acids is made by mixing an equal volume of MEM amino acids solution (×50) (Invitrogen), MEM nonessential amino acids solution (×50) (Invitrogen), and 20 mM L-glutamine solution (Invitrogen). The final concentrations of each amino acid are: 12 mM L-arginine; 6.7 mM L-glutamine, L-isoleucine, L-leucine, L-lysine, L-threonine, and L-valine; 3.3 mM L-alanine, L-asparagine, L-aspartic acid, L-glutamic acid, L-glycine, L-histidine, L-phenylalanine, L-proline, L-serine, and L-tyrosine; 1.7 mM L-cystine, and L-methionine; 0.8 mM L-tryptophan. The mixture is divided into aliquots and stored at −80°C.
3. Creatine kinase (Roche) is dissolved in 20 mM HEPES-KOH, pH 7.5, 50% glycerol to the final concentration of 30 mg/mL. Store at −30°C.
4. Calf liver tRNA (Novagen): Store at −80°C.
5. K3L and GADD34 proteins are expressed and purified as reported (2). Store at −80°C.
6. Mixture-1: Mixture-1 consists of 6 μL of GADD34 (0.2 mg/mL), 6 μL of K3L (0.4 mg/mL). Store at −80°C (see Note 9).
7. Mixture-2: Mixture-2 (650 μL) consists of 56 μL of magnesium acetate (63 mM), 354 μL of potassium acetate (258 mM for cap dependent system 157 mM for the EMCV-IRES-dependent system or 564 mM for the HCV-IRES-dependent system), 82.5 μL of DTT (100 mM), and 158 μL of HEPES-KOH (400 mM), pH 7.5. Store at 4°C without DTT. Add DTT just before use.
8. Mixture-3: Mixture-3 (180 μL) consists of 11.3 μL of ATP (200 mM), 1.1 μL of GTP (200 mM), 36 μL of creatine phosphate (1 M), 18 μL of creatine kinase (6 mg/mL), 18 μL of calf liver tRNA (9 mg/mL, Novagen), 18 μL of the mixture of 20 amino acids, and 77.6 μL of H_2O. Store at −80°C (see Note 9).
9. Mixture-4: Mixture-4 consists of the same ingredients as in mixture-2 except that the concentration of magnesium acetate is 98 mM. Store at 4°C without DTT. Add DTT just before use.

10. Mixture-5: Mixture-5 (180 μL) consists of 11.3 μL of ATP (200 mM), 7.5 μL each of GTP, CTP, and UTP (200 mM each), 36 μL of creatine phosphate (1 M), 18 μL of creatine kinase (6 mg/mL), 18 μL of calf liver tRNA (9 mg/mL, Novagen), 18 μL of a mixture of 20 amino acids, 18 μL of spermidine (50 mM), 20 μL of T7 RNA polymerase (1.0 mg/mL) (see Note 12), and 18.2 μL of H_2O. Store at –80°C (see Note 9).
11. Dialysis membrane (SPECTRUM; molecular weight cut-off 50,000, regenerated cellulose).
12. Outer buffer-1: The outer buffer-1 consists of 1,480 μL of H_2O, 12.5 μL of ATP (200 mM), 1.25 μL of GTP (200 mM), 40 μL of creatine phosphate (1 M), 20 μL of the mixture of 20 amino acids, 10 μL of DTT (1 M), 10 μL of EGTA, pH 7.5 (300 mM), 5.4 μL of magnesium acetate (1 M), 0.3 mL of HEPES-KOH (400 mM), pH 7.5, 13.3 μL of KCl (3 M), and 108 μL of potassium acetate (1.87 M for cap dependent system, 1.5 M for the EMCV-IRES-dependent system or 3 M for the HCV-IRES-dependent system). Prepare every time before use.
13. Outer buffer-2: The outer buffer-2 consists of 1,453 μL of H_2O, 12.5 μL of ATP (200 mM), 8.3 μL each of GTP, CTP, and UTP (200 mM each), 40 μL of creatine phosphate (1 M), 20 μL of the mixture of 20 amino acids, 10 μL of DTT (1 M), 10 μL of EGTA, pH 7.5 (300 mM), 1 μL of spermidine (1 M), 7.6 μL of magnesium acetate (1 M), 0.3 mL of HEPES-KOH (400 mM), pH 7.5, 13.3 μL of KCl (3 M), and 108 μL of potassium acetate (1.5 M for the EMCV-IRES-dependent system or 3 M for the HCV-IRES-dependent system). Prepare every time before use.

3. Methods

We describe HeLa cell and hybridoma-derived cell-free protein synthesis systems. For robustness of translation, the HeLa cell-derived system is the primary choice. Although both systems are able to glycosylate proteins, the hybridoma-dependent system is more efficient than the HeLa cell-based one for this protein modification (Fig. 5.1). To enhance protein synthesis in both systems, addition of recombinant K3L and GADD34 proteins is very effective (Figs. 5.1 and 5.2). eIF2α, an essential translation initiation factor, is heavily phosphorylated in mammalian cell extracts, being partially inactivated therein. K3L prevents phosphorylation of eIF2α, and GADD34 enhances dephosphorylation of eIF2α.

One may choose an mRNA-dependent system or a coupled transcription/translation system (Fig. 5.3), although the latter system is recommended for convenience. Also, for a better yield of proteins, the dialysis system, which continuously supplies the substrates and energy source for protein synthesis and removes waste products through a dialysis membrane, is recommended rather than the conventional closed system (batch system) (Fig. 5.1).

3.1. Cell Culture

3.1.1. HeLa Cells

1. Culture HeLa S3 cells at 37°C in S-MEM, SMEM, or JMEM supplemented with 10% heat-inactivated calf serum, penicillin (1 unit/mL), and streptomycin (0.1 mg/mL) and GlutaMAX (2 mM) (see Note 5) using a spinner flask connected with a controlling system Cellmaster Model 1700 (see Note 6) with the control values of temperature (37°C), pH (7.2), oxygen density (6.7 ppm), and stirring speed (50 rpm).
2. Harvest the cells when the cell density reaches $0.8–1.0 \times 10^6$ cells/mL (see Note 7).

3.1.2. HF10B4 Cells

1. Culture HF10B4 cells at 37°C in E-RDF medium supplemented with 10% heat-inactivated fetal calf serum and GlutaMAX (2 mM) (see Note 5) using a spinner flask connected with Cellmaster Model 1700 (see Note 6) with the control values of temperature (37°C), pH (7.0), oxygen density (6.7 ppm), and stirring speed (20 rpm).
2. Harvest the cells when the cell density reaches $1.0–1.5 \times 10^6$ cells/mL (see Note 7).

3.2. Preparation of Cell Extracts

1. Wash the cells three times with the washing buffer and once with the extraction buffer.
2. Resuspend the cell pellet in an equal volume of the extraction buffer (approximately 3.0×10^8 cells/mL).
3. Disrupt cells by nitrogen pressure (1.0 MPa, 30 min) in the Mini-Bomb cell disruption chamber.
4. Mix the cell homogenates with 1/29 volume of the high potassium buffer.
5. Centrifuge twice at $1,200 \times g$ for 5 min at 4°C and recover the supernatant.
6. Divide the supernatant (24–28 mg protein/mL) into aliquots, and freeze them in liquid nitrogen.
7. Store the frozen aliquots at −80°C (see Note 8).

3.3. mRNA and Plasmids

3.3.1. mRNA Preparation

1. Construct a plasmid harboring T7, T3, or SP6 RNA polymerase promoter and the coding region of a protein + or− poly (A) tail.
2. Digest the plasmid with a restriction enzyme 3′ downstream of the coding region or the poly(A) tail.

3. Synthesize RNA using the digested plasmid as the template with RiboMAX large scale RNA production system or mMESSAGEmMACHINE (see Note 13).
4. Purify the synthesized RNAs by using Chroma spin+TE-30 column.
5. Check the integrity of each RNA preparation by electrophoresis on formaldehyde–agarose gels.

3.3.2. Plasmid Preparation for the Coupled System

1. Construct and amplify a plasmid encoding T7 RNA polymerase promoter, EMCV or HCV-IRES, the coding region of a protein and T7 RNA polymerase terminator in this order.
2. Purify the plasmid using any commercially available kit, and dissolve it in 450 μL TE.
3. Add 25 μL SDS (10%), 10 μL EDTA (0.5 M), and 10 μL proteinase K (20 mg/mL) and incubate for 30–60 min at 37–50°C to inactivate possibly contaminating RNases.
4. Purify the plasmid by treating with phenol and chloroform followed by ethanol precipitation.

3.4. Cell-Free mRNA-Dependent Translation

3.4.1. Batch Methods

1. Preincubate the HeLa or HF10B4 cell extract (7.5 μL) with mixture-1 (1.2 μL) and mixture-2 (6.5 μL) at room temperature for 10 min (Fig. 5.2) (see Note 10).
2. Add mixture-3 (1.8 μL) and mRNA (1.0 μL, 36 ng/μL final concentration) (see Note 14).
3. Incubate the mixture (18 μL, total volume) at 32°C for 1.0 h.

3.4.2. Dialysis Methods

1. Inject the incubation mixture (180 μL, total volume) described in Subheading 3.4.1 into a chamber with a dialysis membrane (molecular weight cut-off 50,000, regenerated cellulose).
2. Incubate the chamber at 32°C for 3–12 h being dialyzed against the outer buffer-1 (2 mL) (see Note 11).

3.5. Cell-Free Coupled Transcription/Translation

3.5.1. The Batch Method

1. Preincubate the HeLa or HF10B4 cell extract (7.5 μL) with mixture-1 (1.2 μL) and mixture-4 (6.5 μL) at room temperature for 10 min (Fig. 5.2) (see Note 10).
2. Add mixture-5 (1.8 μL) and a plasmid (1.0 μL) (15 ng/μL, final concentration) (see Note 15).
3. Incubate the mixture (18 μL, total volume) at 32°C for 1–3 h.

3.5.2. Dialysis Method

1. Inject the incubation mixture (180 μL, total volume) described in Subheading 3.5.1 into a chamber with a dialysis membrane (molecular weight cut-off 50,000, regenerated cellulose).
2. Incubate the chamber at 32°C for 3–12 h being dialyzed against the outer buffer-2 (2 mL) (see Note 16).

4. Notes

1. When purchased, cells should be propagated to make many (more than 50) frozen cell stocks. Avoid continued culture for longer than 1 month.
2. It is important to use the serum lot which promotes cells to grow with the doubling time of ~1 day at the concentration of 10%.
3. Capping and poly (A) tailing increase translation efficiency. A capping kit is available from Ambion. A poly (A) sequence (more than 30 A) is incorporated 3′ downstream of the coding region. When mRNA with the encephalomyocarditis virus (EMCV) internal ribosome entry site (IRES) or the hepatitis C virus (HCV) IRES is used, capping is not necessary (Fig. 5.3).
4. EMCV-IRES and HCV-IRES are employed to enhance translation in the transcription-coupled translation system. These plasmids can be also used to synthesize EMCV-IRES and HCV-IRES mRNA for the mRNA-dependent translation (Fig. 5.3).
5. Addition of GlutaMAX to the culture medium helps cells grow, and leads to an enhanced protein synthesis in the cell-free systems, even when L-glutamine is already included.
6. If an appropriate controlling system for culturing cells is not available, culturing cells on Petri dishes (20–25 mL medium per 150 mm dish) is recommended.
7. It is very important to recover cells when they are in the logarithmic growth phase.
8. The extracts can be kept at –80°C at least for 6 months without a loss of translation activity.
9. Repeated freezing–thawing should be avoided.
10. Preincubation with K3L/GADD34 prior to addition of mixture-3 is important, because ATP/creatine phosphate present in mixture-3 enhances phosphorylation of eIF2α (2).
11. Supplementation of the incubation mixture (120 μL) with mRNA (4.3 μg, each time) every 3–6 h increases the yield of the expressed protein.
12. T7 RNA polymerase can be expressed in bacteria and purified to homogeneity with a comparable activity to commercially available T7 RNA polymerase (4).
13. When capped mRNAs are synthesized, m^7GpppG is included at a four to eightfold molar excess relative to GTP in the transcription reaction.

14. The final concentrations of ingredients in the incubation mixture are: ATP, 1.25 mM; GTP, 0.125 mM; creatine phosphate, 20 mM; creatine kinase, 60 μg/mL; calf liver tRNA, 90 μg/mL; 20 amino acids, 8–120 μM; magnesium acetate, 2.7 mM; potassium (potassium acetate plus KCl), 100, 120 or 180 mM; DTT, 5 mM; and HEPES-KOH, 51 mM.
15. The final concentrations of each ingredient in the incubation mixture are the same as those described for the mRNA-dependent translation system (see Note 14) except the following changes: GTP, CTP, and UTP, 0.83 mM each; spermidine, 0.5 mM; T7 RNA polymerase, 11.1 μg/mL; magnesium acetate, 3.8 mM; and HEPES-KOH, 54 mM.
16. For a preparative purpose, 1.0 mL of the reaction mixture in dialysis tube is dialyzed against 20–40 mL of the outer buffer-2.

References

1. Kobayashi T, Mikami S, Yokoyama S, Imataka H. (2007) An improved cell-free system for picornavirus synthesis. *J. Virol. Methods* **142**, 182–188.
2. Mikami S, Kobayashi T, Yokoyama S, Imataka H. (2006) A hybridoma-based in vitro translation system that efficiently synthesizes glycoproteins. *J. Biotechnol.* **127**, 65–78.
3. Mikami S, Masutani M, Sonenberg N, Yokoyama S, Imataka H. (2006) An efficient mammalian cell-free translation system supplemented with translation factors. *Protein Expr. Purif.* **46**, 348–357.
4. Mikami S, Kobayashi T, Masutani M, Yokoyama S, Imataka H. (2008) A human cell-derived in vitro coupled transcription/translation system optimized for production of recombinant proteins. *Protein Expr. Purif.* **62**, 190–198.

Chapter 6

Analysis of Protein Functions Through a Bacterial Cell-Free Protein Expression System

Takanori Kigawa

Abstract

Cell-free protein synthesis is a suitable protein expression method for the high-throughput use because a PCR-amplified linear DNA fragment is utilized as a template for protein synthesis without any cloning procedures. We have developed a two-step PCR method for high-throughput and robust production of linear templates ready for cell-free protein synthesis. A high-throughput protein expression method has been established by combining the batch-mode cell-free protein synthesis with the two-step PCR, which is performed on multiwell plates, and is thus adapted for robotics. In this chapter, our two-step PCR method and the batch-mode cell-free protein synthesis are described.

Key words: Prokaryotic coupled transcription–translation, Two-step PCR, T7 RNA polymerase, High-throughput protein expression, Automation

1. Introduction

In contemporary genome scale researches, the demand of high-throughput and flexible protein expression has increased. The simplicity and robustness of experimental procedures are especially important for massive, high-throughput experiments using robotics. One of the keys to achieve these specifications is the protein expression method itself, and the other is the template DNA preparation method. For the protein expression method, cell-free protein synthesis is most suitable for high-throughput use.

First, proteins can be produced from a linear template DNA (1, 2). Second, the reaction condition can be optimized for a target protein by adding or depleting components. For example, stable-isotope-labeled proteins were produced and subjected to structural analysis using NMR (e.g., (3–5)). Selenomethionine-substituted

Y. Endo et al. (eds.), *Cell-Free Protein Production: Methods and Protocols,* Methods in Molecular Biology, vol. 607
DOI 10.1007/978-1-60327-331-2_6,

proteins for X-ray crystallography were synthesized and structures were solved (e.g., (6–8)). By the addition of chaperons or protein disulfide isomerases, some proteins were expressed in soluble and active forms (9, 10). Metal-binding proteins can be produced in the native form simply by adding the suitable ligand (11).

For the preparation of a template DNA for protein expression, complicated procedures such as cloning, including ligation, transformation, and culture, are used in many cases. Gateway cloning or ligation-independent cloning can simplify these procedures; however, these methods still require transformation and time-consuming culture steps. We have developed a simple and robust PCR protocol to prepare template DNA for high-throughput cell-free protein synthesis, which does not include any DNA purification step ((1), Fig. 6.1). First, the open reading frame (ORF) or a domain fragment of the ORF is amplified by PCR using gene-specific primers and/or a universal primer. Second, a

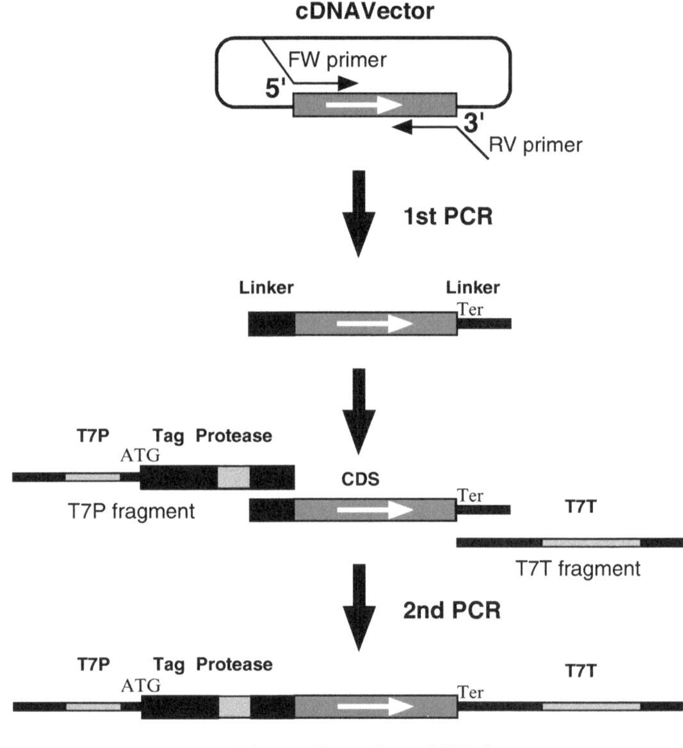

Fig. 6.1. Schematic of the two-step PCR method. First, the open reading frame (ORF) or a domain fragment of ORF is amplified by PCR using gene-specific primers and/or an universal primer. Second, the first PCR product, a T7 promoter fragment with the tag-coding sequence, a T7 terminator fragment, and the universal primer are subjected to overlapping PCR, and the construct that expresses a fusion protein under the control of the T7 promoter is produced.

T7 promoter fragment with the tag-coding sequence, a T7 terminator fragment, and the universal primer, which are non gene-specific and thus in common with all the genes, are subjected to an overlapping PCR with the first PCR product as the template. This produces a construct that expresses a fusion protein under the control of the T7 promoter. This method has an advantage that different tags can be easily attached to the same first PCR product just by changing the tag-coding fragment in the second PCR step.

Both the two-step PCR and cell-free protein synthesis are suitable for robotics, because they are constructed only from simple procedures and are free from the cloning process. Thus, we established a practical platform to produce and analyze proteins on a genome scale in a high-throughput manner (12–14).

2. Materials

2.1. First-Step PCR Reaction (for 96 Reactions in a 96-Well Plate)

1. Expand-HiFi enzyme and Expand Hi-Fi buffer (Roche) (see Note 1).
2. 2 mM each of dNTPs mixture (TOYOBO, etc.).
3. 0.5 mM gene-specific forward (FW) primer (see Note 2).
4. 0.5 mM gene-specific RV primer (see Note 2).
5. cDNA clones (see Note 3).
6. 96-Well PCR plates (e.g., AB gene AB-0900).
7. Cap stripes for PCR plate (e.g., AB gene AB-0265).
8. A PCR machine.
9. Aluminum blocks for 96-well PCR plate.
10. Reservoirs.
11. Multichannel pipette (recommended).

2.2. Second PCR Reaction (for 96 Reactions in a 96-Well Plate)

1. Expand-HiFi enzyme and Expand Hi-Fi buffer (Roche).
2. 2 mM each of dNTPs mixture (TOYOBO, etc.).
3. 2 nM T7 promoter (T7P) fragment (see Note 4).
4. 2 nM T7 terminator (T7T) fragment (see Note 4).
5. 0.1 mM U2 universal primer (see Note 5).
6. 0.1 × TE: 1 mM Tris buffer (pH 7.5) containing 0.1 mM EDTA.
7. First-step PCR reaction product.
8. 96-Well PCR plates (e.g., AB gene AB-0900).

9. Cap stripes for PCR plate (e.g., AB gene AB-0265).
10. A PCR machine.
11. Aluminum blocks for 96-well PCR plate.
12. Reservoirs.
13. Multichannel pipette (dispensable but highly recommended).

2.3. Cell-Free Reaction (for 96 Reactions in a 96-Well Plate)

1. *E. coli* S30 extract (see Chapter 1).
2. LMCP: 160 mM HEPES-KOH buffer (pH 7.5) containing 10.7% (w/v) PEG 8000, 534 mM potassium glutamate, 5 mM DTT, 3.47 mM ATP (pH 7.0), 2.4 mM GTP (pH 7.0), 2.4 mM CTP (pH 7.0), 2.4 mM UTP (pH 7.0), 96 μg/mL Folinic acid•Ca, 1.78 mM cAMP•Na, 74 mM NH_4OAc, and 214 mM creatine phosphate: Store at –20°C.
3. 17.5 mg/mL *E. coli* total tRNA (Roche): Store at –20°C.
4. 1.6 M $Mg(OAc)_2$. Store at –20°C.
5. 20 mM amino acid mixture/10 mM DTT: 20 mM each of glutamine, asparagine, arginine, tryptophan, lysine, histidine, phenylalanine, isoleucine, glutamic acid, proline, asparatic acid, glycine, valine, serine, alanine, threonine, cystaeine, tyrosine, leucine, methionine and 10 mM DTT. Store at –20°C.
6. 3.75 mg/mL creatine kinase. Store at –80°C.
7. 10 mg/mL T7 RNA polymerase (see Chapter 1). Store at –20°C.
8. Second-step PCR reaction product.
9. 96-Well PCR plates (e.g., AB gene AB-0900).
10. Cap stripes for PCR plate (e.g., AB gene AB-0265).
11. An incubator. PCR machine is recommended.
12. Aluminum blocks for 96-well PCR plate.
13. Reservoirs.
14. Multichannel pipette (dispensable but highly recommended).

3. Methods

The protocol described in the following is for 96 reactions in a 96-well plate.

3.1. First-Step PCR Reaction

1. Chill the aluminum block on ice.
2. Thaw all of the reagents on ice.

3. Dispense 3 μL of a cDNA clone to a well of a 96-well PCR plate.
4. Dispense 2 μL each of FW and RV primers to corresponding well.
5. Repeat **steps 3** and **4** for each cDNA clone and corresponding primers.
6. Prepare Premixed solution 1 on ice according to Table 6.1. Do NOT vortex.
7. Dispense 13 μL of Premixed solution 1 to every well of a PCR plate. The use of a reservoir is recommended.
8. Cap the PCR plate with Cap stripes.
9. Place the PCR plate on the PCR machine.
10. Run the program as follows: 94°C for 2 min, 40 cycles of 94°C for 30 s (denaturation), 60°C for 30 s (annealing), and 72°C for 1 min (extension; after the 20th cycle, the duration is prolonged for 5 s per cycle), 72°C for 7 min.
11. An example result is shown in Fig. 6.2.

3.2. Second-Step PCR Reaction

1. Chill the aluminum block on ice.
2. Thaw all of the reagents on ice.
3. Mix 5 μL of the first-step PCR reaction product and 20 μL of 0.1 × TE to prepare 5× dilution of the first-step PCR products in a PCR plate on the aluminum block.
4. Dispense 5 μL of the 5× dilution of the first-step PCR products to another PCR plate on the aluminum block.
5. Prepare Premixed solution 2 on ice according to Table 6.2. Do NOT vortex.
6. Dispense 15 μL of the Premixed solution 2 to every well of the PCR plate. The use of a reservoir is recommended.
7. Cap the PCR plate with Cap stripes.

Table 6.1
Composition of the premixed solution 1 for the first-step PCR (96 reactions)

2 mM dNTPs	192 μL
Expand Hi-Fi buffer	192 μL
Expand-HiFi enzyme (3.5 U/μL)	14.4 μL
Water	849.6 μL
Total	1,248 μL

8. Place the PCR plate on the PCR machine.
9. Run the program as follows: 94°C for 2 min, 30 cycles of 94°C for 30 s (denaturation), 60°C for 30 s (annealing), and 72°C for 1.5 min (extension; after the tenth cycle, the duration is prolonged for 5 s per cycle), 72°C for 7 min.
10. An example result is shown in Fig. 6.2.

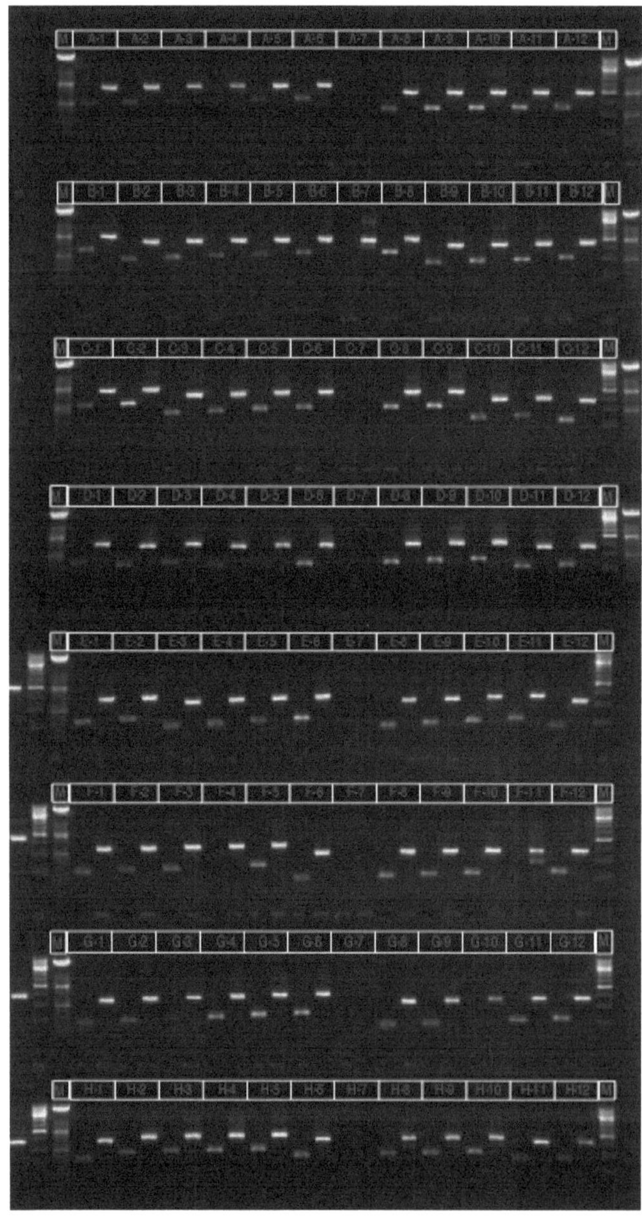

Fig. 6.2. Agarose electrophoresis of the two-step PCR products. Two lanes are paired for each cDNA (*left lane*: the First PCR product, *right lane*; the Second PCR product).

Table 6.2
Composition of the premixed solution 2 for the second-step PCR (96 reactions)

2 nM T7P fragment	48 μL
2 nM T7T fragment	48 μL
0.1 mM U2 primer	19.2 μL
2 mM dNTPs	192 μL
Expand Hi-Fi buffer	192 μL
Expand-HiFi enzyme (3.5 U/μL)	14.4 μL
Water	926.4 μL
Total	1,440 μL

Table 6.3
Composition of the premixed solution 3 for the cell-free reaction (96 reactions)

LMCP	660 μL
17.5 mg/mL tRNA	19.2 μL
1.6 M $Mg(OAc)_2$	12.8 μL
20 mM amino acid mixture/10 mM DTT	96 μL
3.75 mg/mL creatine kinase	128 μL
10 mg/mL T7 RNA polymerase	17.9 μL
Water	45.3 μL
S30 extract	460.8 μL
Total	1,440 μL

3.3. Cell-Free Reaction

1. Chill an aluminum block on ice.
2. Thaw all of the reagents on ice-cold water.
3. Dispense 1.5 μL each of the second-step PCR products in a 96-well PCR plate on the aluminum block on ice.
4. Prepare Premixed solution 3 on ice according to Table 6.3. Mix the S30 extract last. After mixing the S30 extract, step to the next as soon as possible.
5. Dispense 18.5 μL of the Premixed solution 3 to the every well on the PCR plate.

6. Incubate the PCR plate at 37°C for 1 h.
7. Stop the reaction by placing the PCR plate on ice.
8. An example result is shown in Fig. 6.3.

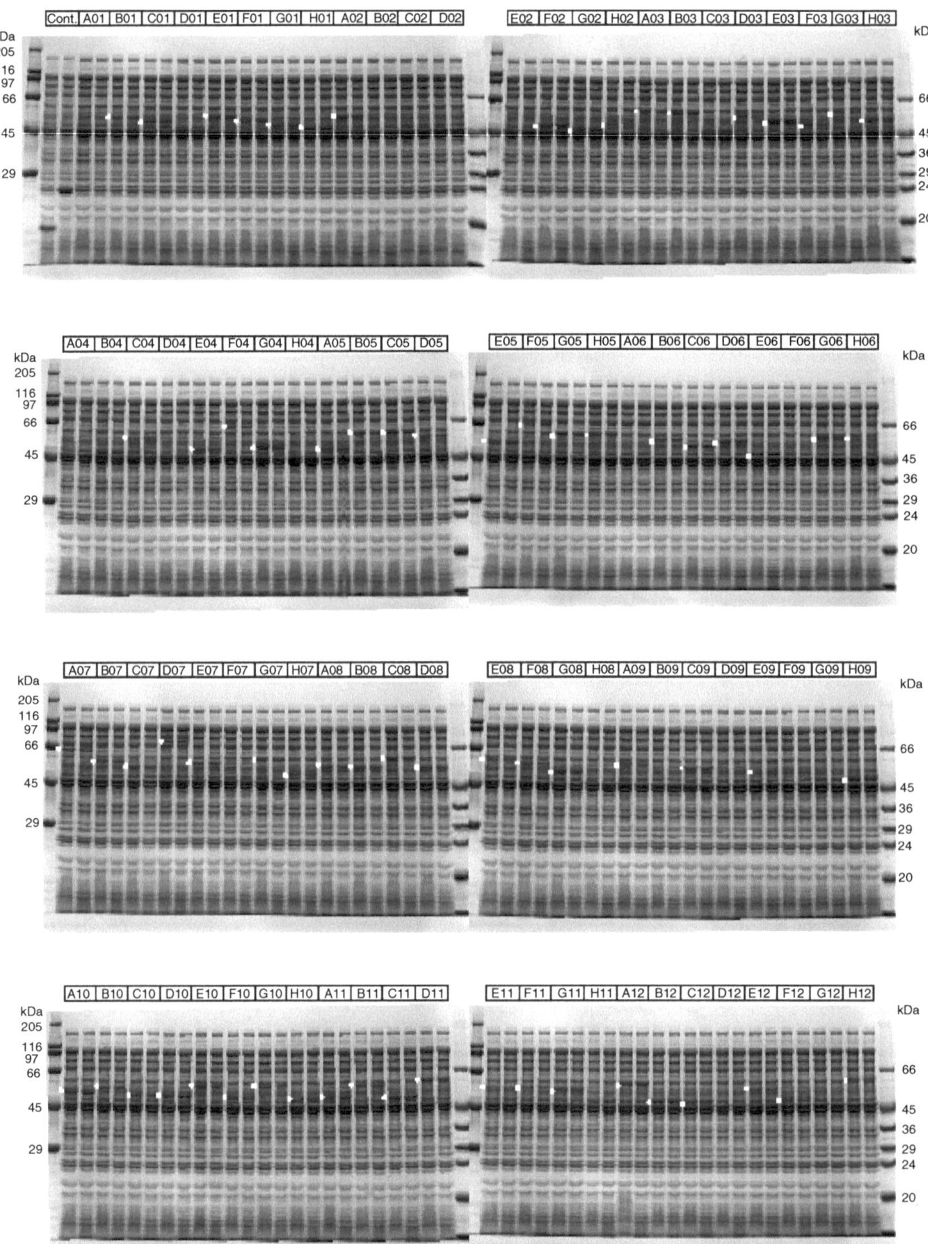

Fig. 6.3. SDS–PAGE analysis of the batch-mode cell-free protein synthesis reaction for MBP-fused *Arabidopsis* proteins. The template DNAs were constructed as MBP-fused form from 96 RAFL *Arabidopsis thaliana* cDNA clones (15). Two lanes are paired for each cDNA (*left lane*: total fraction, *right lane*; soluble fraction). A portion (3 μL) of the reaction (30 μL) was analyzed on 12.5% Perfect NT gel (DRC, Japan) and then stained with Coomassie Brilliant Blue. *White dot* indicates the position of the product.

4. Notes

1. The protocol for PCR reaction described in this chapter is highly optimized for Expand-HiFi enzyme (Roche). Other PCR enzymes, especially enzymes with higher fidelity such as iProof Hi-Fi (Bio-Rad) can be also used if the parameters are reoptimized (1).
2. The sequence of the FW primer is 5′-CCAGCGGCTCCTC-GGGA-X_n-3′ where X_n is identical to the 5′-terminal sequence of the target coding sequence. The sequence of the RV primer is 5′-CCTGACGAGGGCCCCG-Y_n-3′ where Y_n is complementary to the 3′-terminal sequence of the target coding sequence. The lengths of the unique sequences (X_n and Y_n) should be designed to be 14 nt or longer, to provide a T_m of at least 46°C.
3. This protocol is applicable to both purified plasmid cDNA and cultured cell harboring cDNA (1).
4. The T7P fragment contains the promoter sequence for T7 RNA polymerase, the optional N-terminal tag-coding sequence, and linker sequence (CCAGCGGCTCCTCGGGA), which is the same one on the FW primer. The T7T fragments contains the terminator sequence for T7 RNA polymerase, the optional C-terminal sequence, and linker sequence (TCGGGGCCCTCGTCAGG), which is the same as that on the RV primer. A wide variety of the optional sequences are available, for example, GST-fused type, MBP-fused type, and so on (1).
5. The sequence of the U2 primer is 5′-GCTCTTGTCATTGT GCTTCG-3′.

Acknowledgments

The author would like to thank Takashi Yabuki, Satoru Watanabe, Yoko Motoda, and Kazuharu Hanada for developing our two-step PCR method.

References

1. Yabuki, T., Motoda, Y., Hanada, K., Nunokawa, E., Saito, M., Seki, E., Inoue, M., Kigawa, T., and Yokoyama, S. (2007) A robust two-step PCR method of template DNA production for high-throughput cell-free protein synthesis. *J. Struct. Funct. Genomics* **8**, 173–191.
2. Seki, E., Matsuda, N., Yokoyama, S., and Kigawa, T. (2008) Cell-free protein synthesis system from Escherichia coli cells cultured at decreased temperatures improves productivity by decreasing DNA template degradation. *Anal. Biochem.* **377**, 156–161.

3. Yabuki, T., Kigawa, T., Dohmae, N., Takio, K., Terada, T., Ito, Y., Laue, E. D., Cooper, J. A., Kainosho, M., and Yokoyama, S. (1998) Dual amino acid-selective and site-directed stable-isotope labeling of the human c-Ha-Ras protein by cell-free synthesis. *J. Biomol. NMR* **11,** 295–306.

4. Tochio, N., Umehara, T., Koshiba, S., Inoue, M., Yabuki, T., Aoki, M., Seki, E., Watanabe, S., Tomo, Y., Hanada, M., Ikari, M., Sato, M., Terada, T., Nagase, T., Ohara, O., Shirouzu, M., Tanaka, A., Kigawa, T., and Yokoyama, S. (2006) Solution structure of the SWIRM domain of human histone demethylase LSD1. *Structure* **14,** 457–468.

5. Li, H., Koshiba, S., Hayashi, F., Tochio, N., Tomozawa, T., Kasai, T., Yabuki, T., Motoda, Y., Harada, T., Watanabe, S., Inoue, M., Hayashizaki, Y., Tanaka, A., Kigawa, T., and Yokoyama, S. (2008) Structure of the C-terminal PID domain of Fe65L1 complexed with the cytoplasmic tail of APP reveals a novel peptide binding mode. *J. Biol. Chem.* **283,** 27165–27178.

6. Kurimoto, K., Muto, Y., Obayashi, N., Terada, T., Shirouzu, M., Yabuki, T., Aoki, M., Seki, E., Matsuda, T., Kigawa, T., Okumura, H., Tanaka, A., Shibata, N., Kashikawa, M., Agata, K., and Yokoyama, S. (2005) Crystal structure of the N-terminal RecA-like domain of a DEAD-box RNA helicase, the Dugesia japonica vasa-like gene B protein. *J. Struct. Biol.* **150,** 58–68.

7. Murayama, K., Shirouzu, M., Kawasaki, Y., Kato-Murayama, M., Hanawa-Suetsugu, K., Sakamoto, A., Katsura, Y., Suenaga, A., Toyama, M., Terada, T., Taiji, M., Akiyama, T., and Yokoyama, S. (2007) Crystal structure of the rac activator, Asef, reveals its autoinhibitory mechanism. *J. Biol. Chem.* **282,** 4238–4242.

8. Kigawa, T., Yamaguchi-Nunokawa, E., Kodama, K., Matsuda, T., Yabuki, T., Matsuda, N., Ishitani, R., Nureki, O., and Yokoyama, S. (2002) Selenomethionine incorporation into a protein by cell-free synthesis. *J. Struct. Funct. Genomics* **2,** 29–35.

9. Kang, S. H., Kim, D. M., Kim, H. J., Jun, S. Y., Lee, K. Y., and Kim, H. J. (2005) Cell-free production of aggregation-prone proteins in soluble and active forms. *Biotechnol. Prog.* **21,** 1412–1419.

10. Ryabova, L. A., Desplancq, D., Spirin, A. S., and Pluckthun, A. (1997) Functional antibody production using cell-free translation: effects of protein disulfide isomerase and chaperones. *Nat. Biotechnol.* **15,** 79–84.

11. Matsuda, T., Kigawa, T., Koshiba, S., Inoue, M., Aoki, M., Yamasaki, K., Seki, M., Shinozaki, K., and Yokoyama, S. (2006) Cell-free synthesis of zinc-binding proteins. *J. Struct. Funct. Genomics* **7,** 93–100.

12. Yokoyama, S. (2003) Protein expression systems for structural genomics and proteomics. *Curr. Opin. Chem. Biol.* **7,** 39–43.

13. Kigawa, T., Inoue, M., Aoki, M., Matsuda, T., Yabuki, T., Seki, E., Harada, T., Watanabe, S., and Yokoyama, S. (2007) The use of the *Escherichia coli* cell-free protein synthesis for structural biology and structural proteomics, in *Cell-Free Protein Synthesis. Methods and Protocols* (Spirin, A. S., Swartz, J. R., Eds.), pp. 99–109, Wiley-VCH, Weinheim.

14. Kigawa, T., Matsuda, T., Yabuki, T., and Yokoyama, S. (2007) Bacterial cell-free system for highly efficient protein synthesis, in *Cell-Free Protein Synthesis. Methods and Protocols* (Spirin, A. S., Swartz, J. R., Eds.), pp. 83–97, Wiley-VCH, Weinheim.

15. Seki, M., Narusaka, M., Kamiya, A., Ishida, J., Satou, M., Sakurai, T., Nakajima, M., Enju, A., Akiyama, K., Oono, Y., Muramatsu, M., Hayashizaki, Y., Kawai, J., Carninci, P., Itoh, M., Ishii, Y., Arakawa, T., Shibata, K., Shinagawa, A., and Shinozaki, K. (2002) Functional annotation of a full-length Arabidopsis cDNA collection. *Science* **296,** 141–145.

Chapter 7

Cell-Free-Based Protein Microarray Technology Using Agarose/DNA Microplate

Tatsuya Sawasaki and Yaeta Endo

Abstract

Protein microarray is considered to be one of the key analytical tools for high-throughput protein function analysis. We found that Arabidopsis HY5 protein functions as a novel DNA-binding tag (DBtag), and DBtagged proteins are immobilized and purified on a newly designed agarose/DNA microplate. In this chapter, we demonstrate a protocol for making the DBtag-based protein microarray and will provide protocols for two applications using the microarray: (1) detection of autophosphorylation activity of DBtagged human protein kinases and inhibition of their activity by staurosporine, and (2) detection of a protein–protein interaction between the DBtagged UBE2N and UBE2v1.

Key words: Protein microarray, High-throughput functional analysis, DNA-binding tag, Protein kinase, Agarose/DNA microplate

1. Introduction

Currently available protein microarray technology has allowed large-scale screening of biomarker proteins recognized by serum antibodies (1). However, this method is yet to become a commonly used biochemical tool for the analysis of proteins (2). Certainly there is room for further improvement before this technology could become a routinely used laboratory tool. For example, one of the problems is the difficulty in immobilizing a variety of proteins in their functionally active forms. Many proteins needed to be appropriately oriented for proper functioning (3). However, it is not easy to control the orientation of the protein during its mobilization on the surface of the microplate. Another problem is that the high-throughput functional analysis requires freshly produced and purified proteins; however, unlike DNA, many purified proteins are not stable and thus, these cannot

Y. Endo et al. (eds.), *Cell-Free Protein Production: Methods and Protocols,* Methods in Molecular Biology, vol. 607
DOI 10.1007/978-1-60327-331-2_7, © Humana Press, a part of Springer Science+Business Media, LLC 2010

be stored in active condition for long time. The development of functional protein microarrays for practical use, therefore, requires relatively easy methods for the functional immobilization and purification of freshly prepared proteins on the microplate. We recently developed a high-throughput method for protein synthesis using the wheat germ cell-free protein synthesis and an automatic protein synthesizer (4–6) and demonstrated that the automatic synthesizer is very useful for the production of freshly prepared proteins.

To create a new type of protein microarray, we developed a novel tag using a DNA-binding protein and a newly designed microplate consisting of agarose and commercially available genomic DNAs (7). The new tag, named here as DBtag, is the Arabidopsis transcription factor HY5 having a basic leucine-zipper domain (8). We found that the HY5 protein had high binding affinity to commercially available salmon sperm and calf DNAs. Here, we used this DNA-binding ability of HY5 to immobilize and purify the fusion protein on the microplate.

2. Materials

2.1. DBtag Protein Production

1. PCR thermocycler MP (Takarabio Inc., Otsu, Japan).
2. ExTaq DNA polymerase (Takarabio Inc.).
3. pEU-DBtag (accession no. AB369281) or Arabidopsis *HY5* gene (GenBANK accession no. NM_121164).
4. cDNA for synthesized protein.
5. Oligonucleotide primers.
6. Exonuclease I (1 U/10 µL reaction mixture, GE Healthcare, Little Chalfont, UK).
7. PCR product purification kit (Gel-M®Extraction, Viogene, Shijr, Taipei).
8. 5× Transcription buffer (TB): 400 mM HEPES-KOH, pH 7.8, 80 mM magnesium acetate, 10 mM spermidine, and 50 mM DTT.
9. Nucleotide tri-phosphates (NTPs) mix: a solution containing 25 mM each of ATP, GTP, CTP, and UTP.
10. SP6 RNA polymerase (80 units/µL, Promega; Madison, WI).
11. RNasin (80 units/µL, Promega; Madison, WI).
12. Microcon (YM-50; Millipore, Bedford, USA).

13. 10 mg/mL creatine kinase (Roche Diagnostics K. K.).
14. 4× Translational substrate buffer (TSB): 120 mM HEPES-KOH, pH 7.8, 400 mM potassium acetate, 10.8 mM magnesium acetate, 1.6 mM spermidine, 16 mM DTT, 1.2 mM amino acid mix, 4.8 mM ATP, 1 mM GTP, and 64 mM creatine phosphate. Or, 4× SUB-AMIX® (CellFree Sciences Co., Ltd., Matsuyama, Japan).
15. 1× TSB: 30 mM HEPES-KOH, pH 7.8, 100 mM potassium acetate, 2.7 mM magnesium acetate, 0.4 mM spermidine, 4 mM DTT, 0.3 mM amino acid mix, 1.2 mM ATP, 0.25 mM GTP, and 16 mM creatine phosphate. Or, 1× SUB-AMIX® (CellFree Sciences Co., Ltd.).
16. Wheat embryo extract (240 OD/mL, described in Chapter 3) or WEPRO®1240 (CellFree Sciences Co., Ltd.).
17. *Escherichia coli* biotin protein ligase (BirA, Genbank Accession no. NP_312927).
18. Biotin (Nakalai Tesque, Inc., Kyoto, Japan).

2.2. Protein Microarray Using Garose/DNA Microplate

1. Agarose (SeaKem Gold, Takarabio, Inc.).
2. Salmon sperm DNA (Sigma-Aldrich Corp, MO).
3. 1× TMD buffer: 20 mM Tris–HCl at pH 7.8, 2 mM $MgCl_2$, and 1 mM DTT.
4. Slide glass (Asahi glass, Japan).
5. Lab-Tek II Chamber slide (one-well, Nalge Nunc International Co., Naperville, IL).
6. Pin-type spotter like MultiSPRinter™ spotter (Toyobo Bio Instruments, Tsuruga, Japan) or comparable product.
7. Wash buffer: 20 mM Tris–HCl at pH 7.8, 200 mM NaCl, 2 mM $MgCl_2$, and 1 mM DTT.
8. Tupperware box.

2.3. Functional Protein Analysis Using the Protein Microarray

1. cDNAs.
2. 1× PBS buffer: 137 mM Nacl, 2.7 mM Kcl, 10 mM Na_2HPO_4, 1.8 mM KH_2PO_4, pH 7.4.
3. Kination solution: 50 mM Tris–HCl at pH 7.8, 100 mM NaCl, 10 mM $MgCl_2$, and 0.1 mM DTT, 10,000 Ci/μl [γ-^{32}P]-ATP, 0.05% DMSO (plus protein kinase inhibitor for inhibitor solution, if need be).
4. Caspase-3 solution: 20 mM Tris–HCl at pH 7.8, 200 mM NaCl, 2 mM $MgCl_2$, 1 mM DTT, and 17.4 ng/μL caspase 3 (human, Sigma-Aldrich Corp)].
5. 10 μg/ml Alexa488-streptavidin (STA) (Invitrogen Corp., Carlsbad, CA).

6. Typhoon 9400 imaging system (GE Healthcare, Little Chalfont, UK) or comparable product.

3. Methods

3.1. Construction of DNA Template for DBtagged Protein Production

Arabidopsis *HY5* gene was inserted into pEU vector (4) as pEU-DBtag (Fig. 7.1a). The DBtag fragment (accession no. AB369281) was isolated (Subheading 3.1.1.) and fused on N-terminal of purpose gene by split-PCR (Subheading 3.1.2.). We have optimized a PCR-based linear template DNA generation by designing a set of universal primers for fusions of SP6 promoter and translational enhancer (E02) (9). For production of biotinylated protein, biotin ligation site (amino acid sequence: GLNDIFEAQKIEWHE)

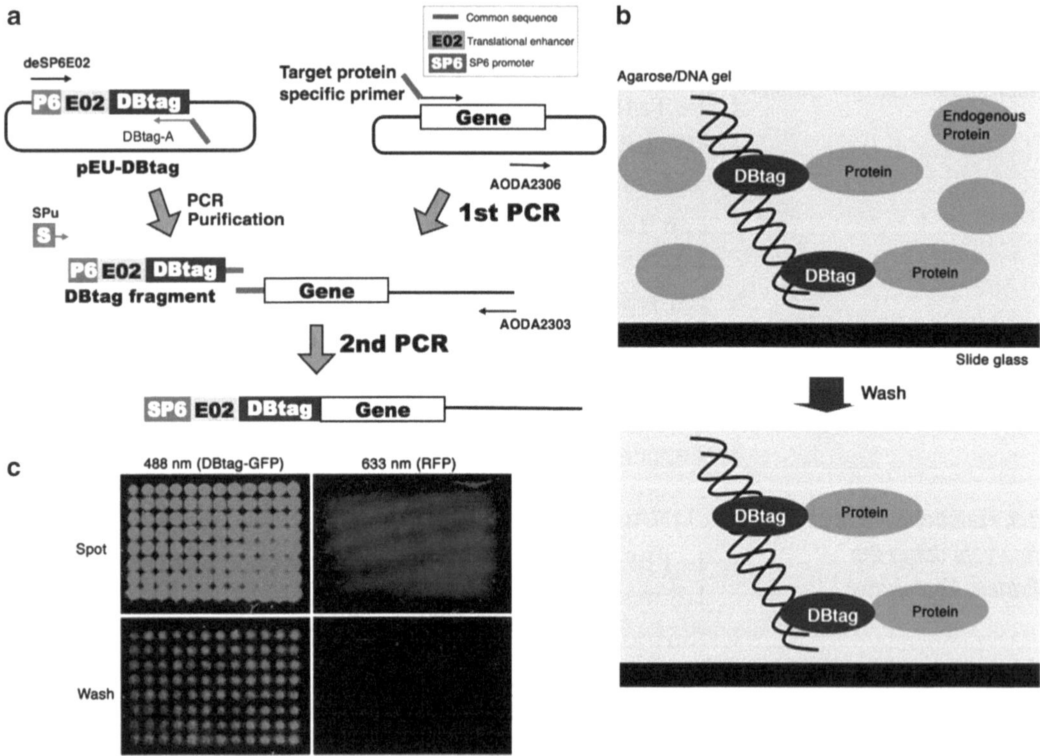

Fig. 7.1. (**a**) Scheme for DBtag-fusion method by the "split-primer" PCR. The DBtag fragment includes partial SP6 promoter. DNA amplification by SPu primer makes full-length SP6 promoter. (**b**) Schematic showing immobilized DBtag-fusion proteins on the microplate before and after washing. (**c**) DBtag-GFP and RFP (untagged) were mixed together and the mixture was spotted on the microplate. The DBtag-GFP was immobilized on the microplate, while the RFP was washed out and no RFP fluorescence was observed. Reproduced from ref. (7) with permission.

is inserted into N- or C-terminal of purpose gene by PCR or general recombinant method.

3.1.1. Isolation of DBtag Fragment

1. Prepare 100 µL of PCR reaction mixture as the manufacturer's instructions (Takarabio Inc.) by mixing 100 pg/µL of plasmid (pEU-DBtag) and 200 nM of deSP6E02 (5′-GGTGACAC TATAGAACTCACCTATCTCTCTACACA) and DBtag-A (5′-TGGTGGTGGTGGGTGGAAGCCCTGGAAG TACAGGTTCTC), 200 µM of each dNTP, 1.25 units of ExTaq DNA polymerase.
2. Set the mixture to PCR thermocycler for 1 min denaturation at 96°C followed by 25 cycles of amplification: 98°C for 10 s, 55°C for 30 s, and 72°C for 1 min.
3. Incubate the amplified product with exonuclease I (1 U/10 µL reaction mixture) for 30 min at 37°C, and then treat for 30 min at 80°C (see Note 1).
4. Purify them with a PCR product purification kit.
5. Check concentration of the DBtag fragment.

3.1.2. Construction of DNA Template for DBtag Protein by Split-PCR

1. Prepare 50 µL of a first PCR reaction mixture by mixing 3 ng of plasmid or 3 µL of *Escherichia coli* (overnight culture) as template, 200 µM each of dNTP, 1.5 units of ExTaq DNA polymerase, 10 nM of target protein-specific primer (5′-CCACCCACCACCACCAatgnnnnnnnnnnnnnnnn; uppercase and lowercase indicate common sequence and the 5′-coding region of the target gene, respectively) and AODA2306 primer (5′-AGCGTCAGACCCCGTAGAAA, this primer is designed on pUC Ori sequence), and the buffer supplied by the manufacturer.
2. Set the first mixture to PCR thermocycler for 4 min denaturation at 94°C followed by 30 cycles of amplification: 98°C for 10 s, 55°C for 30 s, and 72°C for 5 min depending on the length of gene (1 kb per min).
3. Prepare 50 µL of a second PCR reaction mixture by mixing 5 µL of the first PCR product (without any purification) above as template, 200 µM each of dNTP, 1.5 units of ExTaq DNA polymerase, 100 nM SPu primer (5′-GCGTAG CATTTAGGTGACACT), 100 nM AODA2303 primer (5′-GTCAGACCCCGTAGAAAAGA, this primer is designed on pUC Ori sequence), and 2 nM DBtag fragment and the buffer supplied by the manufacturer.
4. Set the second PCR mixture to PCR thermocycler for 1 min denaturation at 98°C followed by 30 cycles of amplification: 98°C for 10 s, 55°C for 30 s, and 72°C for 5 min depending on the length of gene (1 kb per min).

5. Check the DNA template production by agarose electrophoresis (see Note 2).

3.2. DBtag Protein Production by Wheat Cell-Free System

We use bilayer method for protein production by the cell-free system (10). In principle, exchange of substrates and dilution of by-products take place between translation mix and TSB by expansion of translation mix.

3.2.1. Preparation of mRNA

1. Prepare 30 μL of transcription reaction by mixing 3 μL of the PCR mix (without any purification) as template, 6 μL of 5× TB, 3 μL of NTPs mix, 25 units of SP6 RNA polymerase, and 25 units of RNasin with Milli-Q water.
2. Incubate the reaction mixture at 37°C for 3 h.
3. Add 3 volumes of Milli-Q water, 1/7.5 vol. of 7.5 M ammonium acetate, and 2.5 vol. of ~99% EtOH, and keep on ice for 10 min after mixing well.
4. Centrifuge the mixture (20,000*g*; 5 min) and discard the supernatant.
5. Wash the pellet with 500 μL of 70% ethanol and then spin for 5 min at 20,000*g*.
6. Dissolve the mRNA pellet in 10 μL of Milli-Q water.

3.2.2. Preparation of DBtagged Proteins by Wheat Cell-Free Protein Production

1. Prepare 25 μL of translational reaction mixture by combining 6.3 μL of wheat embryo extract (final conc. 60 OD/mL), 1 μL of 10 mg/mL creatine kinase, 4.5 μL of 4× TSB, and 10 μL of the dissolved mRNA above.
2. Add 125 μL of 1× TSB in the U-shaped titer plate well.
3. Carefully place 25 μL of the reaction mixture at the bottom of titer plate well.
4. Place a coverlet on the plate and then wrap with the Saran Wrap to avoid evaporation.
5. Keep the plate in an incubator at 17°C for 16 h without shaking.

3.3. Generation of Protein Microarray Using DBtag Protein

We designed a new microplate that carries a thin layer of agarose gel containing DNA (agarose/DNA microplate) to immobilize and purify the DBtagged proteins on the microplate (Fig. 7.1b). Non-DBtagged proteins are washed out (Fig. 7.1c).

3.3.1. Preparation of Agarose/DNA Microplate

1. Melt 0.2% agarose gel in 1× TMD buffer, and subsequently add 1 mg/mL (final concentration) salmon sperm DNA.
2. Before the gel is solidified, spread 600 or 400 μL of the agarose gel/DNA mixture above on a slide glass or a Lab-Tek II Chamber slide, respectively, to form a thin layer (0.5–0.6 mm) (see Note 3).

3.3.2. Preparation of Protein Microarray on Agarose/DNA Microplate

1. Spot approximately 10 nL of each translational mixture per 0.2 mm^2 (~500 µm in diameter) on the agarose/DNA-coated glass plate by using pin-type spotter according to the instruction manual (see Note 4).
2. Soak the microplate in the wash buffer for 15 min.
3. Use immediately for assay (see Note 5).

3.4. Functional Analysis of DBtagged Protein on Agarose/DNA Microplate (see Note 5)

Using this new protein microarray, we demonstrate (a) the autophosphorylation activity of the fusion human protein kinases (Subheading 3.4.1. and Fig. 7.2a), (b) a protein–protein interaction between UBE2N (Accession no. NM_003348, MGC5063) and UBE2v1 (NM_022442, MGC8586) (Subheading 3.4.2. and Fig. 7.2b), and (c) specific cleavage of the fusion proteins (PAK2, NM_002577) by caspase 3 (Subheading 3.4.3. and Fig. 7.2c).

3.4.1. Inhibition Assay of DBtagged Protein Kinases on Agarose/DNA Microplate

1. Prepare the protein microarray spotting DBtag-fusion protein kinases on the agarose/DNA-coated glass plate as described above.
2. Cover the microplate with a kination solution or inhibitor solution.
3. Incubate for 30 min at 37°C.
4. Soak the microplate in washing buffer for 10 min.
5. Analyze the microplate by Typhoon 9400 imaging system.

3.4.2. Protein–Protein Interaction Analysis Between DBtagged Protein and Biotinylated Protein on Agarose/DNA Microplate

1. Prepare DBtag-UBE2N and biotinylated bls-UBE2V1 proteins by the cell-free system.
2. Mix both proteins with Alexa488-STA (final 10 µg/mL).
3. Incubate for 30 min at 26°C.
4. Spot the mixture on the agarose/DNA-coated glass plate as described above (Subheading 3.3.).
5. Soak the microplate in 1× PBS buffer for 10 min (see Note 6).
6. Analyze the microplate by Typhoon 9400 imaging system.

3.4.3. Caspase Cleavage Assay of DBtagged Proteins on Agarose/DNA Microplate

1. Prepare biotinylated substrate proteins (DBtag-PAK2-bls) by the cell-free system.
2. Mix the biotinylated proteins with Alexa488-STA (final 10 µg/ml).
3. Incubate for 30 min at 26°C.
4. Spot the mixture on the agarose/DNA-coated glass plate as described above (Subheading 3.3.).
5. Cover the microplate with caspase-3 solution.
6. Incubate for 30 min at 26°C.

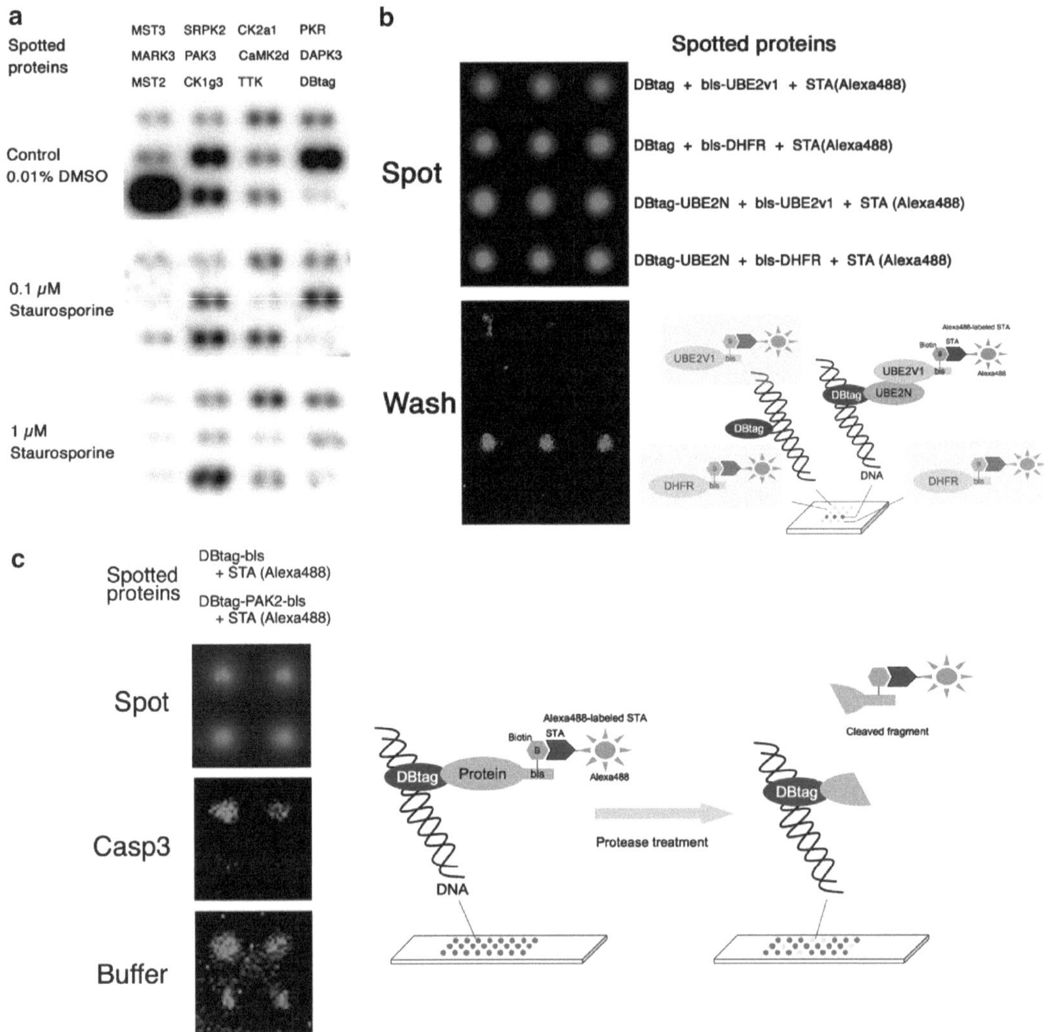

Fig. 7.2. (**a**) Detection and specific inhibition of the autophosphorylation activity of 11 human protein kinases on agarose/DNA microplate. The DBtagged human protein kinases were analyzed in duplicates. "DBtag" indicates the tag protein by itself (control). (**b**) Detection of a protein–protein interaction between UBE2N and UBE2v1 on the agarose/DNA microplate. Four samples (DBtag + biotin-labeled bls-UBE2v1 + STA(Alexa488), DBtag + biotin-labeled bls-DHFR + STA(Alexa488), DBtag-UBE2N + the bls-UBE2v1 + STA(Alexa488) and DBtag-UBE2N + the bls-DHFR + STA(Alexa488) from the top in "spot" panel] were spotted with three spots on the microplate. The first, second, and fourth samples lost the fluorescence after washing ("wash" panel). The third sample showed the fluorescence in "wash" panel, indicating a heterodimer as a protein–protein interaction between UBE2N and UBE2v1 proteins. An image from this result was shown in the *right panel*. (**c**) Two proteins including PAK2 (*lower spots* in each panel: DBtag-PAK2-bls labeled with Alexa488-STA) lost the fluorescence after the caspase 3 (casp3) treatment ("Casp3" panel). DBtag-bls proteins (*upper spots* in each panel: DBtag-bls labeled with Alexa488-STA) and buffer without the casp3 ("buffer" panel) showed fluorescence because they were not recognized by the casp3. *Right panel* indicates the schematic that the fluorescence-labeled region is released by the protease cleavage, and as a result, the fluorescence was lost upon washing.

7. Soak the microplate in 1× PBS buffer for 10 min (see Note 6).
8. Analyze the microplate by Typhoon 9400 imaging system.

4. Notes

1. Exonuclease I degrades extra primers after the PCR reaction. We often found primer contaminant in samples after Purification kit. The exonuclease I is very convenient to remove the extra primers because it is inactivated by simple heat treatment.
2. If DNA template is of low production, increase of primer concentration could provide higher production. In our case, the concentration increases to 100 nM from 10 nM as final concentration.
3. The agarose/slide-coated glass should be used within 1 day.
4. During spotting process, humidity inside the spotter should be kept. MultiSPRinter™ spotter can control high-humidity condition.
5. During making of the microplate and the assay, the microplate should be kept in a Tupperware box containing wet papers to prevent it from drying. Dried microplate could not be used for the functional analysis of proteins.
6. If soaking time is too long, assay sensitivity is decreased.

Acknowledgements

We greatly thank Satoko Matsunaga, Yuko Matsubara, Yoshiko Kodani, Yoshinori Tanaka, and two researchers (Mihoro Saeki and Dr. Ryo Morishita) working in CellFree Sciences Co., Ltd for their expert technical assistance. This work was partially supported by the Special Coordination Funds for Promoting Science and Technology by the Ministry of Education, Culture, Sports, Science and Technology, Japan.

References

1. Qin, S., Qiu, W., Ehrlich, J.R., Ferdinand, A.S., Richie, J.P., O'Leary M.P., Lee, M.L. and Liu, B.C. (2006) Development of a "reverse capture" autoantibody microarray for studies of antigen-autoantibody profiling. *Proteomics* **6,** 3199–3209.
2. Sheridan, C. (2005) Protein chip companies turn to biomarkers. *Nat. Biotechnol.* **23,** 3–4.
3. Zhu, H. and Snyder, M. (2003) Protein chip technology. *Curr. Opin. Chem. Biol.* **7,** 55–63.
4. Sawasaki, T., Ogasawara, T., Morishita, R. and Endo, Y. (2002) A cell-free protein

synthesis system for high-throughput proteomics. *Proc. Natl. Acad. Sci. USA.* **99,** 14652–14657.

5. Endo, Y. and Sawasaki, T. (2006) Cell-free expression systems for eukaryotic protein production. *Curr. Opin. Biotechnol.* **17,** 373–380.
6. Sawasaki, T., Gouda, M.D., Kawasaki, T., Tsuboi, T., Tozawa, Y., Takai, K. and Endo, Y. (2005) The wheat germ cell-free expression system: methods for high-throughput materialization of genetic information. *Methods Mol. Biol.* **310,** 131–144.
7. Sawasaki, T., Kamura, N., Matsunaga, S., Saeki, M., Tsuchimochi, M., Morishita, R. and Endo, Y. (2008) Arabidopsis HY5 protein functions as a DNA-binding tag for purification and functional immobilization of proteins on agarose/DNA microplate. *FEBS Lett.* **582,** 221–228.
8. Oyama, T., Shimura, Y. and Okada, K. (1997) The Arabidopsis HY5 gene encodes a bZIP protein that regulates stimulus-induced development of root and hypocotyl. *Genes Dev.* **11,** 2983–2995.
9. Kamura, N., Sawasaki, T., Kasahara, Y., Takai, K. and Endo, Y. (2005) Selection of 5′-untranslated sequences that enhance initiation of translation in a cell-free protein synthesis system from wheat embryos. *Bioorg. Med. Chem. Lett.* **15,** 5402–5406.
10. Sawasaki, T., Hasegawa, Y., Tsuchimochi, M., Kamura, N., Ogasawara, T. and Endo, Y (2002) A bilayer cell-free protein synthesis system for high-throughput screening of gene products. *FEBS Lett.* **514,** 102–105.

Chapter 8

An Efficient Approach to the Production of Vaccines Against the Malaria Parasite

Takafumi Tsuboi, Satoru Takeo, Tatsuya Sawasaki, Motomi Torii, and Yaeta Endo

Abstract

In malaria vaccine research, one of the major obstacles has been the difficulty of expressing recombinant malarial proteins and it is mainly due to the lack of an efficient methodology for the synthesis of sufficient quantity of quality proteins. We demonstrate that the wheat germ cell-free protein synthesis system can be applied for the successful production of leading malaria vaccine candidate antigens and, thus, prove that it may be a key tool for malaria vaccine research.

Key words: Cell-free protein synthesis, Parasite, Malaria, Vaccine

1. Introduction

Malaria, a serious infectious disease that challenges the global health, causes millions of deaths annually, as well as illness in hundreds of millions of people. The most deadly form of the disease is caused by the inoculation of the malaria parasite, *Plasmodium falciparum*, by infected mosquito bites. The disease is re-emerging mainly due to the emergence of multidrug-resistant parasites and insecticide-resistant mosquitoes (1). Therefore development of malaria vaccine has been considered as one of the essential components for the malaria eradication (2). However, efforts to develop a successful vaccine have not yet accomplished (3). Since we need multiple vaccine candidate antigens to succeed in controlling malaria, post-genome malaria vaccine candidate discovery is necessary. One of the obstacles in this process is at the malaria protein production step and is mainly due to the lack of an efficient methodology to prepare quality proteins. *P. falciparum* genes have a very high A/T content (average 76% per coding

Y. Endo et al. (eds.), *Cell-Free Protein Production: Methods and Protocols,* Methods in Molecular Biology, vol. 607
DOI 10.1007/978-1-60327-331-2_8, © Humana Press, a part of Springer Science+Business Media, LLC 2010

sequence throughout the genome) and a number of them encode repeated stretches of amino acid sequences (4), and these features have been proposed as the major factors limiting *P. falciparum* protein expression in conventional cell-based systems (5–7). Moreover, the presence of glycosylation machinery in eukaryotic cell-based protein expression systems can produce inappropriately glycosylated recombinant malaria proteins, resulting in incorrect immune responses (8–10). There are also constraints such as requirement of disulfide bond formation if the target protein requires it for its bioactivity and requirement of preparation of large quantities of antigen for immunization. The above limiting factors are impediments completely insurmountable in the case of eukaryotic cell-based expression systems such as yeast, baculovirus, or Chinese hamster ovary cell. But fortunately we found that the wheat germ cell-free system can surmount most of the above impediments in the way of finding and developing malaria vaccine candidates (11).

In this chapter, we describe how the wheat germ cell-free system is effective (1) in producing properly folded good quality protein (2) in sufficient quantities (3) that too without the need for codon optimization, using the leading malaria vaccine candidate Pfs25 as an example. And we also describe that (1) proteins produced by the wheat germ cell-free system can be easily purified using simple affinity chromatography, (2) they can be directly used for immunization, and (3) the antibody raised against the proteins are functional in our biochemical, immunocytochemical, and biological analyses. Therefore we hope the wheat germ cell-free system may have dramatic impact on malaria vaccine research.

2. Materials

2.1. Parasite cDNA Preparation, PCR Amplification, and Construction of the DNA Template for Transcription

1. Malaria parasite pellet, *P. falciparum* 3D7 strain (available from Malaria Research and Reference Reagent Resource Center managed by ATCC, Manassas, VA, see Note 1) stored at −80°C in the presence of Complete Protease Inhibitor Cocktail (Roche, Basel, Switzerland).
2. RNeasy Micro Kit (QIAGEN, Valencia, CA).
3. SuperScript III RT First-Strand Synthesis System (Invitrogen, Carlsbad, CA).
4. Oligonucleotide primers (see **Note 2**). Design the 5′ primers for the target to be amplified as follows: desired restriction site followed by a 30-mer of unique sequence covering the 5′ region of the open reading frame containing the start codon.

Design the 3′ primers, desired restriction site followed by a 30-mer of unique sequences covering the 3′ region of the open reading frame upstream of the termination codon. Either 5′ or 3′ primer may contain nucleotide sequence coding hexa-histidine tag for Nickel affinity purification.

5. Phusion™ High-Fidelity DNA Polymerase (New England Biolabs, Ipswich, MA).
6. Thermal cycler (MJ Research, Waltham, MA) (see Note 3).
7. Plasmid of Ehime University (pEU)-E01 protein expression vector specialized for the wheat germ cell-free system.
8. Ligation high ligation reagent (Toyobo, Osaka, Japan).

2.2. Antigen Scale Cell-Free Protein Synthesis and Affinity Purification

1. pEU-E01 plasmid that contains target cDNA.
2. 5× transcription buffer (TB): 400 mM HEPES-KOH (pH 7.8), 80 mM magnesium acetate, 10 mM spermidine, and 50 mM DTT.
3. Nucleotide tri-phosphates (NTPs) mix: a solution containing 25 mM each of ATP, GTP, CTP, and UTP.
4. SP6 RNA polymerase and RNasin (80 U/mL, Promega, Madison, WI).
5. 40 mg/mL creatine kinase.
6. 240 OD/mL WEPRO®1240H (CellFree Sciences, Matsuyama, Japan).
7. 1× Translational substrate buffer (SUB-AMIX): 30 mM HEPES-KOH (pH 7.8), 100 mM potassium acetate, 2.7 mM magnesium acetate, 0.4 mM spermidine, 4.0 mM DTT, 0.3 mM amino acid mix, 1.2 mM ATP, 0.25 mM GTP, and 16 mM creatine phosphate.
8. 6-well microplate (Greiner, Frickenhausen, Germany).
9. Nickel-nitrilotriacetic acid agarose beads (Qiagen).
10. Poly-Prep chromatography column (Bio-Rad, Hercules, CA, USA).
11. Phosphate buffered saline (PBS): Prepare 10× stock with 1.37 M NaCl, 27 mM KCl, 100 mM Na_2HPO_4, 18 mM KH_2PO_4 (adjust to pH 7.4 with HCl if necessary) and autoclave before storage at room temperature. Prepare working solution by diluting one part with nine parts of water.
12. 1× Wash buffer: 1× PBS supplemented with 30 mM of imidazole and 300 mM NaCl.
13. 1× Elution buffer: 1× PBS supplemented with 500 mM of imidazole and 300 mM NaCl.

2.3. Characterization of the Target Molecules by Confocal Immunofluorescence Microscopy

1. Female BALB/c mice 6–8 weeks of age (Kitayama Labes, Ina, Japan).
2. Freund complete and incomplete adjuvant (Wako Pure Chemical, Osaka, Japan)
3. CF11 cellulose powder (Whatman, Maidstone, UK)
4. 50% Percoll: Percoll (GE Healthcare Bio-Sciences, Piscataway, NJ) diluted 1:1 (v/v) with 2× PBS
5. Ookinete culture medium: RPMI1640 medium (Invitrogen) supplemented with 50 μg of hypoxanthine per mL, 25 mM HEPES, 20% heat-inactivated fetal calf serum (Invitrogen), 24 mM NaHCO3, %U of penicillin per mL, and 5 μg of streptomycin per mL (pH 8.4).
6. Eight-well Multitest slides (Flow Laboratories, McLean, VA).
7. Blocking and dilution buffer: 5% (w/v) nonfat dry milk in PBS prepared freshly.
8. Secondary antibody: Antimouse IgG conjugated to Alexa488 (Invitrogen) (see Note 4).
9. Nuclear stain: 2 mg/mL 1,000× DAPI (4,6-diamidino-2-phenylindole) stock solution in 100% methanol.
10. Mounting medium: ProLong Gold Antifade Reagent (Invitrogen) (see Note 5).

2.4. Malaria Transmission-Blocking Vaccine Efficacy Assay

1. Malaria parasite-infected blood specimen collected from malaria patient, under informed consent, at the clinics in Thailand.
2. *Anopheles dirus* mosquitoes reared in the insectary, Armed Forces Research Institute of Medical Sciences, Bangkok, Thailand (see Note 6).
3. Anti-Pfs25 sera produced in BALB/c mice.
4. Malaria naïve human AB-type serum obtained from volunteers lived in Bangkok.
5. Custom made water-jacketed membrane-feeding apparatus.
6. Parafilm M (Alcan Packaging, Neenah, WI)

3. Methods

One of the well-known difficulties in malaria research is the PCR amplification of the target sequence because nucleotide sequences of *P. falciparum* genes are highly A/T rich. This feature requires longer primers than those for the G/C rich organism to amplify the target sequences by PCR. Moreover, there are a number of malaria proteins that contain stretches of amino acid repeat motifs and also a number of malaria genes that contain multiple adenine-rich nucleotide sequences (A-islands) in the coding region. This makes the cDNA cloning dramatically difficult, because these A-islands sometimes cause different number of A-nucleotides in each plasmid clone that causes frameshift. Therefore, the initial hurdle for the malaria protein production is the cDNA-cloning; these factors also require consideration on the plasmid construct design. The following methods explain (1) PCR amplification and cloning of the parasite ORF into the pEU expression vector, (2) synthesis of mRNA using expression vector, (3) translation of prepared mRNA in bilayer system, and (4) biological qualification of the synthesized proteins, which includes the recognition of the parasite molecule with the antibody raised against the recombinant malaria protein. This can be accomplished through indirect immunofluorescence microscopy and vaccine efficacy assay.

3.1. Parasite cDNA Preparation and PCR amplification of DNA-Template Construction for Transcription

1. Extract total RNA from cultured *P. falciparum* 3D7 parasites, then reverse transcribe the mRNA into cDNA by using SuperScript III RT First-Strand Synthesis System.
2. PCR amplify the desired target using Phusion™ High-Fidelity DNA Polymerase.
3. Clone the PCR products (see Fig. 8.1) into the pEU-E01 plasmid using Ligation high ligation reagent.
4. Confirm the sequence of the cloned insert (see Note 7).
5. Purify the plasmid.

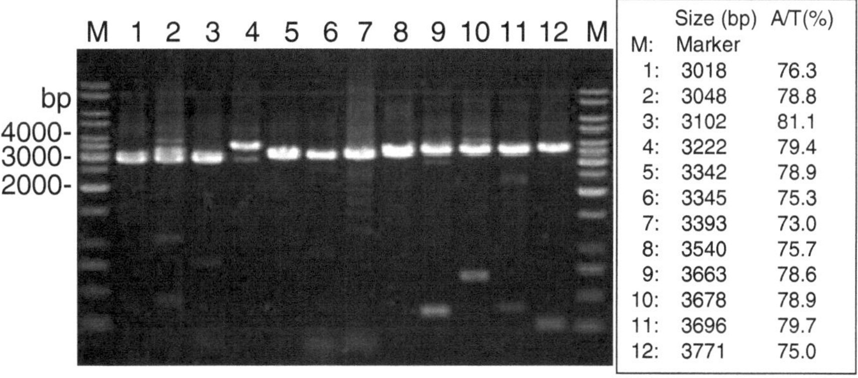

Fig. 8.1. Efficient PCR amplification of the *Plasmodium falciparum* cDNAs. Twelve cDNA targets were selected from *P. falciparum* 3D7, randomly. The cDNA targets were PCR amplified using Phusion™ High-Fidelity DNA Polymerase, under the following conditions: 98°C for 2 min and then 40 cycles at 98°C for 15 s, 55°C for 20 s, and 68°C for 2 min, followed by a final extension at 68°C for 10 min. The amplified products were visualized in a 0.8% agarose gel stained with ethidium bromide. All of the targets ranging from 3,018 bp (*lane 1*) to 3,771 bp (*lane 12*) were successfully amplified even in the presence of very high A/T contents (i.e., 73.0–81.1%).

3.2. Antigen Scale Cell-Free Protein Syntheses of Malaria Protein and Affinity Purification

An example of the methods and the results for Pfs25 is described.

1. Prepare a plasmid clone which has an insert as shown in Fig. 8.2A.
2. Incubate 200 μL of transcription mixture containing 20 μg of the plasmid DNA, 1× TB, 2.5 mM each of NTPs, 200 U of SP6 RNA polymerase, and 200 U of RNasin for 6 h at 37°C.

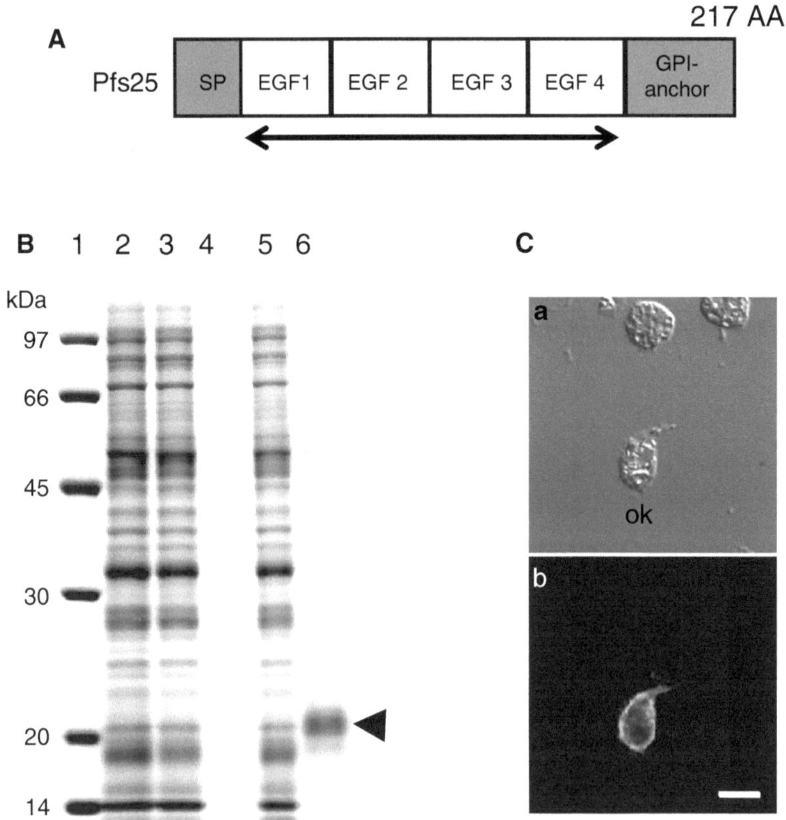

Fig. 8.2. Cloning, expression, and immunolocalization of Pfs25, a leading malaria transmission-blocking vaccine candidate. (**A**) Schematic representation of Pfs25. Pfs25 was expressed without the nucleotide sequences coding for the signal peptide (SP) and the GPI anchor (*arrow*). This Pfs25 gene was amplified by PCR from the *P. falciparum* 3D7 strain using antisense primer that contains nucleotide sequence encoding for hexa-histidine tag at the C terminus and subcloned into pEU-E01 plasmid at the EcoRV site. The A/T content of the *pfs25* insert was 70%. (**B**) The expression, affinity purification, and SDS-PAGE of recombinant Pfs25. Recombinant Pfs25 was expressed using the wheat germ cell-free method, affinity purified by Nickel affinity chromatography, and size-fractionated by 12.5% SDS-PAGE under reducing condition and stained with CBB. *Lane 1*, molecular mass markers in kDa; *lane 2*, total translation mixture; *lane 3*, supernatant fraction of the translation mixture; *lane 4*, pellet fraction of the translation mixture; *lane 5*, flow-through fraction after the affinity purification, *lane 6*; affinity-purified recombinant protein (*arrow head*). (**C**) The immune serum raised against recombinant Pfs25 specifically recognized native Pfs25 proteins expressed on the surface of *P. falciparum* immature ookinetes (ok). (**a**) Differential interference contrast and (**b**) fluorescence confocal images were obtained on a LSM5 PASCAL microscopy (Carl Zeiss MicroImaging, Thornwood, NY) using a 63× oil-immersion lens. *Bar* = 5 μm.

3. Mix the transcription solution containing transcribed mRNA (see Subheading 3.2, step 1) with 200 μL of WEPRO®1240H (240 OD/mL) supplemented with 0.4 μL of creatine kinase (40 mg/mL). Add 4.4-mL of 1× SUB-AMIX into a single well of a 6-well plate and then underlay the above transcription mixture and incubate at 17°C for 16 h.
4. Add imidazole (pH 8.0) in the translation reaction mixture (final concentration, 20 mM) and then add 80 μL of 50% slurry of Ni-NTA beads.
5. Incubate the tube for 16 h on a continuous rotator, at 4°C, for the binding of proteins on to the beads.
6. Transfer the solution with the beads into a Poly-Prep column.
7. Wash the beads by 0.4 mL of wash buffer three times and then elute the recombinant protein with 80 μL of elution buffer five times.
8. Analyze the purified protein by sodium dodecyl sulfate-polyacrylamide gel electrophoresis (SDS-PAGE) under reducing condition, and the bands were visualized with Coomassie brilliant blue (see Fig. 8.2B).

3.3. Preparation of Antiserum and Parasite Antigen for Indirect Immunofluorescence Assay

3.3.1. Antiserum Preparation

1. Immunize female BALB/c mice (6–8 weeks of age) by subcutaneous injection with affinity purified recombinant proteins emulsified with the same volume of Freund's complete adjuvant as a priming dose and then administer two additional booster doses with Freund's incomplete adjuvant at 3-weeks interval.
2. Collect blood by cardiac puncture a week after the final boost under anesthesia and then separate sera after the coagulation of the blood (see Note 8).

3.3.2. Preparation of Parasite Antigens from the Cultured Ookinetes of *Plasmodium falciparum*

1. Collect peripheral blood with heparinized syringes from malaria patients under written informed consent. Purify gametocytes by passing 5–10 mL of blood through CF-11 column to remove leukocytes followed by 50% Percoll density gradient centrifugation (350 × *g* for 25 min at room temperature).
2. Collect the interface rich in gametocytes on the Percoll cushion into a new tube. Wash them twice with PBS and then culture this parasite pellet to ookinete in 1 mL of ookinete medium for 24 h at 24°C in air.
3. Wash the cultured parasite preparations rich in ookinetes twice with PBS and then spot them on 8-well Multitest slides and fix them with ice-cold acetone for 5 min.
4. Store the slides at –80°C until use.

3.3.3. Staining Procedure for Immunofluorescence Assay

1. Take the desired number of antigen slides out from the freezer and then place them quickly in a desiccator until they are brought to room temperature.
2. Block the slides with blocking buffer for 30 min at 37°C in the humidified chamber.
3. Incubate with anti-Pfs25 immune sera (1:100 dilution with blocking buffer) for 1 h at 37°C (see Note 9).
4. Wash the slides with ice-cold PBS for 5 min and incubate with secondary antibody (1:500) and DAPI (1:1,000) diluted with blocking buffer for 30 min at 37°C, followed by washing with ice-cold PBS for 5 min. Mount the slides with Prolong Gold Antifade Reagent and incubate the slides for overnight at room temperature to allow complete solidification of the mounting medium.
5. View the slides under confocal microscopy. Excitation at 488 nm induces the Alexa Fluor 488 fluorescence (green emission) for the Pfs25 (Fig. 8.2Cb), while the differential interference contrast image (Fig. 8.2Ca) is also captured. For example, the fluorescent signal for Pfs25 on the surface of ookinete is shown in Fig. 8.2Cb.

3.4. Malaria Transmission-Blocking Vaccine Efficacy Assay

1. Collect peripheral blood into heparinized syringe from a volunteer patient.
2. Aliquot the collected blood into tubes (300 μL/tube) and remove plasma by a brief centrifugation.
3. Dilute mouse immune sera into 1:2, 1:8, and 1:32 (v/v) with heat-inactivated normal human AB serum prepared from Thai malaria naïve donors.
4. Mix each diluted test serum with *P. falciparum*-infected blood cells (1:1, v/v) and incubate for 15 min at room temperature.
5. Place the mixture into a water-jacketed membrane feeding apparatus (see Fig. 8.3a) whose bottom is sealed by Parafilm M kept at 37°C with circulating water outside of the feeder to allow starved *A. dirus* mosquitoes (Armed Forces Research Institute of Medical Sciences) to feed on the blood meals for 30 min (see Fig. 8.3b).
6. Remove unfed mosquitoes by manual aspiration and maintain only fully engorged mosquitoes for a week by giving 10% sucrose water in the insectary.
7. Dissect 20 mosquitoes for each mouse test immune serum and count the number of oocysts developed within the mosquito midgut under the microscope by staining with 0.5% mercurochrome. The transmission-blocking vaccine efficacy is accessed by the number of oocysts per mosquito (see Fig. 8.3c).

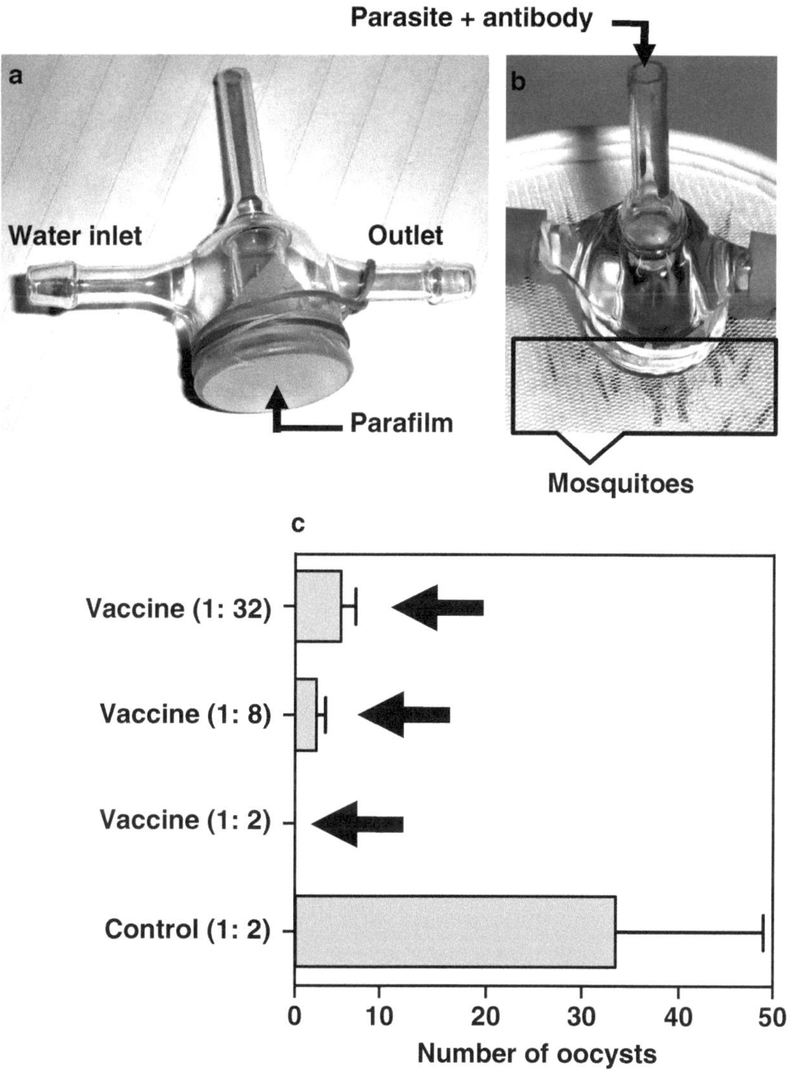

Fig. 8.3. Transmission-blocking efficacy of antibodies against *Plasmodium falciparum* parasites. (**a**) A membrane feeding apparatus with its bottom sealed by Parafilm M. (**b**) The apparatus is kept at 37°C with circulating water outside of the feeder to allow mosquitoes to feed on the infected blood mixed with test and control sera. (**c**) The median numbers of oocysts per mosquito ($n = 20$) (*bars*) with interquartile ranges (*lines on top of the bars*) were compared among groups of mosquitoes fed on either anti-Pfs25 serum serially diluted or control mouse serum. Dilution range of test immune serum used is shown as 1:2 to 1:32. *Arrows* indicate statistically significant differences compared to the control group analyzed using Kruskal–Wallis test ($P < 0.05$).

4. Notes

1. The Malaria Research and Reference Reagent Resource Center (MR4) is a central source of quality reagents to the malaria research community. Materials available to registered users include parasites, mosquito vectors, antibodies, antigens, gene

libraries, etc. MR4 Web site (http://www.mr4.org/Home/tabid/93/Default.aspx) also provides a lot of useful information such as online protocol book named *Methods In Malaria Research* (version 5.2). This book includes protocols that are useful to malaria research, for example, parasite culture methods, immunological assays, or molecular biological techniques.

2. For the efficient PCR amplification, the melting temperature of oligonucleotide primer should be higher than 70°C and it can be roughly calculated by the following formula; 2°C×(A+T)+4°C×(G+C). This is one of the essential factors for the efficient PCR amplification of *P. falciparum* because the *P. falciparum* genome is A/T rich.
3. Thermal cycler, which has gradient temperature function, is useful when optimizing the PCR condition, such as testing a lot of different annealing temperatures.
4. For the multiple labeling experiment, check the quality of the secondary antibodies in advance. For example, some of the commercially available secondary antibodies recognizing mouse IgG are highly cross-adsorbed using bovine, goat, human, rabbit, and rat IgG and also human serum. This secondary antibody can be useful for the double labeling with both mouse and rat primary antibodies.
5. This antifade reagent will solidify, and the sample can be saved for months after mounting, and it also offers enhanced resistance to photobleaching. This criterion is very important for the high-quality imaging because this reagent allows the repeated scanning with the confocal laser microscope until the satisfactory images are obtained.
6. *A. dirus* is a major malaria vector mosquito species in Thailand. This mosquito line is established and adapted as one of the laboratory lines.
7. Pay special attention to confirm the number of adenine nucleotides in the A-islands present in the target region. Otherwise wrong number of adenine nucleotides cause frameshift. If a plasmid clone contains wrong number of adenine nucleotides, verify another plasmid clone which sometimes might have a correct number of adenine nucleotides in the A-island.
8. The antisera for immunofluorescence assay can be saved by addition of sodium azide (final concentration, 0.02%) (Caution highly toxic) and stored at 4°C to prevent repeated freeze thaw cycles which damage the antibodies. However, the antisera for the biological assay such as membrane feeding should not contain any preservatives. In this case, all the antisera are kept frozen at –80°C.

9. One well in the 8-well Multitest slide can hold up to 20 μL of the blocking or antibody solution. In the case of 24-well Multitest slide, one well can hold up to 10 μL of the solution.

Acknowledgments

This work was supported in part by Grants-in-Aid for Scientific Research 18390129 and 19406009; Scientific Research on Priority Areas 21022034 from the Ministry of Education, Culture, Sports, Science, and Technology, Japan; and, in part, by a Grant-in-Aid of the Ministry of Health, Labour, and Welfare (H21-Chikyukibo-ippan-005), Japan.

References

1. Greenwood, B. and Mutabingwa T. (2002) Malaria in 2002. *Nature* **415,** 670-672.
2. Greenwood B. (2009) Can malaria be eliminated? *Trans. R. Soc. Trop. Med. Hyg.* **103S,** S2-S5.
3. Richie, T. L. and Saul, A. (2002) Progress and challenges for malaria vaccines. *Nature* **415,** 694-701.
4. Gardner, M. J., Hall, N., Fung, E., White, O., Berriman, M., Hyman, R. W. et al., (2002) Genome sequence of the human malaria parasite *Plasmodium falciparum*. *Nature* **419,** 498-511.
5. Aguiar, J. C., LaBaer, J., Blair P. L., Shamailova, V. Y., Koundinya, M., Russell, J. A., et al. (2004) High-throughput generation of P. *falciparum* functional molecules by recombinational cloning. *Genome Res.* **14,** 2076-2082.
6. Mehlin, C., Boni, E., Buckner, F. S., Engel, L., Feist, T., Gelb M. H., et al. (2006) Heterologous expression of proteins from *Plasmodium falciparum*: results from 1000 genes. *Mol. Biochem. Parasitol.* **148,** 144-160.
7. Vedadi, M., Lew, J., Artz, J., Amani M., Zhao, Y., Dong, A., et al. (2007) Genome-scale protein expression and structural biology of *Plasmodium falciparum* and related Apicomplexan organisms. *Mol. Biochem. Parasitol.* **151,** 100-110.
8. Gowda, D. C., and Davidson, E. A. (1999) Protein glycosylation in the malaria parasite. *Parasitol. Today* **15,** 147-152.
9. Kedees, M. H., Azzouz, N., Gerold, P., Shams-Eldin, H., Iqbal, J., Eckert, V., et al. (2002) *Plasmodium falciparum*: glycosylation status of *Plasmodium falciparum* circumsporozoite protein expressed in the baculovirus system. *Exp. Parasitol.* **101,** 64-68.
10. Samuelson, J., Banerjee, S., Magnelli, P., Cui, J., Kelleher, D. J., Gilmore, R., et al. (2005) The diversity of dolichol-linked precursors to Asn-linked glycans likely results from secondary loss of sets of glycosyltransferases. *Proc. Natl. Acad. Sci. USA* **102,** 1548-1553.
11. Tsuboi, T., Takeo, S., Iriko, H., Jin, L., Tsuchimochi, M., Matsuda, S., et al. (2008) Wheat germ cell-free system-based production of malaria proteins for discovery of novel vaccine candidates. *Infect. Immun.* **76,** 1702-1708.

Chapter 9

Protein Engineering Accelerated by Cell-Free Technology

Takuya Kanno and Yuzuru Tozawa

Abstract

Utilization of structural information from homologous proteins to design novel enzymes is one of the practical applications of structural biology. Structure-based protein engineering is a more reasonable strategy compared with general random mutagenesis. Here, we describe a useful method for production of a series of mutant enzymes based on a cell-free translation system. We employed PCR-mediated in vitro site-directed mutagenesis in combination with wheat-embryo cell-free protein synthesis to establish a high-throughput system. The efficient generation of a series of mutant enzymes facilitates high-throughput screening of functionally improved enzymes.

Key words: Feedback inhibition, Metabolic enzyme, Site-directed mutagenesis, Wheat-embryo cell-free protein synthesis system

1. Introduction

In the case of engineering usual metabolic enzymes in vitro, there are limited ways for selecting engineered protein from a pool of mutated proteins. Functionally improved enzymes are hardly selectable from the pool by using their substrate-binding affinities. Enzymes differ from antibodies that exhibit binding affinity to specific ligands. Therefore, the selection of functionally improved enzymes still largely relies on bacterial recombinant systems. However, this in vivo approach also has some limitations, for instance, limited utility of appropriate mutant strains as hosts for mutant plasmid screening or disturbance of host metabolism often caused by expression of engineered enzymes.

On the other hand, structure-based protein engineering has been becoming a reasonable strategy because there are plenty of informative data for three-dimensional structures of many

Y. Endo et al. (eds.), *Cell-Free Protein Production: Methods and Protocols,* Methods in Molecular Biology, vol. 607
DOI 10.1007/978-1-60327-331-2_9, © Humana Press, a part of Springer Science+Business Media, LLC 2010

enzymes in high resolution. Moreover, proteins are now producible by a totally cell-free system. We therefore employed PCR for in vitro site-directed mutagenesis based on information of protein structure, and wheat-embryo cell-free protein synthesis for producing each mutant protein in vitro. We selected rice (*Oryza sativa*) anthranilate synthase, which is a key enzyme in tryptophan biosynthetic pathway, as a test enzyme, and we have established a high-throughput in vitro enzyme engineering system. The utilization of this system allows us to save time by skipping the bacterial recombination steps in the process. For instance, one is able to accomplish the procedure from the preparation of the DNA template by PCR to start the in vitro translation within a day. Although it was a special case, we did not need to purify the synthesized product from the reaction mixture. Therefore, the whole process, from template DNA preparation to protein synthesis, was completed within 2 days by the all-PCR system, whereas the Kunkel mutagenesis method (1) requires more than a week, because the system included a bacterial recombination process. Thus, the cell-free mutation scanning system is a potent technology, which is applicable for the engineering of various kinds of functional proteins.

2. Materials

2.1. PCR-Mediated In vitro Site-Directed Mutagenesis

1. 10× Pyrobest Buffer II from TaKaRa (Shiga, Japan). Store at –20°C.
2. 2.5 mM dNTP mixture: 2.5 mM each of dATP, dCTP, dGTP, and dTTP. Store at –20°C.
3. 5 units/µl Pyrobest DNA polymerase from TaKaRa. Store at –20°C.
4. QIAquick PCR purification kit from QIAGEN.
5. Thermal cycler, GeneAmp PCR system 9700, from Applied Biosystems (Tokyo, Japan).

2.2. Split-Primer PCR

1. 10× *EX Taq* buffer from TaKaRa. Store at –20°C.
2. 2.5 mM dNTP mixture: 2.5 mM each of dATP, dCTP, dGTP, and dTTP. Store at –20°C.
3. 5 units/µl *EX Taq* DNA polymerase from TaKaRa. Store at –20°C.

2.3. In Vitro Transcription of mRNA

1. 5× transcription buffer: 400 mM HEPES-KOH, pH 7.8, 80 mM magnesium acetate, 10 mM spermidine, 50 mM DTT. Store at –20°C.
2. 25 mM NTP mix: 25 mM each of ATP, UTP, GTP, and CTP. Store at –20°C.

3. 80 units/μl RNasin ribonuclease inhibitor from Promega (Madison, WI, USA). Store at –20°C.
4. 80 units/μl SP6 RNA polymerase from Promega. Store at –20°C.
5. 7.5 M ammonium acetate. Store at room temperature.
6. 70% (v/v) and >99% ethanol. Store at room temperature.

2.4. Cell-Free Translation with the Wheat-Embryo Extracts

1. Wheat-embryo extracts (kindly provided by T. Shibui at Zoe Gene, Yokohama, Japan) or WEPRO®1240, 240 OD/ml (Cell Free Sciences, Yokohama, Japan). Store at –80°C.
2. 12 MWCO dialysis cup from Biotech International.
3. 1 mg/ml of Creatine kinase from Roche. The final concentration in translation mixture should be 40 ng/μl. Store in single-use aliquots at –80°C.

2.5. Enzyme Assay

1. 0.2 M Tris–HCl (pH 8.3). Store at room temperature.
2. 1 M NH_4Cl. Store at room temperature.
3. 0.1 M $MgCl_2$. Store at room temperature.
4. 0.1 mM and 1 mM tryptophan (Trp). Store at 4°C.
5. 0.5 mM chorismate (Sigma-Aldrich, Saint Louis, MO). Store in single-use aliquots at –80°C.
6. 1 N HCl. Store at room temperature.
7. 100% ethyl acetate. Store at room temperature.
8. Spectrofluorometer (Wallac 1420 ARVOsx Multi-Label Counter) from Perkin Elmer.

3. Methods

The first round of mutation analysis was done for 36 various mutations. Site-directed mutagenesis of the OASA2 gene was carried by in vitro overlap-extension PCR (2). A schematic diagram of this study is shown in Fig. 9.1. Two separate PCRs were performed to amplify two halves of the OASA2 gene, using four primers. An outside-forward primer (P1, 5′-cctcttccagggcccaATGTGCTCCGCGGGGAAGCC-3′, the underlined sequence is an artificially introduced initiation codon and the lowercase letters indicate the linker sequence for the split-primer method) (3) was paired with a middle-reverse mutation primer (P2) to generate the first half of the gene; an outside-reverse primer (P4, the plasmid-specific primer, 5′-CGTCAGACCCCGTAGAAAAGA-3′) was paired with a middle-forward mutation primer (P3) to synthesize the second half. Mutated amino acids and the sequences of the mutagenic primers are listed in Tables 9.1 and 9.2 (see Note 1).

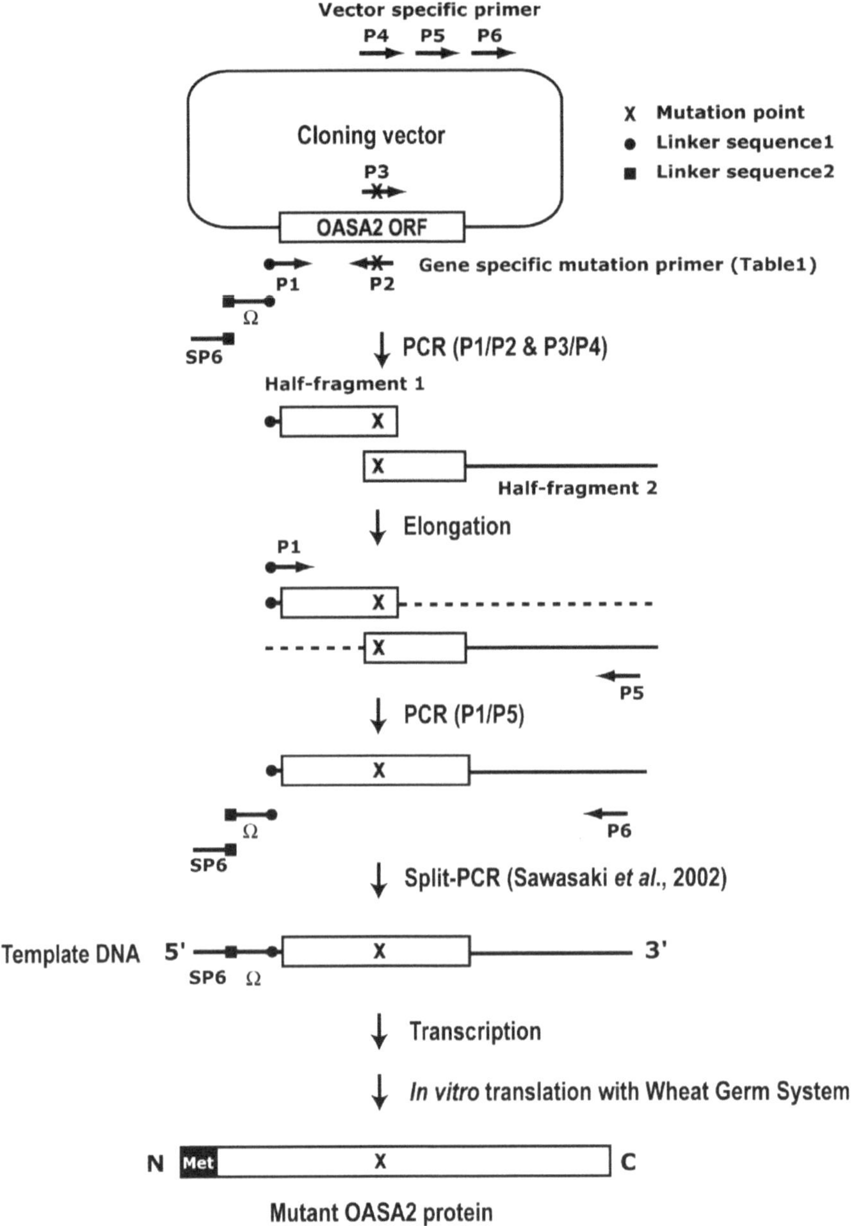

Fig. 9.1. Scheme for site-directed mutagenesis of the OASA2 gene with wheat-embryo cell-free system. P1, an outside-forward primer (S3 linker plus ORF specific sequence); P2, a middle-reverse mutation primer; P3, a middle-forward mutation primer; P4, a plasmid-specific primer, P5 and P6 nested plasmid specific primers; SP6, imperfect SP6 promoter primer; Ω, imperfect SP6 promoter plus OMEGA and S3 linker primer.

Table 9.1
Summary of targeted residues in *OASA2* cDNA

Mutant	Amino acid substitution (codon)	Source and reference
E125A	Glu (GAG)→Ala (GCG)	(7)
E125K	Glu (GAG)→Lys (AAG)	(8)
S126A	Ser (TCC)→Ala (GCC)	(7)
S126F	Ser (TCC)→Phe (TTC)	(8)
Y349A	Tyr (TAC)→Ala (GCC)	(7)
Y349F	Tyr (TAC)→Phe (TTC)	(7)
N351A	Asn (AAT)→Ala (GCA)	(7)
N351D	Asn (AAT)→Asp (GAT)	(7)
N363A	Asn (AAC)→Ala (GCC)	(7)
N363D	Asn (AAC)→Asp (GAC)	(8)
P364A	Pro (CCA)→Ala (GCA)	(7)
P364L	Pro (CCA)→Leu (CTA)	(8)
S365A	Ser (AGT)→Ala (GCT)	(7)
P366A	Pro (CCA)→Ala (GCT)	(7)
Y367A	Tyr (TAC)→Ala (GCC)	(7)
M368A	Met (ATG)→Ala (GCG)	(7)
M368T	Met (ATG)→Thr (ACG)	(8)
A369L	Ala (GCA)→Leu (CTA)	(8)
V371A	Val (GTA)→Ala (GCA)	(7)
V371L	Val (GTA)→Leu (CTA)	(7)
A380S	Ala (GCA)→Ser (AGT)	(8)
Y516A	Tyr (TAC)→Ala (GCC)	(7)
G518A	Gly (GGC)→Ala (GCC)	(7)
L520A	Leu (CTT)→Ala (GCA)	(7)
L520F	Leu (CTT)→Phe (TTT)	(7)
L520V	Leu (CTT)→Val (GTT)	(7)
G521A	Gly (GGA)→Ala (GCA)	(7)
G522A	Gly (GGG)→Ala (GCG)	(7)
G522Y	Gly (GGG)→Tyr (TAC)	(7)

(continued)

Table 9.1 (continued)

Mutant	Amino acid substitution (codon)	Source and reference
G527A	Gly (GGA)→Ala (GCA)	(7)
D528A	Asp (GAC)→Ala (GCC)	(7)
M529A	Met (ATG)→Ala (GCG)	(7)
L530A	Leu (CTT)→Ala (GCT)	(7)
L530D	Leu (CTT)→Asp (GAT)	(7)
I531A	Ile (ATC)→Ala (GCC)	(7)
A532Y	Ala (GCT)→Tyr (TAT)	(8)

Table 9.2
Oligonucleotide primers used for construction of the OASA2 mutants

Primer	Sequence (5′-3′)	Location[a] and orientation
E125A-F	CCAGCTTCCTCTTCgcgTCCGTCGAGCAGGG (31 mer)	359–389, forward
E125A -R	CCCTGCTCGACGGAcgcGAAGAGGAAGCTGG Tm=71.54	359–389, reverse
E125K-F	CCCAGCTTCCTCTTCaagTCCGTCGAGCAGG (31 mer)	358–388, forward
E125K-R	CCTGCTCGACGGActtGAAGAGGAAGCTGGG Tm=68.9	358–388, reverse
S126A-F	AGCTTCCTCTTCGAGgccGTCGAGCAGGGGC (31 mer)	361–391, forward
S126A -R	GCCCCTGCTCGACggcCTCGAAGAGGAAGCT Tm=71.54	361–391, reverse
S126F-F	AGCTTCCTCTTCGAGttcGTCGAGCAGGGGC (31 mer)	361–391, forward
S126F-R	GCCCCTGCTCGACgaaCTCGAAGAGGAAGCT Tm=68.9	361–391, reverse
Y349A-F	TTGAGAGGCGAACAgccGCCAATCCATTTG (30 mer)	1031–1060, forward
Y349A-R	CAAATGGATTGGCggcTGTTCGCCTCTCAA Tm=65.1	1031–1060, reverse
Y349F-F	TTGAGAGGCGAACAttcGCCAATCCATTTG (30 mer)	1031–1060, forward
Y349F-R	CAAATGGATTGGCgaaTGTTCGCCTCTCAA Tm=79.2	1031–1060, reverse
N351A-F	GAGGCGAACATACGCTgcaCCATTTGAAGTCTAT (34 mer)	1035–1068, forward
N351A-R	ATAGACTTCAAATGGtgcAGCGTATGTTCGCCTC Tm=77.43	1035–1068, reverse

(continued)

Table 9.2 (continued)

Primer	Sequence (5′-3′)	Location[a] and orientation
N351D-F	GTTTGAGAGGCGAACGTACGCCgatCCATTTGAAGTCT (38 mer)	1029–1066, forward
N351D-R	AGACTTCAAATGGatcGGCGTACGTTCGCCTCTCAAAC Tm = 67.24	1035–1068, reverse
N363A-F	CTTTACGAATTGTGgccCCAAGTCCATACA (30 mer)	1073–1102, forward
N363A-R	TGTATGGACTTGGggcCACAATTCGTAAAG Tm = 72.51	1073–1102, reverse
N363D-F	CTTTACGAATTGTGgacCCAAGTCCATACA (30 mer)	1073–1102, forward
N363D-R	TGTATGGACTTGGgtcCACAATTCGTAAAG Tm = 72.51	1073–1102, reverse
P364A-F	TACGAATTGTGAACgcaAGTCCATACATGG (30 mer)	1076–1105, forward
P364A-R	CCATGTATGGACTtgcGTTCACAATTCGTA Tm = 74.1	1076–1105, reverse
P364L-F	TACGAATTGTGAACctaAGTCCATACATGG (30 mer)	1076–1105, forward
P364L-R	CCATGTATGGACTtagGTTCACAATTCGTA Tm = 69.35	1076–1105, reverse
S365A-F	CGAATTGTGAACCCAgctCCATACATGGCA (30 mer)	1078–1107, forward
S365A-R	TGCCATGTATGGagcTGGGTTCACAATTCG Tm = 78.81	1078–1107, reverse
P366A-F	AATTGTGAACCCAAGCgctTACATGGCATATGTA (34 mer)	1080–1113, forward
P366A-R	TACATATGCCATGTAagcGCTTGGGTTCACAATT Tm = 76.42	1080–1113, reverse
Y367A-F	GTGAACCCAAGTCCAgccATGGCATACGTACAGGC AAGAGG (41 mer)	1084–1125, forward
Y367A-R	GCCTCTTGCCTGTACGTATGCCATggcTGGACTTG GGTTCAC Tm = 70.7	1084–1125, reverse
M368A-F	AACCCAAGTCCATACgcgGCATACGTACAGGCAA GAGGC (39 mer)	1087–1125, forward
M368A-R	GCCTCTTGCCTGTACGTATGCcgcGTATGGACTTG GGTT Tm = 70.21	1087–1125, reverse
M368T-F	ACCCAAGTCCATACacgGCATATGTACAGG (30 mer)	1088–1117, forward
M368T-R	CCTGTACATATGCcgtGTATGGACTTGGGT Tm = 74.01	1088–1117, reverse
A369L-F	CAAGTCCATACATGctaTATGTACAGGCAA (30 mer)	1091–1120, forward
A369L-R	TTGCCTGTACATAtagCATGTATGGACTTG Tm = 68.59	1091–1120, reverse
V371A-F	CATACATGGCATATgcaCAGGCAAGAGGCT (30 mer)	1097–1126, forward

(continued)

Table 9.2 (continued)

Primer	Sequence (5′-3′)	Location[a] and orientation
V371A-R	AGCCTCTTGCCTGtgcATATGCCATGTATG Tm = 76.63	1097–1126, reverse
V371L-F	CATACATGGCATATctaCAGGCAAGAGGCT (30 mer)	1097–1126, forward
V371L-R	AGCCTCTTGCCTGtagATATGCCATGTATG Tm = 72.05	1097–1126, reverse
A380S-F	GCTGTGTCCTGGTAagtTCTAGTCCAGAAA (30 mer)	1124–1153, forward
A380S-R	TTTCTGGACTAGAactTACCAGGACACAGC Tm = 70.27	1124–1153, reverse
Y516A-F	CAAGACGAGGACCAgccAGTGGCGGCCTTG (30 mer)	1532–1561, forward
Y516A-R	CAAGGCCGCCACTggcTGGTCCTCGTCTTG Tm = 70.56	1532–1561, reverse
G518A-F	GAGGACCATACAGTgccGGCCTTGGAGGGA (30 mer)	1538–1567, forward
G518A-R	TCCCTCCAAGGCCggcACTGTATGGTCCTC Tm = 69.2	1538–1567, reverse
L520A-F	ACCATACAGTGGCGGTgcaGGAGGGATATCATTT (34 mer)	1542–1575, forward
L520A-R	AAATGATATCCCTCCtgcACCGCCACTGTATGGT Tm = 79.43	1542–1575, reverse
L520F-F	CATACAGTGGCGGCtttGGAGGGATATCAT (30 mer)	1544–1573, forward
L520F-R	ATGATATCCCTCCaaaGCCGCCACTGTATG Tm = 76.17	1544–1573, reverse
L520V-F	CATACAGTGGCGGCgttGGAGGGATATCAT (30 mer)	1544–1573, forward
L520V-R	ATGATATCCCTCCaacGCCGCCACTGTATG Tm = 78.35	1544–1573, reverse
G521A-F	ACAGTGGCGGCCTTgcaGGGATATCATTTG (30 mer)	1547–1576, forward
G521A-R	CAAATGATATCCCtgcAAGGCCGCCACTGT Tm = 79.81	1547–1576, reverse
G522A-F	GTGGCGGCCTTGGAgcgATATCATTTGACG (30 mer)	1550–1579, forward
G522A-R	CGTCAAATGATATcgcTCCAAGGCCGCCAC Tm = 66.46	1550–1579, reverse
G522Y-F	AGTGGCGGCCTTGGAtacATATCATTTG (28 mer)	1549–1576, forward
G522Y-R	CAAATGATATgtaTCCAAGGCCGCCACT Tm = 73.48	1549–1576, reverse
G527A-F	GGATATCATTTGACgcaGACATGCTTATCG (30 mer)	1565–1594, forward
G527A-R	CGATAAGCATGTCtgcGTCAAATGATATCC Tm = 73.04	1565–1594, reverse
D528A-F	TATCATTTGACGGAgccATGCTTATCGCTC (30 mer)	1568–1597, forward
D528A-R	GAGCGATAAGCATggcTCCGTCAAATGATA Tm = 75.65	1568–1597, reverse
M529A-F	CATTTGACGGAGACgcgCTTATCGCTCTTG (30 mer)	1571–1600, forward
M529A-R	CAAGAGCGATAAGcgcGTCTCCGTCAAATG Tm = 79.06	1571–1600, reverse

(continued)

Table 9.2 (continued)

Primer	Sequence (5′-3′)	Location[a] and orientation
L530A-F	TTGACGGAGACATGgctATCGCTCTTGCAC (30 mer)	1574–1603, forward
L530A-R	GTGCAAGAGCGATagcCATGTCTCCGTCAA Tm = 78.68	1574–1603, reverse
L530D-F	TTGACGGAGACATGgatATCGCTCTTGCAC (30 mer)	1574–1603, forward
L530D-R	GTGCAAGAGCGATatcCATGTCTCCGTCAA Tm = 77.14	1574–1603, reverse
I531A-F	ACGGAGACATGCTTgccGCTCTTGCACTCC (30 mer)	1577–1606, forward
I531A-R	GGAGTGCAAGAGCggcAAGCATGTCTCCGT Tm = 67.83	1577–1606, reverse
A532Y-F	GAGACATGCTTATCtatCTTGCACTCCGCA (30 mer)	1580–1609, forward
A532Y-R	TGCGGAGTGCAAGataGATAAGCATGTCTC Tm = 73.31	1580–1609, reverse

[a]Nucleotide positions refer to the nucleotide number of the ORF of the *OASA2* gene; the underlined sequence is an artificially introduced restriction enzyme site to check for the introduction of the mutation, and the lowercase letters indicate a mutagenized codon

3.1. PCR-Mediated In Vitro Site-Directed Mutagenesis

1. Add the following reagents in a PCR tube to prepare the first half and the second half of the target gene (OASA2) and mix by pipetting:

   10× Pyrobest buffer II: 5 μl

   2.5 mM dNTP mixture: 4 μl

   10 μM forward primer (P1 or P3): 2.5 μl

   10 μM revese primer (P2 or P4): 2.5 μl

   5 units/μl Pyrobest DNA polymerase: 0.25 μl (see Note 2)

   Template DNA (5 ng/μl pBR-OASA2): 1 μl (see Note 3)

   Nuclease-free water: to 50 μl

2. The PCR conditions are 20 successive cycles of denaturation at 98°C for 15 s, annealing at 60°C for 35 s, and elongation at 72°C for 3 min, followed by a final elongation step at 72°C for 10 min (see Note 4).
3. After checking the PCR product on an agarose gel (see Note 5), purify with a QIAquick PCR purification kit (elute with 50 μl of nuclease-free water) to remove excess primers.
4. These half-fragments bearing overlapping sequences introduced by the two middle primers are then mixed together as follows.

   10× Pyrobest buffer II: 5 μl

   2.5 mM dNTP mixture: 4 μl

Pyrobest DNA polymerase (5 units/μl): 0.25 μl (see Note 2)

Half-fragment 1 (1:25 dilution): 1 μl (see Note 6)

Half-fragment 2 (1:25 dilution): 1 μl (see Note 6)

Nuclease-free water: to 48 μl

5. Subject to three cycles of denaturation at 98°C for 15 s, annealing at 60°C for 30 s, and elongation at 72°C for 3 min.
6. After final step, add 1 μl each of outside-forward primer (P1) and a nested outside-reverse primer (P5, 5′-AGCGTCAGACCCC GTAGAAA-3′) and mix by pipetting (see Note 7).
7. Finally, perform 20 cycles of denaturation at 98°C for 15 s, annealing at 60°C for 35 s, and elongation at 72°C for 3 min, followed by a final elongation step at 72°C for 10 min. These fragments can be used directly as the template DNA for split-primer PCR (3) using 1 μl of a 1:50 dilution of the reaction.
8. OASA2 derivatives bearing double mutations are constructed by the introduction of a second mutation directly to the above-mentioned PCR products, which are single mutants, by using the same PCR-mediated mutation procedure.

3.2. Split-Primer PCR (3)

Nucleotide sequences of the primers used for the split-primer strategy are: SP6 primer, 5′-GCGTAGCATTTAGGTGACACT (the underlined sequence is the 5′-half of the SP6 promoter); Ω primer, 5′-GGTGACACTATAGAAGTATTTTTACAACAATT ACCAACAACAACAACAAACAACAACAACAT TACATTTTACATTCTACAACTACCACCCACCACCA CCAATG (the underlined sequence is the 3′-half of the SP6 promoter and the sequence in italic denotes the complementary region of SP6 primer); P6, 5′-GGAGAAAGGCGGACAGGTAT-3′.

After site-directed mutagenesis by PCR, 1 μl of a 1:50 dilution of mutated DNA fragments were used for a second PCR (50 μl) in the presence of 200 nM SP6 primer and P6 primer and 1 nM Ω primer, 0.2 mM of each dNTP mixture, 0.025 units/μl *ExTaq* DNA polymerase, and the buffer supplied by the manufacturer is set on a GeneAmp PCR system 9700 (2 min denaturation at 95°C followed by 30 cycles of amplification: 98°C for 15 s, 60°C for 35 s, and 72°C for 3 min).

3.3. In Vitro Transcription of mRNA

For the first and second screenings, cDNAs encoding OASA2 mutants were constructed by split-primer PCR (3) with PCR fragments as templates, and the amplified PCR fragments were directly subjected to in vitro transcription for mRNA preparation.

1. Prepare 50 μl of transcription mixture by mixing 10 μl (1/10 volume) of PCR products for template DNA, 10 μl of 5× transcription buffer, 5 μl of 25 mM NTP mix, 0.5 μl of RNasin ribonuclease inhibitor, and 0.625 μl of SP6 RNA polymerase.

2. Gently mix the reaction by pipetting and incubate at 37°C for 3–6 h in a thermal cycler or incubator (see Note 9).
3. After incubation, centrifuge at 22,000×*g* for 3 min to remove a white precipitate of magnesium pyrophosphate in the reaction mixture.
4. Transfer 50 μl of the supernatant to a 1.5-ml centrifuge tube.
5. Add 350 μl of nuclease-free water, 53 μl of 7.5 M ammonium acetate (0.13 vol.), and 1 ml of 100% ethanol (2.5 vol.), mix them well and incubate on ice for 10 min.
6. Pellet the mRNA by centrifugation in a microcentrifuge (22,000×*g*, 20 min, 4°C).
7. After centrifugation, discard the supernatants, and rinse the pellets with 70% ethanol.
8. Allow to air dry and dissolve the dried pellets in 10 μl of nuclease-free water.

3.4. Cell-Free Translation with the Wheat-Embryo Extracts (4)

The translation mixture has a total volume of 50 μl and contains 30 mM HEPES, pH 7.8, 100 mM potassium acetate, 2.7 mM magnesium acetate, 1.2 mM ATP, 0.25 mM GTP, 16 mM creatine phosphate, 0.4 mM spermidine, 0.3 mM of each amino acid, 0.4 mg/mL of creatine kinase, and 0.8 units/μl of RNasin. Wheat germ extract (15 μl) was added from a concentrated commercial preparation to a final OD600 of 50 and the remainder of the total volume was from nuclease-free water. The purified mRNA pellet was dissolved in the translation mixture and the reaction was placed into a 12 MWCO dialysis cup suspended in a buffer reservoir containing all of the above reagents except creatine kinase, RNasin, and wheat germ extract. The reaction was incubated at 26°C for 16–24 h. The OASA2 proteins in the reaction mixture were quantified by immunoblotting, as described previously (5).

3.5. Enzyme Assay (6)

Anthranilate synthase (AS) (E.C. 4.1.3.27) catalyzes the conversion of chorismate to anthranilate, which is a precursor for the biosynthesis of tryptophan as well as secondary metabolites, such as indole 3-acetic acid and various indole alkaloids, in higher plants. In most instances, AS consists of α and β subunits. The α subunit catalyzes the conversion of chorismate to anthranilate with ammonia as the amino donor and is susceptible to feedback inhibition by tryptophan, whereas the β subunit functions as a glutamine amidotransferase, transferring an amido group from glutamine to the α subunit.

The AS activity was assayed as described (5).

1. In a microcentrifuge tube, mix the following:

   Test sample (see Note 10): 5 μl
   0.2 M Tris–HCl (pH 8.3): 10 μl

1 M NH_4Cl: 10 µl
0.1 M $MgCl_2$: 10 µl
Water: to 80 µl

2. To test Trp-feedback inhibition of the enzyme, add 10 µl of H_2O, 0.1 mM or 1 mM Trp to give a final concentration of 0, 10, 100 µM, respectively.
3. Pre-incubate the tube at 32°C for 30 min.
4. Initiate the reactions by adding 10 µl of 5 mM chorismate and incubate at 32°C for 1 h.
5. Terminate the reaction by adding 10 µl of 1 N HCl to each tube.
6. Add 350 µl of 100% ethyl acetate to extract the produced anthranilate and mix thoroughly.
7. Centrifuge at 22,000×*g* for 3 min and transfer 200 µl of the organic phase to a 96-well assay plate.
8. Quantitate with a spectrofluorometer at excitation and emission wavelengths of 340 and 450 nm, respectively.
9. An example of the results produced is shown in Figs. 9.2 and 9.3. The first round of mutation analysis was done for 36 various

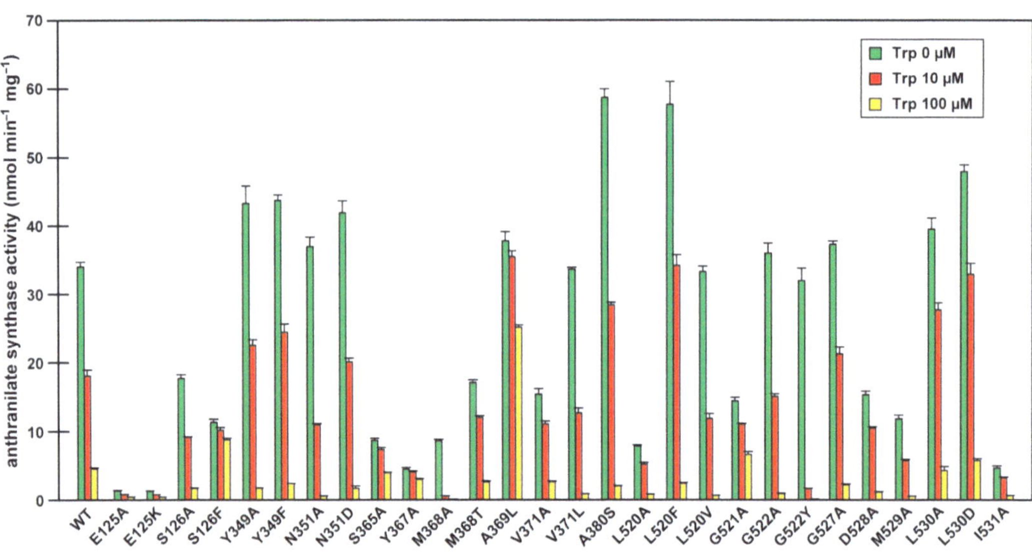

Fig. 9.2. Ammonium-dependent AS activity of the synthesized OASA2 mutant proteins. The AS activities of the OASA2 derivatives, carrying single mutation, synthesized in the translation reaction mixture were determined in the presence of 100 mM NH_4Cl. The assays were performed independently in the presence (10 µM or 100 µM) or absence of Trp. *Letters* under the graph indicate the mutant OASA2 proteins generated by single amino acid substitutions (for example, E125A means the substitution of the Glu residue with Ala at amino acid number 125) or the wild-type protein (wt). The OASA2 proteins in the reaction mixture were quantified by immunoblotting, as described previously (5). The specific activity was calculated based on the amount of OASA2 protein in the reaction mixture. *Bars* indicate standard deviations ($n = 3$). This figure was adapted from Kanno et al. (7), http://www.plantphysiol.org; Copyright American Society of Plant Biologists.

mutations. An ammonium-dependent AS assay was performed at three different concentrations of Trp (0 mM, 10 mM, and 100 mM) for each single AS α-subunit enzyme. This mutation scan revealed several mutation points that affected the enzymatic activity or the Trp-feedback inhibition of the enzyme (Fig. 9.2).

10. The single-mutation analysis identified five feedback insensitive mutations (Fig. 9.2). However, except for A369L, they reduced enzyme catalytic activities (Fig. 9.2). We added another mutation to improve activity of these feedback-insensitive single mutant enzymes. Among these, the addition of the L530D mutation exhibited positive effects on the enzyme catalytic activity, except in the G521 background (Fig. 9.3a–e). Particularly, the combination of Y367A and L530D prominently improved the activity to eightfold of that of the single mutant Y367A (Fig. 9.3c). On the other hand, the performance of the A369L mutant enzyme was not improved by additional mutations (Fig. 9.3d). As for the G521 mutant, there was no positive effect in combinations with other mutations (Fig. 9.3e).

4. Notes

1. Primer design. First, we selected codons in *OASA2* open reading frame sequence, that corresponding to amino residues that are subjects for substitution to other residues. These target codons were changed so that other sequences remain unchanged. For example, Glu (GAG) Ala (GCG). The changed codon sequence was located in center of each primer (30–35 mer). Primers shorter than 40 mer are preferred because that leftover primer could be easily removed by using QIAquick PCR purification kit after PCR reaction. In our case, Tm for each primer was 70–78°C. Estimation of Tm was done by using Nearest Neighbor method (http://www.genosys.jp/whatsnew/tm/tm_syosail.html)
2. Taq DNA polymerase possesses terminal transferase activity, therefore, there might be happening of an excess nucleotide attachment on 3′ end which causes frameshift and unexpected mutation due to its low fidelity. We use high-fidelity α-type enzyme such as Pyrobest.
3. The OASA2 cDNA was previously cloned into pBluescript SK(+), yielding pBROASA2 (6).
4. To void or control generation of an unexpected mutation, cycle number is fixed.

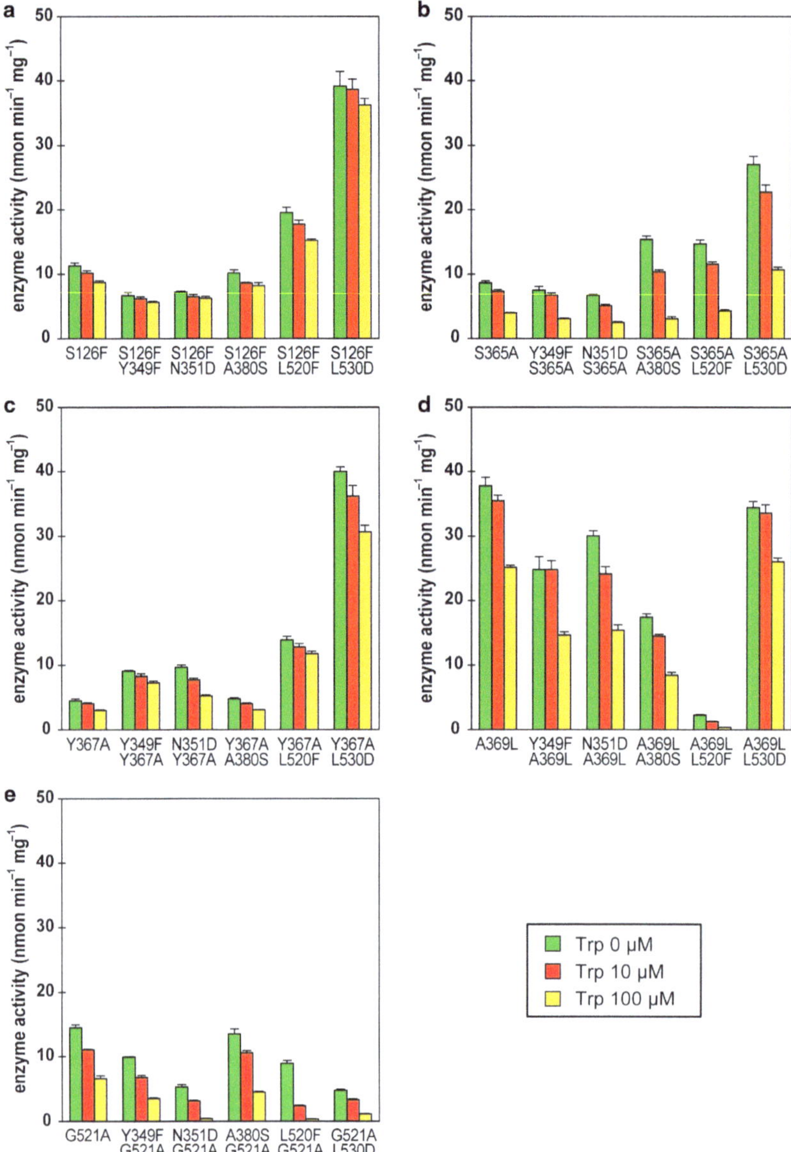

Fig. 9.3. Combined effects of mutations on the enzymatic functions. The second mutation was introduced in the feed-back-insensitive OASA2 derivatives S126F (**a**), S365A (**b**), Y367A (**c**), A369L (**d**), and G521A (**e**). Enzymatic functions of the engineered proteins were examined as described in Fig. 9.2. This figure was adapted from Kanno et al. (7), http://www.plantphysiol.org); Copyright American Society of Plant Biologists.

5. If PCR product resulted in multiple DNA sizes, PCR cycle condition is modified or DNA fragment exhibiting an appropriate size on agarose gel might be excised out and purified for further step.
6. Test between dilution ratios, 1:10–1:50.

7. Soon after the completion of final step, PCR tube was taken out from PCR machine, and primer was added. Prior to the completion of the PCR, we prepare primer mixture (P1/P5) and add them into the mix. The PCR mix was mixed well by gentle pipetting, set on the PCR machine immediately, and start the next PCR program.
8. After this step, we use TaKaRa Ex Taq.
9. White pellet that appears during incubation is magnesium pyrophosphate.
10. Samples are assayed after desalting with a MicroSpin G-25 column (GE Healthcare).

References

1. Kunkel, T.A. (1985) Rapid and efficient site-specific mutagenesis without phenotypic selection. *Proc. Natl. Acad. Sci. U.S.A.* **82**, 488–492.
2. Higuchi, R., Krummel, B., and Saiki, R.K. (1988) A general method of in vitro preparation and specific mutagenesis of DNA fragments: study of protein and DNA interactions. *Nucleic Acids Res.* **16**, 7351–7367.
3. Sawasaki, T., Ogasawara, T., Morishita, R., and Endo, Y. (2002) A cell-free protein synthesis system for high-throughput proteomics. *Proc. Natl. Acad. Sci. USA* **99**, 14652–14657.
4. Madin, K., Sawasaki, T., Ogasawara, T., and Endo, Y. (2000) A highly efficient and robust cell-free protein synthesis system prepared from wheat embryos: Plants apparently contain a suicide system directed at ribosomes. *Proc. Natl. Acad. Sci. USA* **97**, 559–564.
5. Kanno, T., Kasai, K., Ikejiri-Kanno, Y., Wakasa, K., and Tozawa, Y. (2004) In vitro reconstitution of rice anthranilate synthase: distinct functional properties of the alpha subunits OASA1 and OASA2. *Plant Mol. Biol.* **54**, 11–22.
6. Tozawa, Y., Hasegawa, H., Terakawa, T., Wakasa, K. (2001) Characterization of rice anthranilate synthase alpha-subunit genes OASA1 and OASA2. Tryptophan accumulation in transgenic rice expressing a feedback-insensitive mutant of OASA1. *Plant Physiol.* **126**, 1493–1506.
7. Kanno, T., Komatsu, A., Kasai, K., Dubouzet, J.G., Sakurai, M., Ikejiri-Kanno, Y., Wakasa, K., and Tozawa, Y. (2005) Structure-based in vitro engineering of the anthranilate synthase, a metabolic key enzyme in the plant tryptophan pathway. *Plant Physiol.* **138**, 2260–2268.
8. Caligiuri, M.G., and Bauerle, R. (1991) Identification of amino acid residues involved in feedback regulation of the anthranilate synthase complex from *Salmonella typhimurium*. Evidence for an amino-terminal regulatory site. *J. Biol. Chem.* **266**, 8328–8335.

Chapter 10

Cell-Free Protein Production System with the *E. coli* Crude Extract for Determination of Protein Folds

Takanori Kigawa

Abstract

Escherichia coli cell extract-based coupled transcription–translation cell-free system has been developed for large-scale production of protein samples for both X-ray crystallography (selenomethionine substitution) and NMR (stable-isotope labeling). For both cases, higher labeling/substitution efficiency can be achieved compared with the production using cell-based expression system. In addition, as the system is easily adapted to automated and/or high-throughput procedures, it is an especially suitable protein expression method for structural genomics and proteomics project. In this chapter, the procedure for large-scale protein production for structure determination using our *E. coli* cell-free system is presented.

Key words: Prokaryotic coupled transcription–translation, *Escherichia coli* cell extract, Seleomethionine, X-ray crystallography, MAD phasing stable-isotope labeling, NMR spectroscopy

1. Introduction

Protein production process is one of the major bottlenecks in structural biology because the sample is required to be productive, soluble, monodispersed, and so on. Structural genomics and proteomics (SG) is a genome/proteome-scale project of structural biology (1). The process should be scaled up for parallel production of hundreds or thousands of targets for an SG project. It has been widely noticed that the cell-free production is a promising and suitable protein expression method for SG projects (2–5).

We have reported that the cell-free system is useful for preparing selenomethionine-substituted proteins for X-ray crystallography (6) and the application to the structure determination (for example (7, 8)). Incorporation of selenomethionine into

Y. Endo et al. (eds.), *Cell-Free Protein Production: Methods and Protocols,* Methods in Molecular Biology, vol. 607
DOI 10.1007/978-1-60327-331-2_10, © Humana Press, a part of Springer Science+Business Media, LLC 2010

protein can be achieved by simply replacing methionine with selenomethionine and more than 95% of the methionine residues were substituted with selenomethionine (6). Therefore, the cell-free system is certainly a powerful protein expression method for high-throughput structure determinations by X-ray crystallography.

We have also reported that the cell-free production system is extremely useful for three different kinds of stable-isotope labeling of proteins: amino acid-selective (9–11), uniform (10), and site-directed (10, 12–14) labeling. By simply replacing the amino acid(s) of interest in the cell-free reaction mixture with the labeled one(s), amino acid-selective or uniform labeling can be easily achieved. Stereo-array isotope labeling (SAIL) method (15), which is now expected to expand the molecular size limit of NMR spectroscopy, is one of the application of "uniform" stable-isotope labeling using the cell-free system. We developed a novel cell-free system that uses potassium D-glutamate (D-Glu system), enabling highly productive cell-free protein synthesis suitable for stable-isotope labeling (16).

In this chapter, our *Escherichia coli* cell extract-based coupled transcription–translation cell-free system is described. The system is now routinely used for the expression screening of mouse, human, and *Arabidopsis* proteins and protein domains as well as large-scale production such as selenomethionine incorporation for X-ray crystallography and stable isotope labeling for NMR structure determination. We determined more than 200 X-ray structures and 1,300 NMR structures of proteins/protein domains using our *E. coli* cell extract-based system as the protein production method.

2. Materials

2.1. Selenomethionine Incorporation

1. *E. coli* S30 extract (see Chapter 1).
2. LMCP: 160 mM HEPES-KOH buffer (pH 7.5) containing 10.7 % (w/v) PEG 8000, 534 mM potassium glutamate, 5 mM DTT, 3.47 mM ATP (pH 7.0), 2.4 mM GTP (pH 7.0), 2.4 mM CTP (pH 7.0), 2.4 mM UTP (pH 7.0), 96 μg/mL folinic acid·Ca, 1.78 mM cAMP·Na, 74 mM NH_4OAc, and 214 mM creatine phosphate: Store at –20°C.
3. 17.5 mg/mL *E. coli* total tRNA (Roche): Store at –20°C.
4. 5% (w/v) NaN_3: Prepare just before use.
5. 1.6 M $Mg(OAc)_2$. Store at –20°C.

6. 20 mM amino acid mixture (-M)/10 mM DTT: 20 mM each of glutamine, asparagine, arginine, tryptophan, lysine, histidine, phenylalanine, isoleucine, glutamic acid, proline, asparatic acid, glycine, valine, serine, alanine, threonine, cysteine, tyrosine, leucine, and 10 mM DTT. Store at –20°C.
7. 20 mM selenomethionine: Better to prepare just before use.
8. 3.75 mg/mL creatine kinase. Store at –80°C.
9. 10 mg/mL T7 RNA polymerase (see Chapter 1). Store at –20°C.
10. S30 buffer (the same as S30 buffer (B) in Chapter 1): 10 mM Tris-acetate buffer (pH 8.2) containing 14 mM $Mg(OAc)_2$, 60 mM KOAc, 1 mM DTT. Store at –20°C.
11. Dialysis tube (Spectra/Por 7, MWCO: 15 kDa, Spectrum; Cat. 132123): ca. 11 cm for one 3-mL (internal)/30-mL (external) reaction.
12. Dialysis tube closures (Spectra/Por): two pieces for one reaction.
13. A container with enough capacity for the external solution.
14. Template DNA (see Note 1).

2.2. Stable-Isotope Labeling

1. *E. coli* S30 extract (see Chapter 1).
2. LMCP(DGlu): 160 mM HEPES-KOH buffer (pH 7.5) containing 10.7 % (w/v) PEG 8000, 614 mM d-potassium glutamate, 5 mM DTT, 3.47 mM ATP (pH 7.0), 2.4 mM GTP (pH 7.0), 2.4 mM CTP (pH 7.0), 2.4 mM UTP (pH 7.0), 96 µg/mL folinic acid·Ca, 1.78 mM cAMP·Na, 74 mM NH_4OAc, and 214 mM creatine phosphate. Store at –20°C.
3. 17.5 mg/mL *E. coli* total tRNA (Roche). Store at –20°C.
4. 5% (w/v) NaN_3: Prepare just before use.
5. 1.6 M $Mg(OAc)_2$. Store at –20°C.
6. 20 mM amino acid mixture (SI)/10 mM DTT: 20 mM each of stable-isotope labeled glutamine, asparagine, arginine, tryptophan, lysine, histidine, phenylalanine, isoleucine, glutamic acid, proline, asparatic acid, glycine, valine, serine, alanine, threonine, cysteine, tyrosine, leucine, methionine, and 10 mM DTT (see Note 2). Store at –20°C.
7. 3.75 mg/mL creatine kinase. Store at –80°C.
8. 10 mg/mL T7 RNA polymerase (see Chapter 1). Store at –20°C.
9. S30 buffer (the same as S30 buffer (B) in Chapter 1): 10 mM Tris-acetate buffer (pH 8.2) containing 14 mM $Mg(OAc)_2$, 60 mM KOAc, 1 mM DTT. Store at –20°C.

10. Dialysis tube (Spectra/Por 7, MWCO: 15 kDa, Spectrum; Cat. 132123): ca. 11 cm for one 3 mL (internal)/30 mL (external) reaction.
11. Dialysis tube closures (Spectra/Por): two pieces for one reaction.
12. A container with enough capacity for the external solution.
13. Template DNA (see Note 1).

3. Methods

For large-scale protein production in milligram quantities for structure analysis, dialysis-mode cell-free system is usually used. We have developed several formats of the dialysis-mode cell-free system including small-scale and large-scale formats (17). The internal solution in a dialysis tube is dialyzed against the external solution, and the reaction is performed for several hours to overnight (16, 17). In this chapter, the large-scale format of 3-mL internal/30-mL external solutions is described (see Note 3).

3.1. Selenomethionine Incorporation Using Dialysis-Mode Reaction

1. Wash the dialysis tube with water (see Note 4).
2. Thaw all of the reagents on ice-cold water. Do *not* use vortex for mixing creatine kinase and the S30 extract.
3. Prepare the external solution in the container according to Table 10.1.

Table 10.1
Composition of the external solution for selenomethionine incorporation

LMCP	11.2 mL
5 % NaN_3	0.3 mL
S30 buffer	9.0 mL
1.6 M $Mg(OAc)_2$	0.17 mL
20 mM amino acid mixture (-M)/10 mM DTT	2.25 mL
20 mM selenomethionine	2.25 mL
Water	4.83 mL
Total	30 mL

Table 10.2
Composition of the internal solution for selenomethionine incorporation

LMCP	1.12 mL
17.5 mg/mL tRNA	30 µL
5% NaN_3	30 µL
1.6 M $Mg(OAc)_2$	17 µL
20 mM amino acid mixture (-M)/10 mM DTT	225 µL
20 mM selenomethionine	225 µL
3.75 mg/mL creatine kinase.	200 µL
10 mg/mL T7 RNA polymerase	20 µL
S30 extract	0.9 mL
Template DNA	50 µL
Water	183 µL
Total	3 mL

4. Prepare the internal solution except the template DNA according to Table 10.2.
5. Close one end of the dialysis tube with a closure.
6. Add the template DNA to the internal solution and fill it into the dialysis tube.
7. Close another end of the dialysis tube with another closure, and place it into the external solution in the container (Fig. 10.1).
8. Close the container and incubate it at 30°C for 8 h with gentle shaking (see Note 5).
9. Recover the internal solution from the dialysis tube and centrifuge it to precipitate the insoluble fraction.
10. Purify your protein by your conventional method (see Note 6). An example result is shown in Fig. 10.2.

3.2. Stable-Isotope Labeling Using Dialysis-Mode Reaction

1. Wash the dialysis tube with water (see Note 4).
2. Thaw all of the reagents on ice-cold water. Do *not* use vortex for mixing creatine kinase and the S30 extract.
3. Prepare the external solution in the container according to Table 10.3.

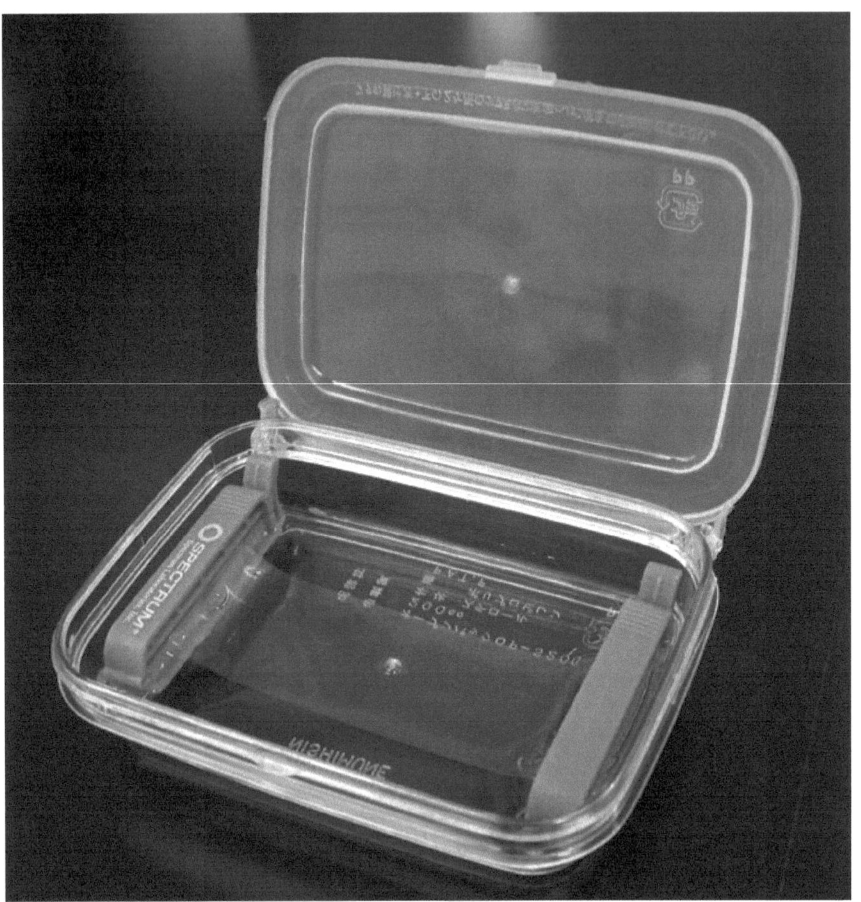

Fig. 10.1. Large-scale dialysis-mode cell-free reaction. Picture of the large-scale dialysis-mode cell-free system (3-mL internal/30-mL external solutions) is shown. A lid of container is closed during the incubation in order to avoid the external solution being dried up.

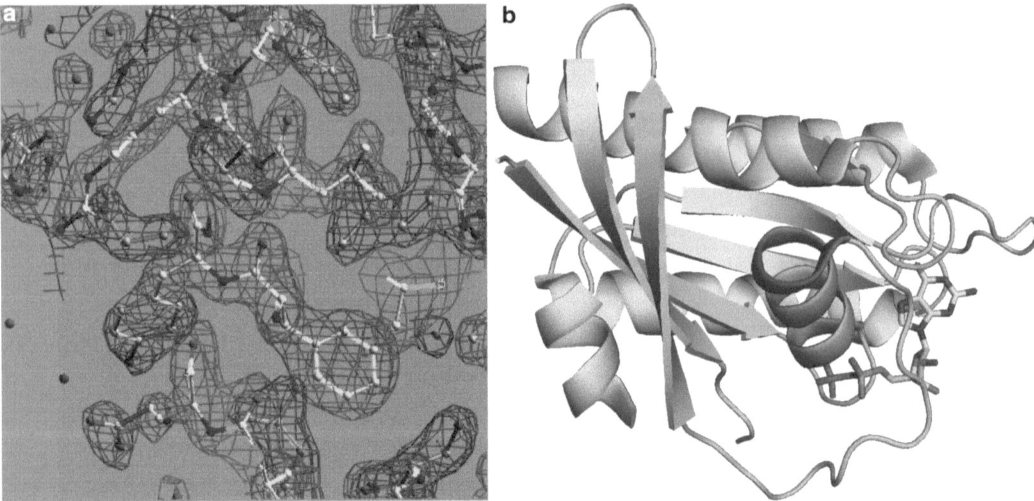

Fig. 10.2. Selenomethionine-substituted Ras protein. (**a**) Electron density map around the Phe156 in helix 5 of the Ras protein. The map was calculated using the experimental MAD phase. (**b**) Ribbon representation of the GDP-bound Ras protein determined by the MAD phasing (PDB code; 1IOZ).

Table 10.3
Composition of the external solution for stable-isotope labeling

LMCP(DGlu)	11.2 mL
5% NaN_3	0.3 mL
S30 buffer	9.0 mL
1.6 M $Mg(OAc)_2$	0.17 mL
20 mM amino acid mixture (SI)/10 mM DTT	2.25 mL
Water	7.08 mL
Total	30 mL

4. Prepare the internal solution except the template DNA according to Table 10.4.
5. Close one end of the dialysis tube with a closure.
6. Add the template DNA to the internal solution and fill it into the dialysis tube.
7. Close another end of the dialysis tube with another closure and place it into the external solution in the container (Fig. 10.1).
8. Close the container and incubate it at 30°C for 8 h with gentle shaking (see Note 5).
9. Recover the internal solution from the dialysis tube and centrifuge it to precipitate the insoluble fraction.
10. Purify your protein by your conventional method. Example results are shown in Figs. 10.3 and 10.4.

4. Notes

1. As previously described (17, 18), the quality of the template DNA is important for successful cell-free protein synthesis. Usually, plasmid DNA purified by the commercially available kits (Qiagen, Promega, etc.) has acceptable quality with little RNase contamination if the plasmid is washed with the pre-wash buffer (40% 1-propanol/4.2 M Guanidine-HCl) just after it is loaded on the column. In addition, the productivity is strongly influenced by the template DNA concentration,

Table 10.4
Composition of the internal solution for stable-isotope labeling

LMCP(DGlu)	1.12 mL
17.5 mg/mL tRNA	30 μL
5% NaN_3	30 μL
1.6 M $Mg(OAc)_2$	17 μL
20 mM amino acid mixture (SI)/10 mM DTT	225 μL
3.75 mg/mL creatine kinase.	200 μL
10 mg/mL T7 RNA polymerase	20 μL
S30 extract	0.9 mL
Template DNA	50 μL
Water	408 μL
Total	3 mL

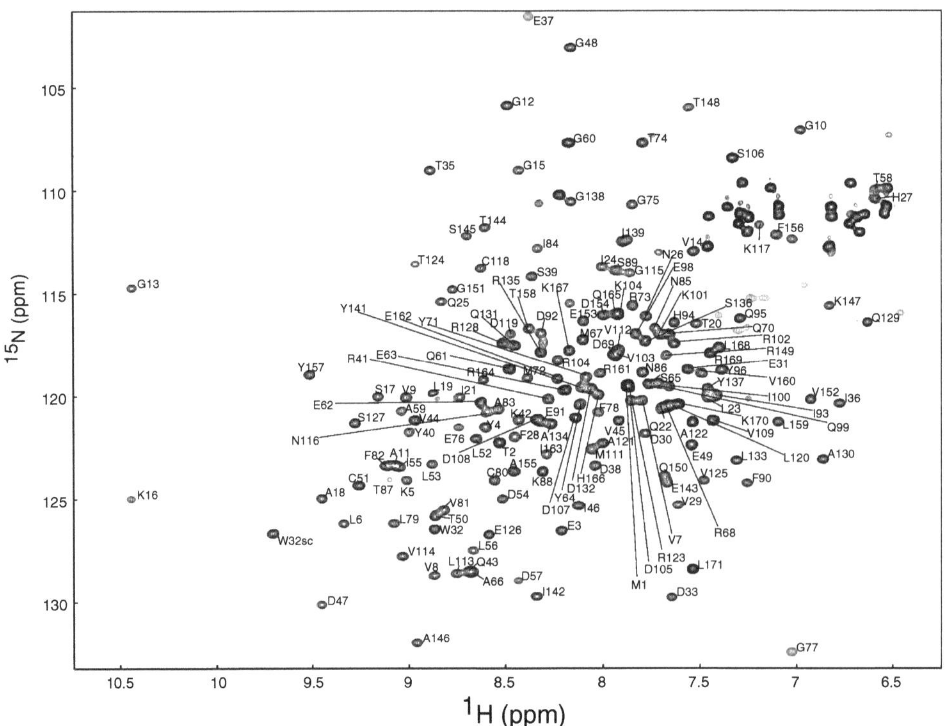

Fig. 10.3. The ^{1}H-^{15}N HSQC spectrum of the uniformly ^{13}C/^{15}N-labeled Ras(Y32W). Assignments are indicated beside the *cross peaks* or defined by *lines*. W32sc denotes the *cross peak* of the side chain amide of Trp32.

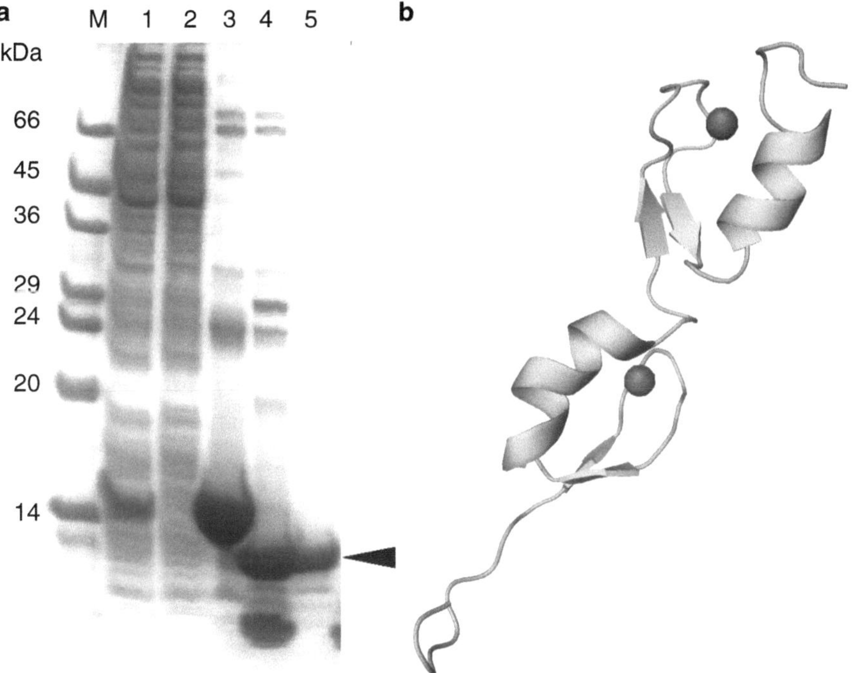

Fig. 10.4. Uniformly $^{13}C/^{15}N$-labeled zf-TRAF domain from human TNF receptor-associated factor 4 protein. (**a**) SDS-PAGE analyses of the supernatant fraction of the cell-free reaction (*lane 1*), the flow-through fraction of the 1st HisTrap chromatography (*lane 2*), the elution fraction of the 1st HisTrap chromatography (*lane 3*), the fraction after TEV protease cleavage (*lane 4*), the flow-through fraction of the second HisTrap chromatography containing purified zf-TRAF domain (*lane 5*), respectively. (**b**) *Ribbon* representation of the structure (PDB code; 2EOD).

suggesting that the optimum concentration should be examined for each protein to obtain the best result. Thanks to the high productivity of our dialysis-mode system and the property of the cell extract we prepared (19) (see Chapter 1). The PCR-amplified linear DNA fragments can be used as a template even in the dialysis-mode reaction though the productivity is approximately 70% of that with the plasmid template (17, 19).

2. If ^{15}N-labeled amino acid(s) is used, ^{15}N labeling can be achieved. If *2*H-labeled one(s) is used, ^{2}H labeling can be done. If SAIL amino acids are used, complicated SAIL labeling can be easily done (15). Instead of expensive purified labeled amino acids, a relatively inexpensive stable-isotope labeled algal amino acid mixture can be used (10). In this case, labeled cysteine, tryptophan, asparagine, and glutamine are still required for uniform labeling, because most of the commercially available algal amino acid mixture lacks these four amino acids. Some vendors are now selling an amino acid mixture consisting of the algal amino acid mixture supplemented with cysteine, tryptophan, asparagine, and glutamine.

3. Reaction scale can be expanded to 9-mL internal/90-mL external format with almost the same productivity per reaction mixture volume if the proportion of the surface area of the dialysis tube to the reaction volume is kept at optimum as previously described (17).
4. In the past, the water used for cell-free protein synthesis was believed to be treated with diethylpyrocarbonate (DEPC) for RNase inactivation. However, in our experience, ultra-pure water prepared by the Milli-Q Synthesis system (Millipore) has sufficient quality for the cell-free synthesis. Thus, this ultra-pure water is routinely used as the RNase-free water for stock solution preparation, apparatus cleaning, and so on and is simply denoted as "water" in this chapter.
5. In general, a longer reaction produces a larger amount of protein. However, a shorter incubation (4–6 h) sometimes gives a better yield because some kind of the protein tends to precipitate during the incubation.
6. Selenomethionine-substituted protein is usually more easily oxidized than the native one. Thus, it is better to keep the sample under a reduced condition in order not to be oxidized.

Acknowledgments

The author would like to thank Drs. Takashi Yabuki, Takayoshi Matsuda, and Makoto Inoue for development of our cell-free system.

References

1. Mittl, P. R., and Grutter, M. G. (2001) Structural genomics: opportunities and challenges. *Curr. Opin. Chem. Biol.* **5,** 402-408.
2. Yokoyama, S., Hirota, H., Kigawa, T., Yabuki, T., Shirouzu, M., Terada, T., Ito, Y., Matsuo, Y., Kuroda, Y., Nishimura, Y., Kyogoku, Y., Miki, K., Masui, R., and Kuramitsu, S. (2000) Structural genomics projects in Japan. *Nat. Struct. Biol.* 7 **Suppl.,** 943-945.
3. Busso, D., Kim, R., and Kim, S. H. (2003) Expression of soluble recombinant proteins in a cell-free system using a 96-well format. *J. Biochem. Biophys. Methods* **55,** 233-240.
4. Sawasaki, T., Ogasawara, T., Morishita, R., and Endo, Y. (2002) A cell-free protein synthesis system for high-throughput proteomics. *Proc. Natl. Acad. Sci. USA* **99,** 14652-14657.
5. Vinarov, D. A., Lytle, B. L., Peterson, F. C., Tyler, E. M., Volkman, B. F., and Markley, J. L. (2004) Cell-free protein production and labeling protocol for NMR-based structural proteomics. *Nat. Methods* **1,** 149-153.
6. Kigawa, T., Yamaguchi-Nunokawa, E., Kodama, K., Matsuda, T., Yabuki, T., Matsuda, N., Ishitani, R., Nureki, O., and Yokoyama, S. (2002) Selenomethionine incorporation into a protein by cell-free synthesis. *J. Struct. Funct. Genomics* **2,** 29-35.
7. Wada, T., Shirouzu, M., Terada, T., Ishizuka, Y., Matsuda, T., Kigawa, T., Kuramitsu, S., Park, S. Y., Tame, J. R., and Yokoyama, S. (2003) Structure of a conserved CoA-binding protein synthesized by a cell-free system. *Acta Crystallogr. D Biol. Crystallogr.* **59,** 1213-1218.

8. Arai, R., Kukimoto-Niino, M., Uda-Tochio, H., Morita, S., Uchikubo-Kamo, T., Akasaka, R., Etou, Y., Hayashizaki, Y., Kigawa, T., Terada, T., Shirouzu, M., and Yokoyama, S. (2005) Crystal structure of an enhancer of rudimentary homolog (ERH) at 2.1 A resolution. *Protein Sci.* **4,** 888-893.
9. Kigawa, T., Muto, Y., and Yokoyama, S. (1995) Cell-free synthesis and amino acid-selective stable isotope labeling of proteins for NMR analysis. *J. Biomol. NMR* **6,** 129-134.
10. Kigawa, T., Yabuki, T., Yoshida, Y., Tsutsui, M., Ito, Y., Shibata, T., and Yokoyama, S. (1999) Cell-free production and stable-isotope labeling of milligram quantities of proteins. *FEBS Lett.* **442,** 15-19.
11. Yabuki, T., Kigawa, T., Dohmae, N., Takio, K., Terada, T., Ito, Y., Laue, E. D., Cooper, J. A., Kainosho, M., and Yokoyama, S. (1998) Dual amino acid-selective and site-directed stable-isotope labeling of the human c-Ha-Ras protein by cell-free synthesis. *J. Biomol. NMR* **11,** 295-306.
12. Hirao, I., Ohtsuki, T., Fujiwara, T., Mitsui, T., Yokogawa, T., Okuni, T., Nakayama, H., Takio, K., Yabuki, T., Kigawa, T., Kodama, K., Nishikawa, K., and Yokoyama, S. (2002) An unnatural base pair for incorporating amino acid analogs into proteins. *Nat. Biotechnol.* **20,** 177-182.
13. Kiga, D., Sakamoto, K., Kodama, K., Kigawa, T., Matsuda, T., Yabuki, T., Shirouzu, M., Harada, Y., Nakayama, H., Takio, K., Hasegawa, Y., Endo, Y., Hirao, I., and Yokoyama, S. (2002) An engineered *Escherichia coli* tyrosyl-tRNA synthetase for site- specific incorporation of an unnatural amino acid into proteins in eukaryotic translation and its application in a wheat germ cell-free system. *Proc. Natl. Acad. Sci. USA* **99,** 9715-9720.
14. Kodama, K., Fukuzawa, S., Nakayama, H., Kigawa, T., Sakamoto, K., Yabuki, T., Matsuda, N., Shirouzu, M., Takio, K., Tachibana, K., and Yokoyama, S. (2006) Regioselective carbon-carbon bond formation in proteins with palladium catalysis; new protein chemistry by organometallic chemistry. *Chembiochem* **7,** 134-139.
15. Kainosho, M., Torizawa, T., Iwashita, Y., Terauchi, T., Mei Ono, A., and Guntert, P. (2006) Optimal isotope labelling for NMR protein structure determinations. *Nature* **440,** 52-57.
16. Matsuda, T., Koshiba, S., Tochio, N., Seki, E., Iwasaki, N., Yabuki, T., Inoue, M., Yokoyama, S., and Kigawa, T. (2007) Improving cell-free protein synthesis for stable-isotope labeling. *J. Biomol. NMR* **37,** 225-229.
17. Kigawa, T., Matsuda, T., Yabuki, T., and Yokoyama, S. (2007) Bacterial cell-free system for highly efficient protein synthesis, in *Cell-free Protein Synthesis Methods and Protocols* (Spirin, A. S., and Swartz, J. R., Eds.), pp. 83-97, Wiley-VCH, Weinheim.
18. Kigawa, T., Yabuki, T., Matsuda, N., Matsuda, T., Nakajima, R., Tanaka, A., and Yokoyama, S. (2004) Preparation of *Escherichia coli* cell extract for highly productive cell-free protein expression. *J. Struct. Funct. Genomics* **5,** 63-68.
19. Seki, E., Matsuda, N., Yokoyama, S., and Kigawa, T. (2008) Cell-free protein synthesis system from *Escherichia coli* cells cultured at decreased temperatures improves productivity by decreasing DNA template degradation. *Anal. Biochem.* **377,** 156-161.

Chapter 11

NMR Assignment Method for Amide Signals with Cell-Free Protein Synthesis System

Toshiyuki Kohno

Abstract

Nuclear magnetic resonance (NMR) methods are widely used to determine the three-dimensional structures of proteins, to estimate protein folding, and to discover high-affinity ligands for proteins. However, one of the problems to apply such NMR methods to proteins is that we should obtain mg quantities of ^{15}N and/or ^{13}C labeled pure proteins of interest.

Here, we describe the method to produce dual amino acid-selective ^{13}C–^{15}N labeled proteins for NMR study using the improved wheat germ cell-free system, which enables sequence-specific assignments of amide signals simply even for very large protein.

Key words: Wheat germ cell-free protein synthesis, NMR, HSQC, HNCO, Dual selective labeling

1. Introduction

NMR methods are widely used to determine the three-dimensional structures of proteins (1–3), to estimate the protein folding (3), and to discover high-affinity ligands for proteins (4). However, one of the problems to apply such NMR methods to proteins is that we should obtain mg quantities of ^{15}N and/or ^{13}C labeled pure proteins of interest. A recently improved wheat germ cell-free system (5, 6) exhibits several attractive features for preparing NMR samples. (1) It is stable over long periods and contains a very low amount of degradation enzymes such as proteases or ribonucleases (6). (2) It can produce proteins in amounts of mg order per mL of reaction volume (6, 7). (3) When it is used for stable-isotope labeling, only the protein of interest can be labeled (7). (4) When it is used with some inhibitors for amino acid metabolic enzymes, the complete amino acid selective labeling may be performed (8). These features are suitable for application to a dual amino

Y. Endo et al. (eds.), *Cell-Free Protein Production: Methods and Protocols,* Methods in Molecular Biology, vol. 607
DOI 10.1007/978-1-60327-331-2_11, © Humana Press, a part of Springer Science+Business Media, LLC 2010

acid-selective ^{13}C–^{15}N labeling method (9–12). This method utilizes protein samples in which the main chain carbonyl carbons of one amino acid type are labeled with ^{13}C and the amide nitrogens of another amino acid type are labeled with ^{15}N. The advantage of this dual labeling method is that sequence-specific assignments of NH signals can be performed even for larger proteins such as IgG (13). However, if in vivo expression system is used for this method, it often suffers from amino acid scrambling problem (9, 11, 14). Wheat germ cell-free protein synthesis system with some inhibitors for amino acid metabolic enzymes can overcome such scrambling problem and give excellent results. In this chapter, I will discuss the procedure for the production of stable isotope labeled proteins for NMR study by using yeast ubiquitin as an example. I will also discuss its application to amide signal assignments with the dual amino acid-selective labeling method.

2. Materials

1. Yeast ubiquitin cDNA.
2. *E. coli* strain JM109.
3. Restriction enzymes.
4. T4 DNA polymerase.
5. T4 DNA ligase.
6. LB medium.
7. Ampicillin.
8. pEU3b vector (see Fig. 11.1. and Note 1) (6).
9. 1 M Hepes-KOH (pH 7.8).
10. Magnesium acetate.
11. Spermidine trihydrochloride.
12. DTT (dithiothreitol).
13. RNase-free water.
14. ATP.
15. UTP.
16. CTP.
17. GTP.
18. SP6 RNA polymerase (Takara, Kyoto, Japan or Promega, Madison, WI).
19. RNase inhibitor (Takara, Kyoto, Japan or Promega, Madison, WI).
20. 7.5 M Ammonium chloride.

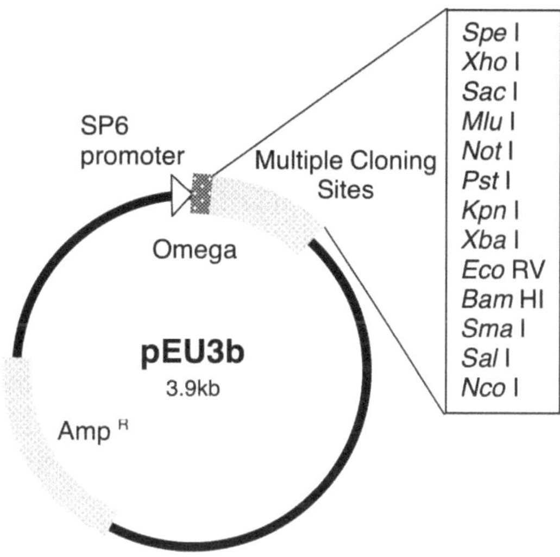

Fig. 11.1. Schematic drawing of the pEU3b cell-free expression plasmid.

21. Ethanol.
22. Quant-iT™ RNA BR assay kit (Invitrogen, Carlsbad, CA).
23. Potassium acetate.
24. Creatine phosphate.
25. Amino acids, ^{15}N-labeled amino acids and ^{13}C′-labeled amino acids (Cambridge Isotope Laboratories, Andover, MA).
26. Wheat germ extract (Cell Free Science, Matsuyama, Japan).
27. Creatine kinase (Roche, Indianapolis, IN).
28. Wheat germ transfer RNA (Sigma-Aldrich, St. Louis, MO) (see Note 2).
29. β-Chloloalanine.
30. Aminooxyacetic acid.
31. L-methioninesulfoximine.
32. Sodium azide.
33. NMR buffer: 50 mM sodium phosphate, pH 6.5, 0.1 M NaCl, and 1 mM 4,4-dimethyl-4-silapentane-1-sulfonic Acid (DSS).

3. Methods

3.1. Plasmid Template for Wheat Germ Cell-free System

3.1.1. Cloning

DNA manipulations were performed by standard recombinant DNA methods (15) to construct the expression plasmid and are not described here in detail due to space limitations. The pEU3b plasmid was digested with *Spe* I and blunted with T4 DNA polymerase. The plasmid was then digested with *Sal* I. The coding region for yeast ubiquitin was prepared as follows. The pET-24a/ubiquitin (16) was digested by *Nde* I, blunted with T4 DNA polymerase, and then digested with *Sal* I. The digested pEU3b plasmid and the coding ubiquitin region were ligated with T4 DNA ligase to produce pEU3b/ubiquitin. The plasmid DNA was then isolated and checked for the presence of the insert and for the correct orientation using restriction enzyme digestions and DNA sequencing.

3.1.2. Plasmid Preparation

The JM109 cells harboring the plasmid pEU3b/ubiquitin were cultured at 37°C overnight in 100 mL of LB medium containing 50 µg/mL of ampicillin. The cells were harvested and the plasmid was extracted according to the standard alkaline lysis with SDS method (15). The extracted plasmid was further purified according to equilibrium centrifugation in CsCl-ethidium bromide gradients method (15). Alternatively, the plasmid pEU3b/ubiquitin may be purified with ion-exchange resin commercially available (Wizard Plus SV, Promega) (15) (see Note 3). Finally, the plasmid was dissolved in RNase-free water. The final concentration of the plasmid was adjusted at 1.0 µg/µL (Tables 11.1 and 11.2).

3.2. mRNA Transcription for Wheat Germ Cell-Free System

3.2.1. mRNA Transcription

1. Prepare a 400 µL of transcription solution for the yeast ubiquitin mRNA by mixing the ingredients listed in Table 11.2 (see also Table 11.1).
2. Incubate the transcription solution for 5 h at 40°C. White precipitant can be seen due to magnesium pyrophosphate after the reaction (see Note 4).
3. Centrifuge the transcription solution (20,000 × *g*; 5 min) at 4°C and remove the white pellet.

Table 11.1
5× Transcription buffer (TB)

Stock	Reagent	Final conc.	
1,000 mM	Hepes-KOH (pH 7.8)	400 mM	4.0 ml
1,000 mM	Magnesium acetate	80 mM	0.8 ml
100 mM	Spermidine	10 mM	1.0 ml
1,000 mM	Dithiothreitol	50 mM	0.5 ml
	RNase-free water		3.7 ml
		Total	10.0 ml

Table 11.2
Transcription solution

Stock	Reagent	Final conc.	
5×	5× TB	1×	80 μL
25 mM	ATP, CTP, GTP, UTP	5.0 mM	80 μL
70 U/μL	RNase inhibitor	1.0 U/μL	6 μL
50 U/μL	SP6 RNA polymerase	1.5 U/μL	12 μL
1.0 μg/μL	pEU3b/ubiquitin	0.1 μg/μL	40 μL
	RNase-free water		182 μL
		Total	400 μL

4. Put the supernatant on ice and add 110 μL of 7.5 M ammonium acetate and 1.0 mL of 100% ethanol (*see* Note 5). Put the tube on ice for 10 min. Centrifuge the solution (20,000 × *g*; 20 min) at 4°C and discard the supernatant. Rinse the pellet with about 300 μL of ice-cold 70% ethanol, then centrifuge (20,000 × *g*; 1 min) again. Remove the supernatant and dry the pellet. Resuspend the pellet with 200 μL of RNase-free water.
5. The synthesized mRNA can be stored at –80°C for several weeks.

3.2.2. Quantitation of Synthesized mRNA

1. Make the Quant-iT™ working solution by diluting the Quant-iT™ Reagent (Invitrogen) 1:200 in Quant-iT™ RNA BR buffer. For example, to prepare enough working solution for two standards and five samples in 200 μL volumes, add 7 μL Quant-iT™ Reagent to 1,393 μL Quant-iT™ RNA BR buffer.
2. Load 190 μL of Quant-iT™ working solution into each of thin-wall, clear 0.5 mL optical-grade real-time PCR tubes.
3. Add 10 μL of each of the two Quant-iT™ RNA BR standards and the transcribed ubiquitin RNA to the tube and mix by vortexing 2–3 s, being careful not to create bubbles. The final volume of each sample tube should be 200 μL. Incubate all tubes at room temperature for 2 min.
4. Calibrate the Qubit fluorometer (Invitrogen) by using two standard RNA solutions prepared above and determine the concentration of the transcribed ubiquitin mRNA.
5. If you want to verify that the mRNA is not degraded by trace contaminations of RNases, analyze the mRNA by agarose gel

Table 11.3
5× Translation solution

Stock	Reagent	Final conc.	
1.0 M	Hepes-KOH (pH 7.8)	100 mM	1.0 ml
4.0 M	Potassium acetate	378 mM	945 μL
1.0 M	Magnesium acetate	8.0 mM	80 μL
100 mM	Spermidine trihydrochloride	2.0 mM	200 μL
1.0 M	Dithiothreitol	10 mM	100 μL
2.5 mM each	Labeled or nonlabeled amino acids	1.2 mM	4.8 mL
100 mM	ATP	6.0 mM	600 μL
20 mM	GTP	1.3 mM	650 μL
500 mM	Creatine phosphate	80 mM	1.6 μL
	RNase-free water		25 μL
		Total	10.0 mL

electrophoresis using standard protocol (15). Usually, ladder bands or smear bands around 1–3 kb are visible when the good mRNA is obtained.

6. Adjust the concentration of mRNA at about 0.5–1.0 μg/μL. The mRNA can be stored for several weeks at –80°C.

3.3. Protein Synthesis by Wheat Germ Cell-free Protein Synthesis System

3.3.1. Preparation of Uniformly Labeled Proteins for NMR Analyses

1. Prepare 5× Translation Solution described in Table 11.3. 5× Translation Solution can be stored at –20°C.
2. Mix 200 μL of 5× Translation Solution, 325 μL of wheat germ extract (OD = 150), 10 μL of wheat germ tRNA (20 μg/μL), 10 μL of creatine kinase (40 μg/μL), 14 μL of RNase inhibitor (40 units/μL), and 165 μg of the transcribed yeast ubiquitin mRNA prepared in Subheading 3.2. Adjust the volume to 1.0 mL with RNase-free water.
3. Load the 1 mL Float-A-Lyzer G2 (Spectrum Laboratories) tube with the Translation Solution. Dialyze the solution against about 15 mL of the Dialysis Buffer A (see Table 11.4) in a sterile tube. Incubate the Translation Solution at 26°C for 2 days with stirring of the Dialysis Buffer A.
4. After 2 days, add concentrated yeast ubiquitin mRNA to the Translation Solution and change the Dialysis Buffer for new one. Incubate at 26°C for two more days with stirring of the Dialysis Buffer (*see* Note 6).

Table 11.4
Dialysis buffer A

Stock	Reagent	Final conc.	
1.0 M	Hepes-KOH (pH 7.8)	30 mM	1.5 mL
4.0 M	Potassium acetate	100 mM	1.25 mL
1.0 M	Magnesium acetate	2.7 mM	135 μL
100 mM	Spermidine trihydrochloride	0.4 mM	200 μL
1.0 M	Dithiothreitol	2.5 mM	125 μL
2.5 mM each	Labeled or nonlabeled amino acids	0.3 mM	6.0 mL
100 mM	ATP	1.2 mM	600 μL
20 mM	GTP	0.25 mM	625 μL
500 mM	Creatine phosphate	16 mM	1.6 mL
0.5%	Sodium azide	0.005%	500 μL
	RNase-free water		37.5 mL
		Total	50.0 mL

3.3.2. Preparation of Single or Dual Amino Acid Selectively Labeled Proteins for NMR Analyses

1. In this chapter, ^{15}N-Val selective labeling is shown as an example for single amino acid-selective labeling. All the procedures are the same as those written in Subheading 3.3.1, except that ^{15}N-Val should be used instead of nonlabeled Val in Tables 11.3 and 11.4. In order to label ^{15}N-Ala, ^{15}N-Asp, or ^{15}N-Glu, the labeled amino acids should be used instead of nonlabeled amino acids in Tables 11.3 and 11.4 and the Dialysis Buffer B (see Table 11.5) should be used instead of Dialysis Buffer A (see Note 7). All the procedures are the same as those written in Subheading 3.3.1.
2. In order to prepare dual amino acid-selective labeled proteins, a ^{13}C′-labeled amino acid and a ^{15}N-labled amino acid should be used instead of nonlabeled amino acids and the Dialysis Buffer B (see Table 11.5) should be used instead of Dialysis Buffer A in Table 11.4 (see Note 7). All the other procedures are the same as those written in Subheading 3.3.1. In this chapter, ^{13}C′-Glu/^{15}N-Val labeling and ^{13}C′-Phe/^{15}N-Val labeling are demonstrated.

3.3.3. Analysis of Synthesized Proteins

1. Recover the Translation Solution with long and narrow plastic pipette. Take 5 μL of the protein synthesis solution. Centrifuge it (20,000 × *g*; 20 min) and retain both supernatant and pellet.

Table 11.5
Dialysis buffer B

Stock	Reagent	Final conc.	
1.0 M	Hepes-KOH (pH 7.8)	30 mM	1.5 mL
4.0 M	Potassium acetate	100 mM	1.25 mL
1.0 M	Magnesium acetate	2.7 mM	135 μL
100 mM	Spermidine trihydrochloride	0.4 mM	200 μL
1.0 M	Dithiothreitol	2.5 mM	125 μL
2.5 mM each	Labeled or nonlabeled amino acids	0.3 mM	6.0 mL
100 mM	ATP	1.2 mM	600 μL
20 mM	GTP	0.25 mM	625 μL
500 mM	Creatine phosphate	16 mM	1.6 mL
200 mM	Aminoocyacetic acid	1 mM	250 μL
200 mM	L-Methioninesulfoximine	0.1 mM	25 μL
200 mM	β-Chloro-L-alanine	7 mM	1.8 mL
0.5%	Sodium azide	0.005%	500 μL
	RNase-free water		35.4 mL
		Total	50.0 mL

2. Dissolve the pellet with 5 μL of water. Load both the supernatant and pellet fractions or both the total and the supernatant fractions by Coomassie brilliant blue staining of a 15–25% gradient SDS-PAGE gel (see Fig. 11.2). Verify that yeast ubiquitin is synthesized and is in the supernatant fraction.
3. Estimate the concentration of the synthesized yeast ubiquitin by comparing the band in SDS-PAGE gel with those of molecular weight markers.
4. The synthesized solution can be stored for a week at 4°C or for several months at –20°C.

3.4. NMR Sample Preparation of Yeast Ubiquitin Synthesized by Wheat Germ Cell-free System

1. Put the 1 mL of protein synthesis solution into Centricon YM-3 and centrifuge (5,000 × *g*) at 4°C.
2. Centrifuge until the solution is concentrated to the volume of 200 μL (about a few hours).
3. Equilibrate two MicroSpin G-25 Columns with NMR sample buffer. At first, resuspend the resin in the column by vortexing gently.

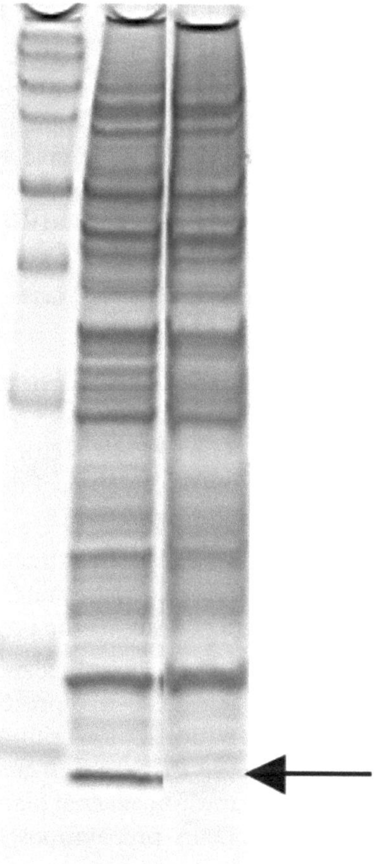

Fig. 11.2. SDS-PAGE analysis of stable-isotope labeled yeast ubiquitin synthesized by wheat germ cell-free protein synthesis system. *Left lane*: molecular weight markers. *Middle lane*: yeast ubiquitin translation solution. *Right lane:* wheat germ alone (control). Yeast ubiquitin is indicated by an *arrow.*

4. Snap off the bottom closure and place the column in 1.5-mL microcentrifuge tubes. Then, prespin the column for 1 min at 735 × *g*.
5. Discard the solution in the microcentrifuge tubes and apply 250 μL of NMR sample buffer to each column. Spin the columns for 1 min at 735 × *g* and discard the buffer in the microcentrifuge tubes. Repeat this procedure eight times. Alternatively, if available, a MicroPlex 24 Vacuum (Amersham Biosciences) system can be useful to equilibrate MicroSpin G-25 Columns.
6. Place the columns in new 1.5-mL microcentrifuge tubes. Apply 100 μL of the synthesized protein to each tube.
7. Spin the columns for 2 min at 735 × *g*. Collect the NMR sample from the bottom of the microcentrifuge tubes (see Note 8).

Adjust the sample volume to 225 μL with NMR sample buffer. Add 25 μL of D_2O.

8. Alternatively, buffer exchange can be done by repeated dilution with NMR sample buffer and concentration using Centricon YM-3. For example, four repeats of fivefold dilution and concentration with NMR sample are sufficient for buffer exchange (see Notes 9 and 10).

4. Notes

1. The pEU3b vector cell-free expression vector is utilized very effectively in producing large amount of proteins with wheat germ system (15).
2. Wheat germ extract is commercially available from Cell Free Science (Matsuyama, Japan). Wheat germ extract can be prepared according to the protocol described (5).
3. In order to obtain a large amount of proteins with this wheat germ cell-free protein synthesis system, even trace contaminations of ribonucleases should be avoided throughout the transcription and translation steps. Therefore, a great care must be taken if plasmid purification kits are used for template DNA preparations because these kits always use a solution including RNase A in the first step of plasmid purification.
4. SP6 RNA polymerase is usually used at 37°C. However, this enzyme is more active at 40°C than 37°C by 30%. Therefore, 40°C is used in this protocol.
5. In the step of ethanol precipitation, the use of sodium acetate instead of ammonium acetate should be avoided because trace sodium salt may interfere the translation reaction.
6. With this wheat germ cell-free system, the translation reaction proceeds over a week, when mRNA is supplied and the Dialysis Buffer A or B is changed every 2 days. If much more protein is needed, continue the translation reaction. Some antibiotics can be added to the Dialysis Solution in order to avoid bacterial contaminations.
7. Some transaminases and glutamine synthase are active in wheat germ extract (8), selective labeling is not successful without inhibitors in the case of Ala, Asp, and Glu. In the case of Ala selective labeling, β-chloroalanine can be used to inhibit the activity of alanine transaminase. In the case of Asp selective labeling, aminooxyacetic acid can be used to inhibit the activity of aspartate transaminase. In the case of Glu selective labeling, both aminoocyacetic acid and L-methionine sulfoximine

can be used to inhibit the activities of aspartate transaminase and glutamine synthase. These inhibitors do not inhibit the protein synthesis activity of the wheat germ extract (8).

8. White precipitation of proteins from wheat germ extract can be seen when the buffer of the translation solution is exchanged for that with lower pH. Remove the precipitation by centrifuge (20,000 × *g*, 15 min).
9. Figure 11.3 shows the ^{1}H–^{15}N HSQC spectrum of nonpurified yeast ubiqutin prepared by wheat germ in vitro system. All the signals visible in Fig. 11.3 match to those visible in the spectrum of purified yeast ubiquitin prepared by *E. coli* cells (data not shown) and no other signals derived from some contaminations of other proteins are visible. This indicates that yeast ubiquitin prepared by wheat germ system assumes a correct fold and that no purification is necessary to measure

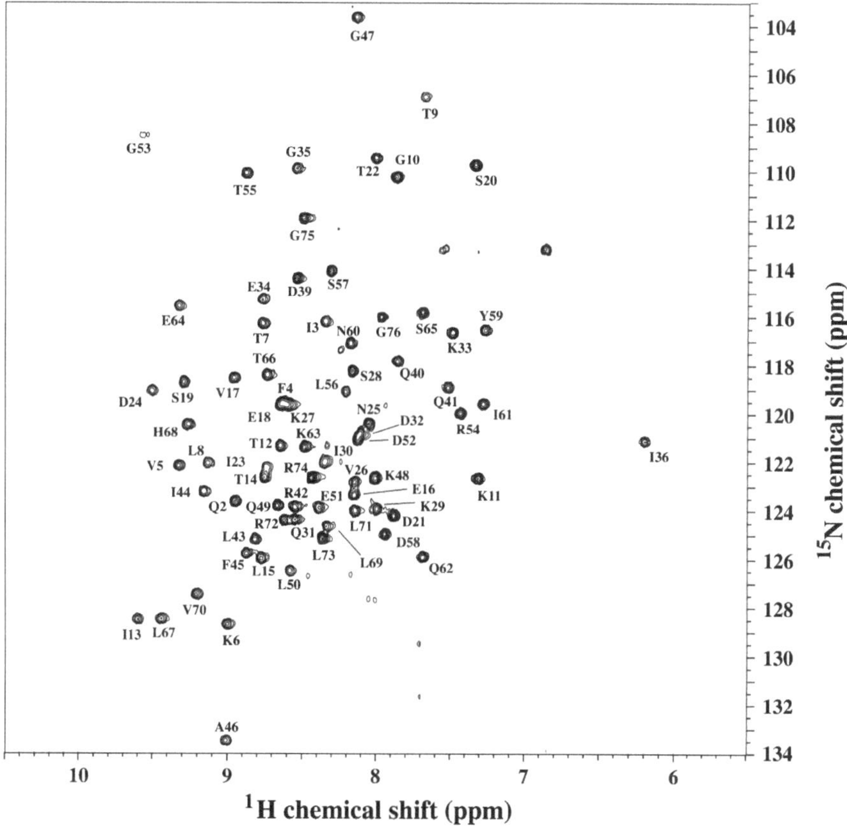

Fig. 11.3. ^{1}H–^{15}N HSQC spectrum of yeast ubiquitin synthesized by wheat germ cell-free protein synthesis system, not purified. All NMR measurements were performed on a Bruker Avance500 spectrometer at 30°C. A 2D ^{1}H–^{15}N HSQC spectrum (17) was acquired with 64 (t_1) × 1,024 (t_2) complex points. The ^{1}H shifts were referenced to the methyl resonance of 4,4-dimethyl-4-silapentane-1-sulfonic acid (DSS), used as an internal standard. The ^{15}N chemical shifts were indirectly referenced using the ratio (0.101329118) of the zero-point frequencies to ^{1}H (18). Spectral widths are 1,600 and 6,250 Hz in *F1* and *F2*, respectively.

MQIFVKTLTGKTITLEVESSDTIDN 25
VKSKIQDKEGIPPDQQRLIFAGKQL 50
EDGRTLSDYNIQKESTLHLVLRLRG 75
G

Fig. 11.4. The amino acid sequence of yeast ubiquitin. There are four Val residues in yeast ubiquitin and they are *underlined*.

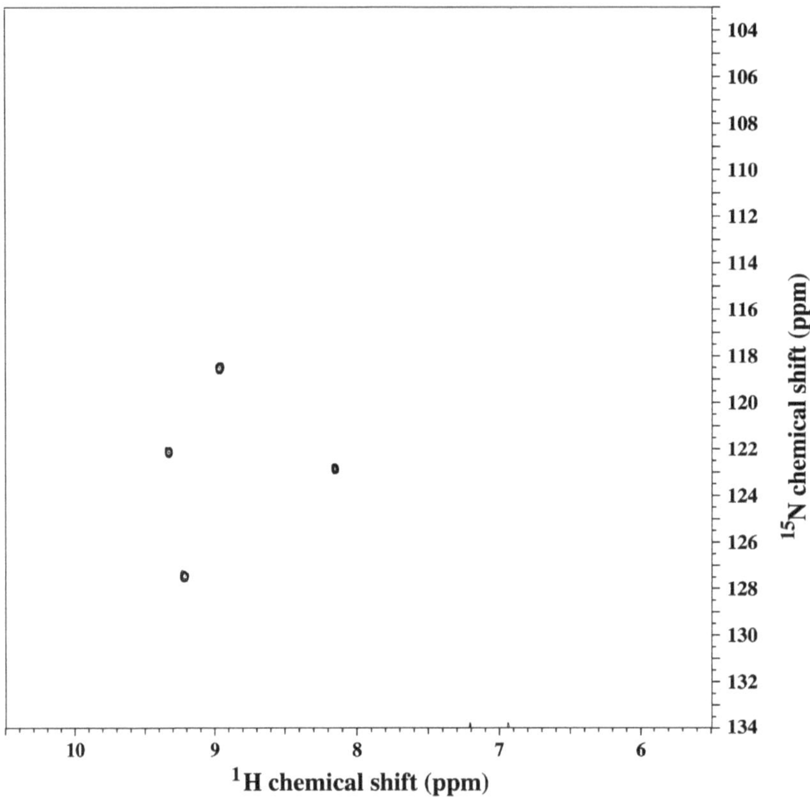

Fig. 11.5. ^{1}H–^{15}N HSQC spectra of ^{15}N-Val selectively labeled yeast ubiquitin synthesized by wheat germ cell-free protein synthesis system. All amide signals derived from Val were observed. The conditions for spectral measurements are same as that in Fig. 11.3.

^{1}H–^{15}N HSQC spectrum of proteins prepared by wheat germ system. This feature may be useful to screen the "foldedness" (3) of proteins of interest very rapidly by NMR.

10. Under optimized conditions, amino acid metabolic enzymes can be completely inhibited. Therefore, the wheat germ cell-free system is very useful for amino acid-selective monitoring in the ^{1}H–^{15}N HSQC spectra of proteins (Figs. 11.4 and 11.5) and also dual amino acid-selective labeling method (Fig. 11.6).

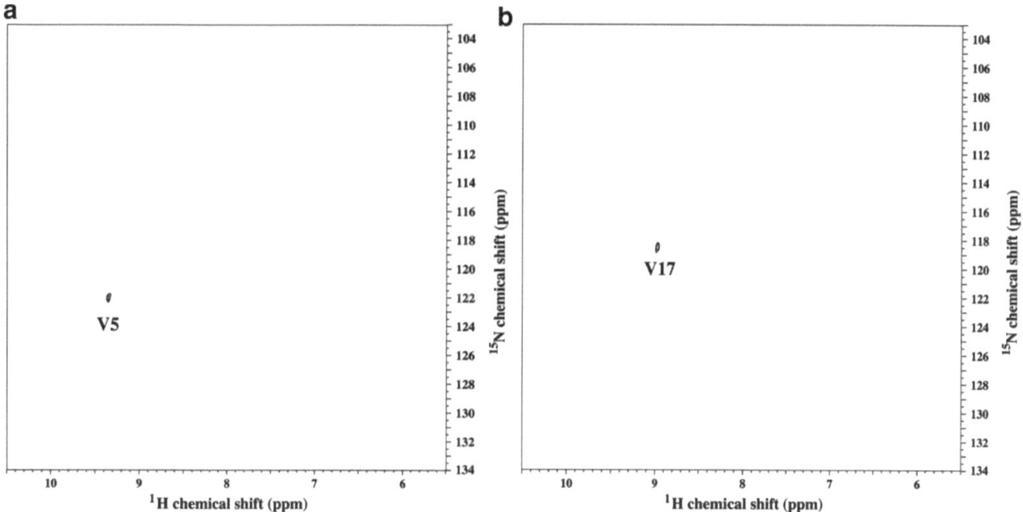

Fig. 11.6. 2D ^{1}H–^{15}N HNCO (19) spectra of yeast ubiquitin synthesized with dual amino acid-selective ^{13}C–^{15}N labeling method [(**a**) ^{13}C′-Phe/^{15}N-Val and (**b**) ^{13}C′-Glu/^{15}N-Val]. The conditions for spectral measurements are same as that in Fig. 11.3. There is only one Phe–Val sequence in the yeast ubiquitin: Phe^{4}–Val^{5}, and one *cross peak* derived from Val^{5} amide was observed in (**a**). And there is only one Glu–Val sequence in the yeast ubiquitin: Glu^{16}–Val^{17}, and one *cross peak* derived from Val^{17} amide was observed in (**b**).

References

1. Wüthrich, K. (2000). Protein recognition by NMR. *Nat. Struct. Biol.* 7, 188-189.
2. Yokoyama, S. et al. (2000). Structural genomics projects in Japan. *Nat. Struct. Biol.* 7 **Suppl,** 943-945.
3. Montelione, G.T., Zheng, D., Huang, Y.J., Gunsalus, K.C. and Szyperski, T. (2000). Protein NMR spectroscopy in structural genomics. *Nat. Struct. Biol.* 7 **Suppl,** 982-985.
4. Shuker, S.B., Hajduk, P.J., Meadows, R.P. and Fesik, S.W. (1996). Discovering high-affinity ligands for proteins: SAR by NMR. *Science* **274,** 1531-1534.
5. Madin, K., Sawasaki, T., Ogasawara, T. and Endo, Y. (2000). A highly efficient and robust cell-free protein synthesis system prepared from wheat embryos: plants apparently contain a suicide system directed at ribosomes. *Proc. Natl. Acad. Sci. USA* **97,** 559-564.
6. Sawasaki, T., Ogasawara, T., Morishita, R. and Endo, Y. (2002). A cell-free protein synthesis system for high-throughput proteomics. *Proc. Natl. Acad. Sci. USA* **99,** 14652-14657.
7. Morita, E.H., Sawasaki, T., Tanaka, R., Endo, Y. and Kohno, T. (2003). A wheat germ cell-free system is a novel way to screen protein folding and function. *Protein Sci.* **12,** 1216-1221.
8. Morita, E.H., Shimizu, M., Ogasawara, T., Endo, Y., Tanaka, R. and Kohno, T. (2004). A novel way of amino acid-specific assignment in ^{1}H-^{15}N HSQC spectra with a wheat germ cell-free protein synthesis system. *J. Biomol. NMR* **30,** 37-45.
9. Kainosho, M. and Tsuji, T. (1982). Assignment of the three methionyl carbonyl carbon resonances in Streptomyces subtilisin inhibitor by a carbon-13 and nitrogen-15 double-labeling technique. A new strategy for structural studies of proteins in solution. *Biochemistry* **21,** 6273-6279.
10. Kainosho, M., Nagao, H., Imamura, Y., Uchida, K., Tomonaga, N., Nakamura, Y. and Tsuji, T. (1985). Structural studies of a protein using the assigned back-bone carbonyl carbon-13 NMR resonances. *J. Mol. Struct.* **126,** 549-562.
11. Kainosho, M., Nagao, H. and Tsuji, T. (1987). Local structural features around the C-terminal segment of Streptomyces subtilisin inhibitor studied by carbonyl carbon nuclear magnetic resonances three phenylalanyl residues. *Biochemistry* **26,** 1068-1075.
12. Yabuki, T. et al. (1998). Dual amino acid-selective and site-directed stable-isotope labeling of the human c-Ha-Ras protein by cell-free synthesis. *J. Biomol. NMR* **11,** 295-306.

13. Kato, K., Matsunaga, C., Nishimura, Y., Waelchli, M., Kainosho, M. and Arata, Y. (1989). Application of ^{13}C nuclear magnetic resonance spectroscopy to molecular structural analyses of antibody molecules. *J. Biochem.* **105,** 867-869.
14. McIntosh, L.P. and Dahlquist, F.W. (1990). Biosynthetic incorporation of ^{15}N and ^{13}C for assignment and interpretation of nuclear magnetic resonance spectra of proteins. *Q. Rev. Biophys.* **23,** 1-38.
15. Sambrook, J. and Russell, D.W. (2001) Molecular Cloning, A Laboratory Manual, 3rd Ed., Cold Spring Harbor Laboratory Press. New York.
16. Kohno, T., Kusunoki, H., Sato, K. and Wakamatsu, K. (1998). A new general method for the biosynthesis of stable isotope-enriched peptides using a decahistidine-tagged ubiquitin fusion system: an application to the production of mastoparan-X uniformly enriched with ^{15}N and $^{15}N/^{13}C$. *J. Biomol. NMR* **12,** 109-121.
17. Schleucher, J., Schwendinger, M., Sattler, M., Schmidt, P., Schedletzky, O., Glaser, S.J., Sorensen, O.W. and Griesinger, C. (1994). A general enhancement scheme in heteronuclear multidimensional NMR employing pulsed field gradients. *J. Biomol. NMR* **4,** 301-306.
18. Wishart, D.S., Bigam, C.G., Yao, J., Abildgaard, F., Dyson, H.J., Oldfield, E., Markley, J.L. and Sykes, B.D. (1995). ^{1}H, ^{13}C, ^{15}N chemical shift referencing in biomolecular NMR. *J. Biomol. NMR* **6,** 135-140.
19. Ikura, M., Kay, L.E. and Bax, A. (1990). A novel approach for sequential assignment of ^{1}H, ^{13}C, and ^{15}N spectra of proteins: heteronuclear triple-resonance three-dimensional NMR spectroscopy. Application to calmodulin. *Biochemistry* **29,** 4659-4667.

Chapter 12

Cell-Free Protein Synthesis Technology in NMR High-Throughput Structure Determination

Shin-ichi Makino, Michael A. Goren, Brian G. Fox, and John L. Markley

Abstract

This chapter describes the current implementation of the cell-free translation platform developed at the Center for Eukaryotic Structural Genomics (CESG) and practical aspects of the production of stable isotope-labeled eukaryotic proteins for NMR structure determination. Protocols are reported for the use of wheat germ cell-free translation in small-scale screening for the level of total protein expression, the solubility of the expressed protein, and the success in purification as predictive indicators of the likelihood that a protein may be obtained in sufficient quantity and quality to initiate structural studies. In most circumstances, the small-scale reactions also produce sufficient protein to permit bioanalytical and functional characterizations. The protocols incorporate the use of robots specialized for small-scale cell-free translation, large-scale protein production, and automated purification of soluble, His_6-tagged proteins. The integration of isotopically labeled proteins into the sequence of experiments required for NMR structure determination is outlined, and additional protocols for production of integral membrane proteins in the presence of either detergents or unilamellar liposomes are presented.

Key words: Eukaryotic proteins, Human proteins, Membrane proteins, Wheat germ extract, Transcription, Translation, Screening for protein production and solubility

1. Introduction

The major bottleneck in high-throughput protein structure determination remains the production of protein that is properly folded and soluble. This is particularly true for eukaryotic proteins, which are often considered to be more difficult than prokaryotic proteins to prepare in the quality and quantity needed for structure determination. The Center for Eukaryotic Structural Genomics (CESG), a specialized research center of the NIH-funded Protein Structure Initiative has developed one platform for eukaryotic protein production that relies on wheat germ cell-free

Y. Endo et al. (eds.), *Cell-Free Protein Production: Methods and Protocols,* Methods in Molecular Biology, vol. 607
DOI 10.1007/978-1-60327-331-2_12, © Humana Press, a part of Springer Science+Business Media, LLC 2010

technology (1–4). Results have shown that CESG's cell-free platform is economically competitive with *Escherichia coli* cell-based methods for protein production (5). Importantly, on average, more structural genomics targets have been prepared successfully for NMR structure analysis by the cell-free platform than by the cell-based platform, although both platforms are useful, because some targets that fail by cell-free translation are successful when produced from *E. coli*. This chapter describes the current CESG platform for cell-free translation and practical aspects of the production of stable isotope-labeled proteins for NMR structure determination. Although all our applications have been to solution state NMR, the same protein production methods are applicable for the production of proteins for solid state NMR spectroscopy, which is not yet considered "high-throughput" but is progressing rapidly in this direction.

1.1. Role of NMR in High-Throughput Investigations of Proteins

In our laboratory, the workhorse NMR spectrometers are ones operating at 600 MHz and equipped with triple-resonance (^{1}H, ^{13}C, ^{15}N plus ^{2}H lock) cryogenic probes. Such spectrometers are adequate generally for high-throughput investigations of proteins or protein complexes of molecular weight 25 kDa or less. It can be useful to have access to spectrometers operating at higher fields for larger proteins or proteins that yield less than ideal spectra.

We first prepare a protein sample labeled uniformly with ^{15}N by translating the mRNA in the presence of a complete mixture of ^{15}N-labeled amino acids. The ^{1}H-^{15}N two-dimensional NMR spectrum of this sample is used as the basis for evaluating both the protein and solution conditions for their suitability for a structure determination. The criteria are that the number of peaks observed should be >80% of the theoretical number and that the peaks should be of roughly uniform intensity. If these criteria are not met, the solution conditions or the amino acid sequence are changed and the test is repeated. If the criteria are met, then the sample is kept for 1 week, and the spectrum is acquired again to test for the stability of the preparation. The second spectrum should match the first for this test to be positive. If the initial screening and stability tests are positive, then we prepare a sample labeled uniformly with ^{13}C and ^{15}N by carrying out the translation reaction in the presence of ^{13}C, ^{15}N-labeled amino acids.

A number of approaches are available for rapid NMR data collection: the most widely used methods are GFT (6), APSY (7), and HIFI-NMR (8), which was developed in our laboratory. These methods can reduce the NMR measurement time by several-fold. Peak lists extracted either from conventional or fast NMR data collection are then used as input to the PINE server (9), which provides robust automatic resonance assignments for the backbone and side chains. Following the collection of NOESY data used to generate structural constraints, the next steps are automated NOE peak picking and assignments using the programs ATNOS/

CANDID (10) and semiautomatic protein structure refinement using CYANA (11). The quality of the protein structure is assessed by using the protein structure validation suite (PSVS) (12).

1.2. Protein Requirements for NMR

Ideally the protein solubility will permit the preparation of a sample of concentration >1 mM, which will enable rapid NMR data collection. The practical lower limit for protein solubility is on the order of 0.2 mM. The minimum sample volume is 300 μL for a 5-mm probe or 30 μL for a 1.7 mm probe.

As noted above, our standard labeling patterns are uniform ^{15}N for screening for sample suitability and ^{13}C, ^{15}N double labeling for structure determination. However, other labeling patterns may be desirable, particularly for larger proteins or membrane proteins. Stereo array isotope labeling (SAIL) is a relatively new approach that makes use of amino acids labeled stereospecifically with ^{2}H, ^{13}C, and ^{15}N so as to provide the optimal pattern for structure determination or functional investigations (13, 14).

Other labeling patterns can be prepared by supplying the appropriate labeled amino acids in the cell-free translation reaction. Larger proteins can be investigated by incorporating ^{2}H, ^{13}C, ^{15}N-labeled amino acids. The incorporation of perdeuterated (^{2}H) amino acids is a useful strategy for preparing a protein that is NMR invisible for use, for example, in the investigation of a protein–protein complex. Selective labeling is easily achieved by supplying the desired mixture of amino acids as substrates for the cell-free translation reaction.

1.3. Overview of the Cell-Free Platform used at CESG

The wheat germ cell-free pipeline protocol developed at the CESG for NMR structure determinations (1, 3) is presented as a workflow diagram in Fig. 12.1. The approach consists of four steps: (1) creation of a plasmid used for in vitro transcription, (2) small-scale (25 μL) screening to assay the level of protein production and solubility, (3) larger scale (4–12 mL) production of [U-^{15}N] protein used to evaluate whether solution conditions render the target suitable for NMR structure determination (soluble, monodisperse, folded, and stable), and (4) production of sufficient [U-^{13}C,^{15}N]-labeled protein for multidimensional, multinuclear magnetic resonance data collection.

2. Materials

2.1. Equipment

1. GenDecoder robot (CellFree Sciences, Ltd., Yokohama, Japan) for analytical scale protein productions.
2. Protemist DT-II benchtop cell-free robotic system (CellFree Sciences, Ltd., Yokohama, Japan) for analytical and preparative protein production.

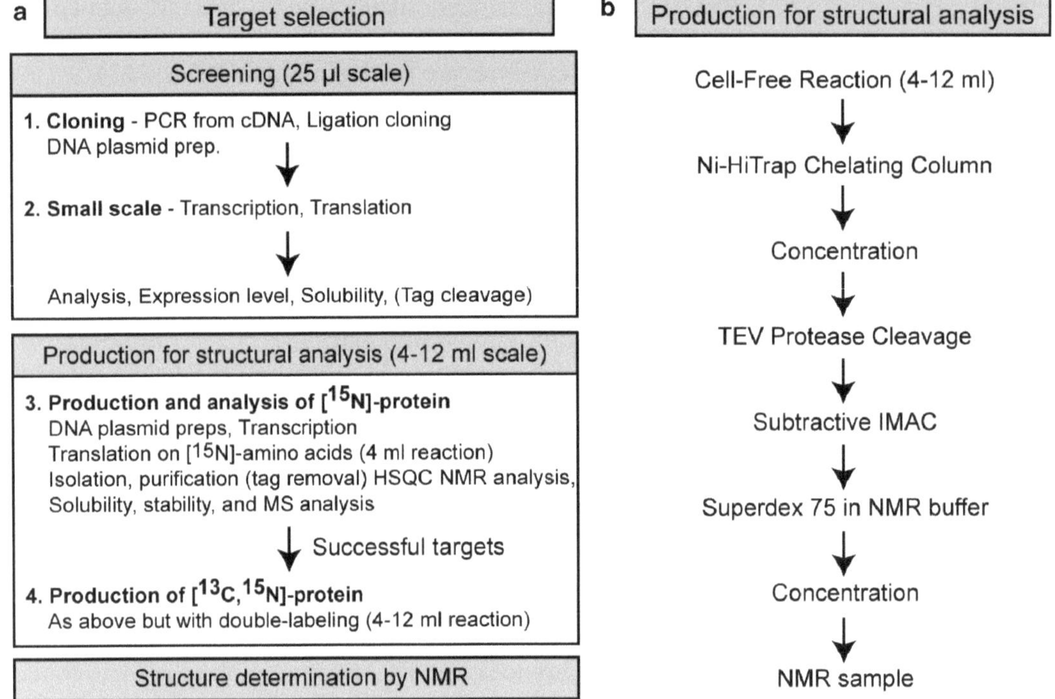

Fig. 12.1. (**a**) Workflow diagram showing how wheat germ cell-free translation is used to screen constructs for the expression of soluble protein, to produce ^{15}N-labeled protein for NMR screening for suitability as a structural candidate, and to produce double-labeled (^{13}C,^{15}N) protein suitable for structure determination. (**b**) Schematic of the steps involved in isolating and purifying proteins produced by wheat germ cell-free translation.

3. Protemist10 and Protemist100 robots (CellFree Sciences, Ltd., Yokohama, Japan) for preparative scale protein productions.
4. Mini-Extruder (Avanti Polar Lipids, Alabaster AL, USA) for preparation of unilamellar liposomes.

2.2. Reagents

1. Unless otherwise stated, 18 MΩ water (Milli-Q water, Millipore, Billerica MA, USA) is used to prepare all reagents. Diethylpyrocarbonate-treated water should not be used in these protocols.
2. All reagents used for in vitro transcription and translation must be RNase-free (see Note 1).
3. Pfu Ultra II fusion hotstart DNA polymerase (Stratagene, La Jolla CA, USA).
4. 10× Flexi Enzyme Blend containing restriction endonucleases *Sgf*I and *Pme*I (Promega, Madison WI, USA).
5. Proteinase K (Sigma-Aldrich, St. Louis, MO). 10× buffer used with proteinase K: 100 mM Tris–HCl, pH 8.0, containing 50 mM EDTA and 1% (w/v) SDS.

6. High-purity Maxiprep system (Marligen Bioscience, Ijamsville MD, USA).
7. Brij-35 detergent, 10% (w/v) (Anatrace, Maumee OH, USA).
8. Soybean total extract (20% lecithin) (Avanti Polar Lipids).
9. Transcription buffer (TB+Mg, 5×): 400 mM HEPES-KOH, pH 7.8, containing 100 mM magnesium acetate, 10 mM spermidine hydrochloride, and 50 mM DTT. Store this buffer at –20°C.
10. NTP solution containing 25 mM each of ATP, GTP, CTP, and UTP: prepared from 100 mM solutions of each NTP in Milli-Q water. NTP solutions are 0.2 μm filter-sterilized and stored at –80°C.
11. SP6 RNA polymerase and RNase inhibitor (RNasin) (Promega).
12. Transcription mixture (TB, 2×): 2× TB+Mg, 8 mM NTPs, 3.2 unit/μL of SP6 RNA polymerase, and 1.6 unit/μL of RNasin. This solution is prepared immediately before use.
13. Dialysis buffer (DB, 5×): 120 mM HEPES-KOH, pH 7.8, containing 500 mM potassium acetate, 12.5 mM magnesium acetate, 2 mM spermidine hydrochloride, 20 mM DTT, 6 mM ATP, 1.25 mM GTP, 80 mM creatine phosphate, and 0.025% (w/v) sodium azide. Store this buffer at –80°C.
14. Unlabeled amino acids (Advanced ChemTech, Louisville KY, USA). ^{15}N and ^{13}C, ^{15}N-labeled amino acids (Cambridge Isotope Laboratories, Andover MA, USA).
15. Mixture of the 20 unlabeled amino acids: weigh out and solubilize each amino acid in Milli-Q water so that the concentration of each amino acid is 2 mM. Mixture of ^{15}N-labeled amino acids: weigh out and solubilize each amino acid in Milli-Q water so that the concentration of each 15N amino acid is 8 mM. Mixture of ^{13}C, ^{15}N-labeled amino acids: weigh out and solubilize each amino acid in Milli-Q water so that the concentration of each amino acid is 5 mM. Do not filter these preparations, because some amino acids are not fully dissolved.
16. 1× DB containing amino acids: dilute 5× DB with Milli-Q water and add the appropriate amino acid mixture to achieve 0.3 mM in each amino acid; sonicate the mixture for 5 min, and then pass the solution through a 0.2-μm filter.
17. Wheat germ cell-free extract (WEPRO2240, CellFree Sciences, Ltd.). This preparation has 240 OD_{260} per mL and does not contain amino acids. Store the extract at –80°C. Note that unused WG extract should be flash frozen.
18. Creatine kinase (Roche Applied Sciences, Indianapolis IN, USA). Dissolve in Milli-Q water to make 50 mg/mL and store at –80°C. Dilute the stock solution to 1 mg/mL prior to use.

19. Lipid rehydration buffer: 25 mM HEPES, pH 7.5, containing 100 mM NaCl.
20. Track-etch polycarbonate membranes: 0.4 μm and 0.1 μm (Nucleopore, Pleasanton CA, USA).
21. Accudenz (Accurate Chemical and Scientific, Westbury NY, USA). Prepare 80% (w/v) and 35% (w/v) solutions in 25 mM HEPES, pH 7.5, containing 100 mM NaCl and 10% (w/v) glycerol. Store these solutions at room temperature.
22. Ultra-clear centrifuge tubes (5 mm id × 41 mm h) (Beckman-Coulter, Fullerton CA, USA).
23. U-bottom 96-well plates (Greiner Bio-One, Monroe NC, USA).
24. Ni-Sepharose high-performance chromatography resin (GE Healthcare, Piscataway NJ, USA).
25. Ni binding/washing buffer: 50 mM sodium phosphate, pH 8.0, containing 300 mM NaCl, and 50 mM imidazole. Store at room temperature.
26. Ni elution buffer: 50 mM sodium phosphate, pH 8.0, containing 300 mM NaCl, and 500 mM imidazole. Store at room temperature.
27. 96-Well multi-screen HTS, HV, 0.45-μm filter plate (Millipore).
28. Centrifuge rotors: JS 5.9, C0650, F241.5P, F2402H, and SW 50.1 (Beckman-Coulter); FA45-30-11 (Eppendorf, Hamburg, Germany).
29. Criterion 10–20% Tris–HCl, 26-well, 1.0 mm SDS-PAGE gels (Bio-Rad, Hercules CA, USA).
30. 3× SDS sample buffer: 150 mM Tris–HCl, pH 6.8, containing 37.5 mM EDTA, 6% (w/v) SDS, 0.01% (w/v) bromophenol blue, 6% (v/v) 2-mercaptoethanol, and 30% (v/v) glycerol.
31. Molecular weight standards, Mark12 (Invitrogen, Carlsbad CA, USA). For visual estimation purposes, the 31-kDa band is present at 0.15 μg in the protocol used for SDS-PAGE analysis.
32. Amicon Ultra-15 (10,000 MWCO) concentrators (Millipore).

3. Methods

3.1. Preparation of Plasmid DNA

Standard molecular cloning techniques are used, and detailed instructions for FlexiVector cloning are available elsewhere (15). The pEU-His-FV plasmid is available from the NIH Protein Structure Initiative Material Repository (http://www.hip.

harvard.edu/PSIMR/index.htm). This vector contains 5′-*Sgf*I and 3′-*Pme*I restriction sites for cloning, 5′-IRES and 3′-UTR sequences to enhance the efficiency of in vitro translation, and a toxic selection cassette for FlexiVector cloning in the multicloning site. The TEV cleavage sequence also is introduced on cloning. Genes are amplified by PCR to incorporate the *Sgf*I and *Pme*I sites into the product. All PCR-amplified genes are sequenced after capture into the plasmid to confirm their fidelity prior to expression studies. Expression from pEU-His-FV produces a protein with an N-terminal His_6 purification tag. The tag can be proteolyzed by treatment with TEV protease, which releases a polypeptide containing a Ser residue as the N-terminus. This N-terminal Ser is the only cloning artifact.

1. For each different protein to be tested, up to 96 variants, inoculate 150 mL of 2× YT medium with each different pEU-His-FV transformant, and grow the culture in a shaking incubator overnight at 37°C (see Note **2**).
2. Recover the cells by centrifugation, and separately purify each different plasmid using a Marligen high-purity plasmid Maxiprep kit. Resuspend each separate DNA pellet in 500-μL Milli-Q water, and measure the absorbance at 260 nm to determine the plasmid DNA concentration. A typical yield is 600–900 μg of plasmid DNA.
3. To remove trace RNase contamination, treat each purified plasmid with 50 ng/μL proteinase K for at least 60 min at 37°C in 1× proteinase K buffer (see Note **3**).
4. Add an equal volume of a 1:1 phenol/chloroform solution to each plasmid preparation; vortex it vigorously, and then centrifuge each preparation at 14,000 rpm (18,000 × g) and ambient temperature in an F241.5P rotor and Microfuge-18 centrifuge for 5 min to separate the phases. This centrifugation may also be performed at 14,000 rpm (18000 × g) and 4°C in an F2402H rotor and Allegra 21R centrifuge (Beckman-Coulter). Transfer each aqueous phase to a separate new tube, and repeat the extraction and centrifugation.
5. Add 1/10 volume 3 M sodium acetate, pH 5.2, and 2.5 volumes ethanol to each separate aqueous phase obtained from **step 4** and mix well. To facilitate precipitation of the plasmid DNA, chill the tube at –20°C for 10 min.
6. Centrifuge each tube at 14,000 rpm (20,800 × g) and 4°C in an FA45-30-11 rotor and 5417R centrifuge for 10 min and remove the supernatant. This centrifugation may also be performed at 14,000 rpm (18,000 × g) and 4°C in an F2402H rotor and Allegra 21R centrifuge. Wash the pellet with 500 μL of ice-cold 70% ethanol.
7. Centrifuge each tube again at 14,000 rpm (20,800 × g) and 4°C in an FA45-30-11 rotor and 5417R centrifuge for 5 min,

discard the supernatant, and air-dry the pellet, which contains the desired plasmid DNA. This centrifugation may also be performed at 14,000 rpm (18,000 × g) and 4°C in an F2402H rotor and Allegra 21R centrifuge.

8. Dissolve the plasmid DNA pellet in 400 μL of Milli-Q water and measure the absorbance at 260 nm to determine the concentration of each separate plasmid DNA preparation. Adjust the volume of each preparation with Milli-Q water to obtain a DNA concentration for the solution of 1 μg/μL.

3.2. Small-Scale Translation on the GenDecoder

1. In a 96-well PCR plate, dispense 2.5 μL of the transcription mixture into each well and add 2.5 μL of each separate plasmid DNA prepared in Subheading 3.1 to separate wells of the PCR plate.
2. Tightly cap the wells to avoid concentration of the samples by evaporation. Incubate the transcription reaction at 37°C for 4 h. This RNA solution is used in translation reactions without further purification.
3. Prepare the translation mixture. Each individual translation mixture consists of 6.25 μL water, 2.75 μL 5× DB, 3.75 μL 2 mM unlabeled amino acids, 1 μL 1 mg/mL creatine kinase, and 6.25 μL WEPRO2240 wheat germ extract. Depending on the number of separate reactions to be performed, scale these volumes and include ~10% extra volume in order to fill each translation well and account for handling losses.
4. Set up the GenDecoder1000 for use in translation.
5. Transfer 5 μL of mRNA from each separate transcription reaction to a U-bottom 96-well plate, and place it in the GenDecoder1000.
6. Place reservoirs of 1× DB with amino acids and translation mixture into the GenDecoder1000.
7. Resume the robot cycle, which will execute a bilayer method translation (16) by adding 20 μL of translation mixture to make 25 μL of lower layer and overlaying 125 μL of 1× DB with amino acids. The reaction will continue in the robot for 20 h at 26°C.

3.3. Screening of GenDecoder Results

1. Tightly seal the top of the reaction plate with a sticky film to prevent evaporation from the translation reactions.
2. Centrifuge the reaction plate in the JS 5.9 rotor and Avanti J-30I centrifuge (Beckman-Coulter) for 60 min at 5,500 rpm (5,700 × g) and 4°C.
3. Carefully peel the film off of the plate in order to avoid spilling the solutions or mixing the well contents.
4. Remove the supernatant from each well with a multichannel pipette and transfer it to a new 96-well PCR plate. Endeavor

to remove all the supernatant from the original plate without disturbing the pellet in order to provide the best reliability to the assessment of the translation of insoluble protein. This plate now contains the soluble translation products.

5. Add 150 μL of Milli-Q water to each well of the original translation plate and gently mix until the pellet is fully resuspended. This plate now contains the insoluble translation products.
6. Transfer 10-μL aliquots from the soluble and insoluble translation product plates into new plates in which each well contains 10-μL water and 10 μL 3× SDS sample buffer. Tightly seal the plate and boil for 5 min in a heat block incubator. After boiling, spin the plate in the JS 5.9 rotor and Avanti J-30I centrifuge for 1 min at 3,640 rpm (2,500 × g) and 4°C to get rid of any condensation on the sealing tape. These plates are the denaturing electrophoresis sample plates for the soluble and insoluble translation products, respectively.
7. Carefully peel the film off of the plate in order to avoid spilling the solutions. Load 9 μL from each well of the denaturating electrophoresis plates, corresponding to 3 μL of the original sample, onto a Criterion SDS-PAGE gel. Load 5 μL of the molecular weight standards in outermost lanes of the gel. This provides 0.15 μg of the 31-kDa protein band, which can be used for visual estimation of the amount of translated protein products. Run the gels at 40 mA/gel for 85 min. Stain the gel with Coomassie Brilliant Blue R-250, destain, and wash as recommended by the manufacturer of the Criterion system. Figure 12.2 shows an example of results obtained from using this protocol.
8. Determine the total expression rating according to the following criteria by visual inspection of the SDS-PAGE gel and comparison of the translated protein bands with the known amounts of protein added in the molecular weight marker lanes. A *high* total expression rating (H) corresponds to greater than 2.5 μg of total protein translated in one reaction. A *medium* total expression rating (M) corresponds to more than 1.25 μg but less than 2.5 μg. A *weak* total expression rating (W) corresponds to less than 1.25 μg, but a still identifiable amount, and an *uncertain* total expression rating (U) corresponds to an uncertain determination, possibly because of no detectable expression or because of overlap with endogenous wheat germ protein bands.
9. Determine the solubility ratings according to the following criteria by visual inspection of the SDS-PAGE gel and comparison of the translated protein bands relative to the molecular weight standards. The solubility is assessed by comparison of the intensity of the corresponding protein bands in the

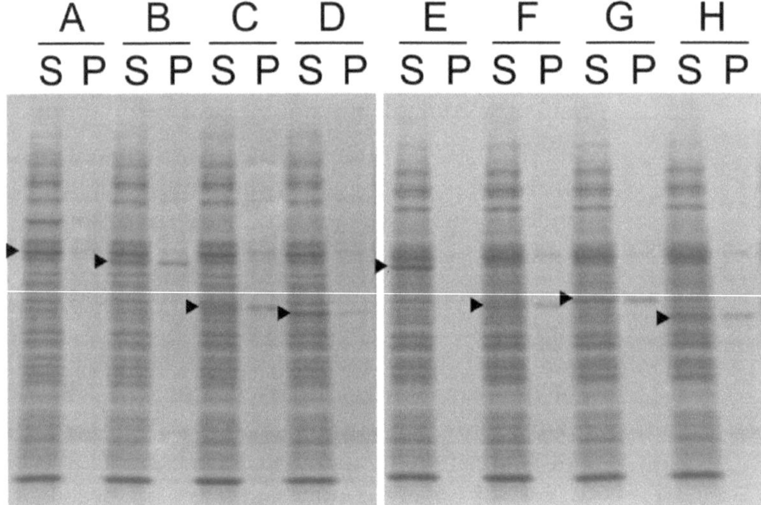

Fig. 12.2. SDS-PAGE analysis of proteins (A–H) soluble (S) and insoluble (P) fractions of expressed proteins prepared as described in Subheading 3.3. The expressed proteins are marked with an *arrowhead*. The amount of total protein expressed is estimated by visual inspection and comparison to the intensity of the 31-kDa marker band in the molecular weight standards, which is present at 0.15 μg. An H total expression rating corresponds to greater than 2.5 μg of protein expressed per translation reaction; an M total expression rating corresponds to more than 1.25 μg but less than 2.5 μg of protein expressed per translation reaction; a W total expression rating corresponds to less than 1.25 μg of protein expressed per translation reaction, but still detectable; and a U rating corresponds to an uncertain determination, possibly because of no detectable expression or overlap with endogenous wheat germ protein bands. The solubility is assessed by comparison of the intensity of the corresponding protein bands in the soluble and insoluble translation products gels. A high-solubility rating (H) is assigned when the soluble translation band has 3× or greater of stained intensity than the insoluble translation band. A medium-solubility rating (M) is assigned when the soluble translation band has approximately equal stained intensity with the insoluble translation band. A weak solubility rating (W) is assigned when the soluble translation band has less stained intensity than the insoluble translation band, and an uncertain solubility rating (U) corresponds to an uncertain determination, possibly because of no detectable expression or because of overlap with endogenous wheat germ protein bands. The proteins investigated, identified by their Gene Ontology numbers, and their purification ratings are as follows: *A*, GO.74329, expression rating H, solubility rating H; *B*, GO.34351, expression rating H, solubility rating W; *C*, GO.70653, expression rating H, solubility rating W; *D*, GO.7312, expression rating H, solubility rating M; *E*, GO.24674, expression rating H, solubility rating H; *F*, GO.79368, expression rating M, solubility rating M; *G*, GO.37540, expression rating H, solubility rating W; *H*, GO.80048, expression rating H, solubility rating M.

soluble and insoluble translation products gels. A *high*-solubility rating (H) is assigned when the soluble translation band has 3× or greater of stained intensity than the insoluble translation band. A *medium*-solubility rating (M) is assigned when the soluble translation band has approximately equal stained intensity with the insoluble translation band. A *weak* solubility rating (W) is assigned when the soluble translation band has less stained intensity than the insoluble translation band, and an *uncertain* solubility rating (U) corresponds to an uncertain determination, possibly because of no detectable expression or because of overlap with endogenous wheat germ protein bands.

3.4. Small-Scale Purification

1. Immediately before use, add 1 M DTT to the Ni binding/washing buffer and Ni elution buffer, respectively, to make the buffers with 2 mM DTT.
2. Equilibrate the Ni-Sepharose resin with water to make a ~50% (v/v) slurry.
3. Place a 96-well filter plate onto a U-bottom 96-well plate.
4. Place 20 μL of the Ni-Sepharose slurry into each well of the filter plate, now retained in the U-bottom 96-well plate. Add 100 μL of the Ni binding/washing buffer supplemented with DTT to each well. This combination of a filter plate and the U-bottom plates is the purification plate.
5. Add 100 μL of the supernatant from the soluble translation products plate to each well of the purification plate.
6. Mix the solution in a plate shaker for 10 min at room temperature, making sure not to spill or cross-contaminate the wells.
7. Centrifuge the plate in the JS 5.9 rotor and Avanti J-30I centrifuge for 1 min at 3,640 rpm (2,500 × g) and ambient temperature.
8. Add 150 μL of the Ni binding/washing buffer supplemented with DTT to wash out nonspecifically bound proteins. Gently mix the solution using a plate shaker for 10 s at room temperature. Centrifuge the plate in the JS 5.9 rotor and Avanti J-30I centrifuge for 1 min at 3,640 rpm (2,500 × g) and ambient temperature. Repeat this step three times. The filtrate should not contain the translated protein, but can be retained for analysis of binding efficiency and product yield if needed.
9. Place the filter plate onto a new U-bottom 96-well plate. Add 50 μL of the Ni elution buffer supplemented with DTT into each well of the purification plate. Use a plate shaker to gently mix the solution for 1 min at room temperature to elute the bound proteins.
10. Centrifuge the plate in the Avanti J-30I centrifuge with JS 5.9 rotor for 1 min at 3,640 rpm (2,500 × g) and ambient temperature, and save the filtrate.
11. Take an 18-μL aliquot of the filtrate from each well, and mix it with 9 μL of 3× SDS sample buffer. Boil the sample for 3 min. After boiling, spin the sample to get rid of any condensation.
12. Load a 9-μL aliquot of each denatured sample, which corresponds to 6 μL of the eluted sample, onto a Criterion SDS-PAGE gel. Run the electrophoresis at 40 mA/gel for 85 min. Stain, destain, and wash the gel as described above in Subheading 3.3. Figure 12.3 shows an example of results obtained from using this protocol.

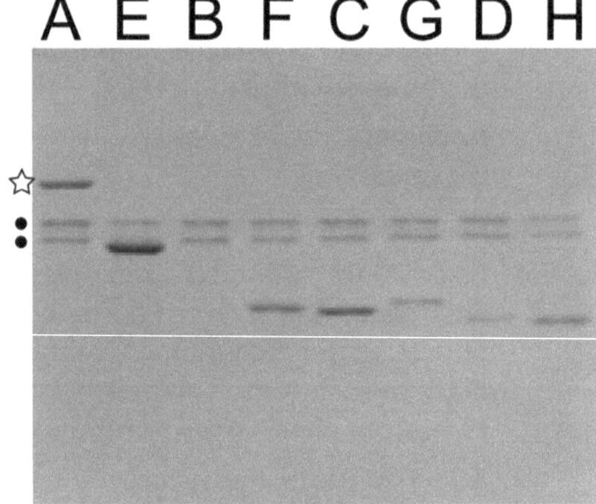

Fig. 12.3. SDS-PAGE analysis of proteins purified from the small-scale cell-free translation reaction used to screen for protein production, solubility, and likelihood of success in large-scale purification. The two proteins observed in all lanes (~50 kDa, marked with a *black dot*) are endogenous wheat germ proteins that copurify with His_6-tagged proteins using IMAC. The successfully purified protein of sample A is marked with a *white star* to show an example. Their Gene Ontology numbers are the same as the legend of Fig. 12.2. Their purification ratings are H, H, U, H, H, M, W, and H (from *left* to *right*). An H rating corresponds to a purified yield of greater than 2.5 μg per small-scale translation reaction; an M rating corresponds to a purified yield of 2.5–1.25 μg per small-scale translation reaction; a W rating corresponds to a purified yield of less than 1.25 μg per small-scale translation reaction; and a U rating corresponds to no purified protein detected (see Note 9).

13. Determine the small-scale purification rating by using the following criteria. A *high* small-scale purification rating (H) corresponds to greater than 2.5 μg of affinity-purified protein recovered per small-scale reaction. A *medium* small-scale purification rating (M) corresponds to more than 1.25 μg but less than 2.5 μg of affinity-purified protein recovered per small-scale reaction. A *weak* small-scale purification rating (W) corresponds to less than 1.25 μg of affinity-purified protein recovered per small-scale reaction, but still detectable, and an *unpurified* rating (U) corresponds to no purified protein detected.

3.5. Protein Production on the Protemist DT-II

1. Place the plasmid DNA preparation, the transcription mixture, the translation mixture, buffers, and the His-tag resin onto the deck of the Protemist DT-II robot as described in the instrument manual. The DT-II robot can hold a maximum of 6 reaction cups containing 6 mL of total reaction volume, with each reaction consisting of 0.5 mL of translation mixture (lower layer of the bilayer) and 5.5 mL of buffer (upper layer). Start the robot (see Note **4**).
2. The translated protein, purified by IMAC chromatography by the robot, is provided after 36 h (see Note **5**).

3.6. Analysis of Protein from the Protemist DT-II

1. The yield of protein provided by a successful Protemist DT-II run may be sufficient for SDS-PAGE analysis, biochemical assays for function, tests of the cleavability of fusion proteins, analytical gel-filtration, MS analysis, or other bioanalytical procedures. Details of these analyses are specific to the individual proteins under study.
2. After appropriate buffer exchange and concentration, the NMR-HSQC spectrum of a ^{15}N-labeled protein can be obtained from proteins prepared by thc DT-II (17).

3.7. Large-Scale Protein Production on the Protemist10 or Protemist100

The need for a large amount of mRNA as a substrate for the large-scale protein production requires first that a large-scale, high-purity plasmid preparation be obtained and second that the transcription reaction be performed before operation of the robot. Once a suitable quantity of mRNA has been obtained, it is added as a reagent to the large-scale reactions, which are carried out by the robot.

1. Prepare the plasmid DNA as described in Subheading 3.1 (see Note **3**).
2. To prepare sufficient mRNA for a large-scale protein production, carry out a transcription reaction in a 50-mL conical tube with a total volume of 4 mL of 1× TB+Mg containing 4 mM NTPs, 0.05 mg/mL plasmid DNA, 0.5 unit/μL of SP6 RNA polymerase, and 0.25 unit/μL of RNasin. Incubate the reaction at 37°C for 3–5 h (see Note **3** and Note **6**).
3. Remove the white precipitate from the transcription reaction by centrifugation in the C0650 rotor and Allegra X-22R centrifuge for 5 min at 6,230 rpm (4,000 × g) and 26°C. Transfer the supernatant to a new tube. This clarified solution is used as the mRNA solution (see Note 7).
4. Add 1,452 μL of 1× DB with amino acids, 48 μL of creatine kinase (50 mg/mL in water), 1,000 μL of WEPRO2240, and 1,500 μL of mRNA solution to an Amicon Ultra-15 (10,000 MWCO) concentrator. Spin the concentrator in the C0650 rotor and Allegra X-22R centrifuge for 8 min at 5,395 rpm (3,000 × g) and 26°C. Add 2 mL of 1× DB with amino acids, mix gently by pipetting, and spin for 5 min. Add 1 mL of 1× DB with amino acids, mix gently by pipetting, and spin for 8 min. If the volume is less than 4 mL, add 1× DB with amino acids to achieve a reaction volume of 4 mL.
5. Prepare an exchanging buffer by mixing 50 mL 1× DB with amino acids and 2.5-mL mRNA solution. Load the concentrators and the exchanging buffer onto the robot, which incubates the concentrators at 26°C. After each reaction cycle, fresh buffer is added to bring the volume to 6.5 mL, and the concentrator tube is centrifuged to decrease the volume to

~4 mL. The robot is programmed to repeat these cycles, each of which takes ~1 h, 18 times.

6. Purify the protein manually (see Note **8** and Note **9**).

3.8. NMR Data Collection and Analysis

1. Dissolve the ^{15}N-labeled protein sample in 90% H_2O/10% D_2O containing 50 mM NaCl and an appropriate buffer (e.g., 10 mM Bis-Tris, pH 7.0, containing 10 mM DTT).
2. Collect a two-dimensional ^{1}H-^{15}N homonuclear single-quantum correlation (HSQC) spectrum and count the peaks resolved. If >80% of the expected peaks are resolved, incubate the sample at 25°C for 1 week and then repeat the HSQC measurement. If there is a decrease in the number of resolved peaks, the sample has failed the stability test and different solution conditions will be required: e.g., different buffer, different pH, higher salt concentration, addition of stabilizing metal ions, cofactors, or substrates, or addition of 50 mM sodium glutamate and 50 mM arginine chloride. Testing of these solution conditions may require preparation of additional ^{15}N-labeled protein samples.
3. When sample conditions that give the appropriate number of expected peaks and stability are identified, a ^{13}C, ^{15}N-labeled sample should be prepared as described in Subheading 3.7.
4. If the HSQC peaks are uniformly sharp, use HIFI-NMR data collection; otherwise use conventional data collection. Collect data sets as needed from the following list:

- 2D ^{1}H-^{15}N heteronuclear NOE
- 3D ^{1}H-^{15}N NOESY-HSQC
- 3D ^{1}H-^{13}C NOESY-HSQC (aliphatic side chain assignments)
- 3D ^{1}H-^{13}C NOESY-HSQC (aromatic side chain assignments)
- 2D ^{1}H-^{13}C HSQC (aromatic)
- 2D ^{1}H-^{13}C CT-HSQC (aliphatic)
- 2D ^{1}H-13C CT-HSQC (aromatic)
- 3D HNCO
- 3D HN(CO)CA
- 3D HNCA
- 3D HN(CA)CO
- 3D HNCACB
- 3D CBCA(CO)NH
- 3D C(CO)NH
- 3D H(CCO)NH
- 3D HBHA(CO)NH
- 3D HN(C)CACB

   - 3D HCCH-TOCSY
5. Use peak lists as input to PINE-NMR (9) to derive backbone and side chain assignments.
6. Complete side chain assignments with NOESY data.
7. Extract torsion angle constraints from chemical shift data with TALOS (18).
8. Use the ATNOS/CANDID (10) package for automated NOE peak picking and assignment and for determining an initial structure.
9. Use the CYANA (11) software package to calculate manually refined structures.
10. If needed, collect residual dipolar coupling data for use in refining the structure further using the XPLOR-NIH (19) package. A final refinement is carried out in explicit water using CNS.
11. Assess the quality of the final structure with the protein structure validation suite (PSVS) (12). The quality criteria to be met include a Z score >–5 for all dihedral angles, discrimination power (DP) score >0.7, at least 85% of backbone dihedral angles in the Ramachandran most favored region, and RDC Q factor <0.2.
12. Submit the results to the Worldwide Protein Data Bank through the ADIT-NMR deposition system (20), which requires submission of coordinates, structural constraints, and assigned chemical shifts (21). In addition, it is strongly suggested that all raw (time domain) NMR data and intermediate and final data related to the structure determination be submitted to the BioMagResBank (BMRB).

3.9. Specialized Methods for Membrane Proteins

3.9.1. Overview

In recent years, in vitro translation systems have emerged as a viable platform for examining integral membrane protein structure and function (22–24). Some membrane proteins can be produced by cell-free translation as insoluble precipitates and refolded by established approaches. For other membrane proteins, the expressed protein can be stabilized through the addition of either detergents or liposomes directly to the translation reactions. In both *E. coli* and wheat germ cell-free expression systems, nonionic polyoxyethylene alkyl-ether detergents are compatible with translation and can solubilize membrane proteins. These detergents include Brij-35 (whose use is described in one protocol below), Brij-56, and Tween-20. Other nonionic detergents, such as alkyl glucosides, zwitterionic detergents, and short-chain phospholipids might also be useful for membrane protein solubilization. These can be examined on a case-by-case basis in order to determine their impact on translation and their ability to solubilize the membrane protein of interest. In order to assess the compatibility of a particular detergent with cell-free translation, a simple control experiment is to compare the yield and fluorescence of green fluorescent protein (GFP) translated in the presence and absence of a specific detergent (25).

In addition to using detergents, the translation of integral membrane proteins can be stabilized by the direct addition of unilamellar liposomes to the cell-free reaction (25). This approach has some advantages over detergent coexpression. First, proteoliposomes can be purified >90% in a single density gradient centrifugation step as described below. Additionally, the use of liposomes obviates the needs for detergents, which can interfere with attaining the native fold and can be difficult to remove from solution for functional studies or proteoliposome formation. Moreover, membrane proteins incorporated into the liposome bilayer during cell-free translation may have a protein fold, membrane insertion topology, and biological activity comparable to naturally produced materials. The incorporation of membrane proteins into liposomes can be monitored by liposome floatation, protease protection assays, and functional studies using established methods. For downstream structural applications, proteoliposomes can be disrupted by the addition of detergents above the critical micelle concentration and purified by established methods. The following methods assume the use of a GenDecoder for protein expression.

3.9.2. Translation in the Presence of Detergent

1. In a 96-well PCR plate, dispense 2.5 μL of the transcription mixture into each well and add 2.5 μL of each separate plasmid DNA prepared in Subheading 3.1 to a separate well.
2. Tightly cap the wells to avoid concentrating the samples by evaporation. Incubate the transcription reaction at 37°C for 4 h. This RNA solution is used in the translation reaction without further purification (see Note **4**).
3. Prepare a single 20-μL translation mixture by mixing 2.5 μL 1% (w/v) Brij-35, 3.75 μL Milli-Q water, 2.75 μL 5× DB, 3.75 μL 2 mM unlabeled amino acids, 1 μL 1 mg/mL creatine kinase, and 6.25 μL WEPRO2240 wheat germ extract. Depending on the number of separate reactions desired, scale these volumes and include ~10% extra volume in order to fill each translation well needed and account for handling losses.
4. Set up the GenDecoder1000 for use in translation.
5. Transfer 5-μL aliquots of each mRNA solution into separate wells of a U-bottom 96-well plate, and put the plate in place on the GenDecoder1000.
6. Place reservoirs of 1× DB with amino acids and translation mixture on the GenDecoder1000.
7. Resume the robot, which will perform a bilayer method translation (16) by adding 20-μL translation mixture and 125 μL

1× DB containing amino acids and 0.1% (w/v) Brij-35. The reaction is incubated for 20 h at 26°C.

8. Analysis of the translation reaction may proceed as described in Subheading 3.3.
9. Membrane proteins expressed in the presence of 0.1% (w/v) Brij-35 can be purified as described in Subheading 3.4 with the inclusion of 0.1% (w/v) Brij-35 in all purification buffers. Detergents can be swapped by performing an extra wash step in the new detergent before elution in the new detergent.

3.9.3. Translation in the Presence of Liposomes

1. In a 96-well PCR plate, dispense 2.5 μL of the transcription mixture into each well and add 2.5 μL of each separate plasmid DNA prepared in Subheading 3.1 to a separate well.
2. Tightly cap the wells to avoid concentrating the samples by evaporation. Incubate the transcription reaction at 37°C for 4 h. This RNA solution is used in the translation reaction without further purification (see Note 4).
3. Prepare a single 20-μL translation mixture by mixing 2 μL of a 15 mg/mL liposome solution prepared as described in Subheading 3.9.4, 4.25 μL Milli-Q water, 2.75 μL 5× DB, 3.75 μL 2 mM unlabeled amino acids, 1 μL 1 mg/mL creatine kinase, and 6.25 μL WEPRO2240 wheat germ extract. Depending on the number of separate reactions desired, scale these volumes and include ~10% extra volume in order to fill each translation well and account for handling losses.
4. Set up the GenDecoder1000 for use in translation.
5. Transfer 5 μL of mRNA from each separate transcription reaction to a U-bottom 96-well plate, and place it in the GenDecoder1000.
6. Place reservoirs of 1× DB with amino acids and translation mixture into the GenDecoder1000.
7. Resume the robot cycle, which will execute a bilayer method translation (16) by adding 20 μL of translation mixture and 125 μL 1× DB containing amino acids and 150-μg liposomes. The reaction will continue for 20 h at 26°C.
8. The purification of proteoliposomes from the translation using density gradient ultracentrifugation is described in Subheading 3.9.5.

3.9.4. Preparation of Liposomes

1. Weigh and dissolve soybean total extract (20% lecithin) in chloroform.
2. Remove the bulk of the organic solvent from the lipid by flushing with a stream of N_2 gas. Further dry the lipid under a vacuum for 30 min.

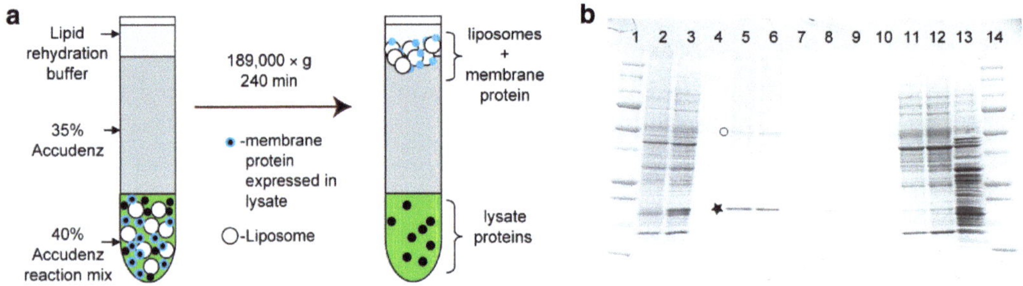

Fig. 12.4. (**a**) A schematic of the use of density gradient centrifugation to purify a membrane protein expressed by wheat germ cell-free translation in the presence of unilamellar liposomes (25). (**b**) An SDS-PAGE analysis of wheat germ cell-free translation and density gradient purification of human full-length cytochrome b_5. *Lanes* 1 and 14, molecular weight standards; *lane* 2, wheat germ extract control; *lane* 2, wheat germ extract after translation of human cytochrome b_5; *lanes* 4–13, fractions obtained from the top to the bottom of the density gradient after ultracentrifugation. A *black star* indicates the presence of full-length cytochrome b_5. The membrane protein incorporated into floated liposomes is contained in *lanes* 5 and 6 at the interface between the 35% Accudenz and lipid hydration buffer layers. A *white circle* indicates a frequent contaminant of the density gradient centrifugation, identified by mass spectrometry to be a homolog of heat shock protein 70, a known chaperone of membrane protein folding.

3. Dissolve the lipid cake in lipid rehydration buffer to a concentration of 15 mg/mL. Vortex the lipid solution for 5 min, followed by three freeze–thaw cycles.
4. Unilamelar liposomes are formed by extrusion through a miniextruder. The liposome solution is passed 11 times through a 0.4-μm track-etch polycarbonate membrane and then passed 11 times through a 0.1-μm membrane.
5. Aliquot the liposomes and flash freeze them. Store liposomes prepared in this way at –80°C.

3.9.5. Protein Purification by Liposome Floatation

Figure 12.4a shows a schematic representation of the purification of proteoliposomes containing membrane proteins translated by wheat germ cell-free translation.

1. Add 150 μL 80% (w/v) Accudenz solution to 150 μL of the translation reaction, mix carefully, and place the mixed sample in an ultraclear centrifuge tube.
2. Carefully layer 650 μL 35% (w/v) Accudenz solution on top of the mixture of 80% (w/v) Accudenz and the translation reaction.
3. Carefully layer 650 μL of lipid rehydration buffer on top of the 35% (w/v) Accudenz solution. This is the density gradient tube.
4. Spin the density gradient tube in an SW 50.1 rotor and L-60 ultracentrifuge for at least 4 h at 45,000 rpm (189,000 × *g*) and 4°C.
5. Fractionate the density gradient by carefully removing 500-μL aliquots from the top to the bottom of the gradient. Store each separate fraction in a separate 1.5-mL microfuge tube.

6. Combine 16 μL of 2× SDS loading dye with 16 μL from each fraction in a new microfuge tube. Seal the tubes and boil them for 5 min in a heat block incubator. After boiling, spin the tubes to recover any liquid condensed on the top of the tube.
7. Run an SDS-PAGE analysis as described in Subheading 3.3, adding 16 μL of sample. Typically, proteoliposomes will migrate at the interface between the 35% (w/v) Accudenz solution and the liposome rehydration buffer. Figure 12.4b shows this analysis applied to the separation of human cytochrome b_5 proteoliposomes from the wheat germ extract. The time required for floatation is dependent on the properties of the proteoliposomes created, so optimization of centrifugation time may be required to obtain the best separation.

4. Notes

1. All glassware must be baked for 3 h at 180°C to eliminate contaminating RNases. In order to prevent RNase contamination from hands or saliva, wear gloves and keep the mouth closed while handling reagents. All the buffers must be sterilized by passage through a 0.2-μm filter and stored at −20°C unless otherwise indicated.
2. For small-scale protein synthesis, a sufficient amount of plasmid DNA can be obtained from a 4-mL culture with a miniprep kit. The required final concentration of the plasmid is at least 0.2 mg/mL (around 0.4 mg/mL).
3. Plasmid DNA prepared by commercially available kits often contains a trace contamination of RNase. This contaminant must be removed for successful transcription and translation.
4. The default protocol for the Protemist DT-II robot uses a commercially available wheat germ extract containing unlabeled amino acids. We make a slight change in the default protocol in order to compensate for the addition of the labeled amino acids solution by decreasing the volume of wheat germ extract.
5. For a protein with ratings of high total expression, high solubility, and high yield of purified protein in the small-scale evaluations, the typical yield of protein from the large-scale reaction and automated purification is 0.1–0.2 mg of purified protein per reaction cup in the Protemist DT-II robot.
6. If the transcription reaction is proceeding correctly, a white precipitate of magnesium pyrophosphate will form and make the transcription solution turbid.

7. In order to avoid coprecipitation of mRNA, the reaction should not be chilled.
8. Typically, His_6-tagged proteins are purified using IMAC. With proteins expressed from pEU-His-FV, the fusion protein can be treated with TEV protease to remove the His_6 tag and can be further purified with a subtractive IMAC in order to remove the liberated His_6 peptide and the TEV protease, which also contains a His_6 tag (26). Gel filtration is often used as a final purification step and can also be used to change the buffer composition. The final yield of soluble protein from the large-scale preparation typically ranges from 0.2 to 0.7 mg of purified protein per 4 mL reaction for a protein rated H in the small-scale purification trial.
9. Contaminating wheat germ extract bands in the first IMAC purification step can be diminished by use of WEPRO2240H wheat germ extract (CellFree Sciences, Yokohama, Japan), which has been pretreated with an IMAC resin to remove these endogenous contaminants.

Acknowledgments

This work was supported by NIGMS Protein Structure Initiative grant 1U54 GM074901 (which supports CESG) and NIGMS grant P41 RR02301 (which supports the National Magnetic Resonance Facility at Madison, where NMR spectroscopy was carried out). Dr. Dmitriy A. Vinarov led initial work at CESG on the development of the wheat germ cell-free system for structural genomics efforts with assistance from Ms. Ejan Tyler and Dr. Carrie Loushin Newman. The authors thank Prof. Yaeta Endo and others at Ehime University for advice and encouragement and also the staff members of CellFree Sciences Ltd. (Yokohama, Japan) who made this approach commercially viable.

References

1. Vinarov, D. A., Lytle, B. L., Peterson, F. C., Tyler, E. M., Volkman, B. F., and Markley, J. L. (2004) Cell-free protein production and labeling protocol for NMR-based structural proteomics. *Nat Methods* **1,** 149–153.
2. Vinarov, D. A. and Markley, J. L. (2005) High-throughput automated platform for nuclear magnetic resonance-based structural proteomics. *Expert Rev Proteomics* **2,** 49-55.
3. Vinarov, D. A., Loushin Newman, C. L., and Markley, J. L. (2006) Wheat germ cell-free platform for eukaryotic protein production. *FEBS J.* **273,** 4160-4169.
4. Vinarov, D. A., Newman, C. L., Tyler, E. M., Markley, J. L., and Shahan, M. N. (2006) Wheat germ cell-free expression system for protein production. *Curr. Protoc. Protein Sci.* Chapter 5:Unit 5 18.

5. Markley, J. L., Aceti, D. J., Bingman, C. A., et al. (2009) The Center for Eukaryotic Structural Genomics. *J. Struct. Funct. Genomics* **10,** 165-179.
6. Szyperski, T., Yeh, D. C., Sukumaran, D. K., Moseley, H. N., and Montelione, G. T. (2002) Reduced-dimensionality NMR spectroscopy for high-throughput protein resonance assignment. *Proc. Natl. Acad. Sci. USA* **99,** 8009-8014.
7. Hiller, S., Fiorito, F., Wüthrich, K., and Wider, G. (2005) Automated projection spectroscopy (APSY). *Proc. Natl. Acad. Sci. USA* **102,** 10876-10881.
8. Eghbalnia, H. R., Bahrami, A., Tonelli, M., Hallenga, K., and Markley, J. L. (2005) High-resolution iterative frequency identification for NMR as a general strategy for multidimensional data collection. *J. Am. Chem. Soc.* **127,** 12528-12536.
9. Bahrami, A., Assadi, A., Markley, J. L., and Eghbalnia, H. (2009) Probabilistic interaction network of evidence algorithm and its application to complete labeling of peak lists from protein NMR spectroscopy. *PLoS Comp. Biol.* **5,** e1000307. doi:10.1371/journal.pcbi.1000307.
10. Herrmann, T., Güntert, P., and Wüthrich, K. (2002) Protein NMR structure determination with automated NOE-identification in the NOESY spectra using the new software ATNOS. *J. Biomol. NMR* **24,** 171-189.
11. Güntert, P. (2004) Automated NMR structure calculation with CYANA. *Methods Mol. Biol.* **278,** 353-378.
12. Bhattacharya, A., Tejero, R., and Montelione, G. T. (2007) Evaluating protein structures determined by structural genomics consortia. *Proteins* **66,** 778-795.
13. Kainosho, M., Torizawa, T., Iwashita, Y., Terauchi, T., Mei Ono, A., and Güntert, P. (2006) Optimal isotope labelling for NMR protein structure determinations. *Nature* **440,** 52-57.
14. Takeda, M., Sugimori, N., Torizawa, T., et al. (2008) Structure of the putative 32 kDa myrosinase-binding protein from *Arabidopsis* (At3g16450.1) determined by SAIL-NMR. *FEBS J.* **275,** 5873-5884.
15. Blommel, P. G., Martin, P. A., Wrobel, R. L., Steffen, E., and Fox, B. G. (2006) High efficiency single step production of expression plasmids from cDNA clones using the Flexi Vector cloning system. *Protein Expr. Purif.* **47,** 562-570.
16. Sawasaki, T., Hasegawa, Y., Tsuchimochi, M., et al. (2002) A bilayer cell-free protein synthesis system for high-throughput screening of gene products. *FEBS Lett.* **514,** 102-105.
17. Frederick, R. O., Bergeman, L., Blommel, P. G., et al. (2007) Small-scale, semi-automated purification of eukaryotic proteins for structure determination. *J. Struct. Funct. Genomics* **8,** 153-166.
18. Cornilescu, G., Delaglio, F., and Bax, A. (1999) Protein Backbone Angle Restraints From Searching a Database for Chemical Shift and Sequence Homology. *J. Biomol. NMR* **13,** 289-302.
19. Schwieters, C. D., Kuszewski, J. J., Tjandra, N., and Clore, G. M. (2003) The Xplor-NIH NMR molecular structure determination package. *J. Magn. Reson.* **160,** 65-73.
20. Ulrich, E. L., Akutsu, H., Doreleijers. J. F., et al. (2008) BioMagResBank. *Nucleic Acids Res.* **36**(Database issue), D402-D408.
21. Markley, J. L., Ulrich, E. L., Berman, H. M., Henrick, K., Nakamura, H., and Akutsu, H. (2008) BioMagResBank (BMRB) as a partner in the Worldwide Protein Data Bank (wwPDB): new policies affecting biomolecular NMR depositions. *J. Biomol. NMR* **40,** 153-155.
22. Nozawa, A., Nanamiya, H., Miyata, T., et al. (2007) A cell-free translation and proteoliposome reconstitution system for functional analysis of plant solute transporters. *Plant Cell Physiol.* 48, 1815-1820.
23. Klammt, C., Schwarz, D., Dötsch, V., and Bernhard, F. (2007) Cell-free production of integral membrane proteins on a preparative scale. *Methods Mol. Biol.* **375**, 57-78.
24. Wuu, J. J. and Swartz, J. R. (2008) High yield cell-free production of integral membrane proteins without refolding or detergents. *Biochim. Biophys. Acta* **1778,** 1237-1250.
25. Goren, M. A. and Fox, B. G. (2008) Wheat germ cell-free translation, purification, and assembly of a functional human stearoyl-CoA desaturase complex. *Protein Expr. Purif.* **62,** 171-178.
26. Blommel, P. G. and Fox, B. G. (2007) A combined approach to improving large-scale production of tobacco etch virus protease. *Protein Expr. Purif.* **55,** 53-68.

Chapter 13

Cell-Free Protein Synthesis for Structure Determination by X-ray Crystallography

Miki Watanabe, Ken-ichi Miyazono, Masaru Tanokura, Tatsuya Sawasaki, Yaeta Endo, and Ichizo Kobayashi

Abstract

Structure determination has been difficult for those proteins that are toxic to the cells and cannot be prepared in a large amount in vivo. These proteins, even when biologically very interesting, tend to be left uncharacterized in the structural genomics projects. Their cell-free synthesis can bypass the toxicity problem. Among the various cell-free systems, the wheat-germ-based system is of special interest due to the following points: (1) Because the gene is placed under a plant translational signal, its toxic expression in a bacterial host is reduced. (2) It has only little codon preference and, especially, little discrimination between methionine and selenomethionine (SeMet), which allows easy preparation of selenomethionylated proteins for crystal structure determination by SAD and MAD methods. (3) Translation is uncoupled from transcription, so that the toxicity of the translation product on DNA and its transcription, if any, can be bypassed. We have shown that the wheat-germ-based cell-free protein synthesis is useful for X-ray crystallography of one of the 4-bp cutter restriction enzymes, which are expected to be very toxic to all forms of cells retaining the genome. Our report on its structure represents the first report of structure determination by X-ray crystallography using protein overexpressed with the wheat-germ-based cell-free protein expression system. This will be a method of choice for cytotoxic proteins when its cost is not a problem. Its use will become popular when the crystal structure determination technology has evolved to require only a tiny amount of protein.

Key words: Structural genomics, Toxic protein, Wheat germ, Restriction enzyme, Selenomethionine, SAD

1. Introduction

In general, it is difficult to determine crystal structure of proteins toxic to cells because the toxicity does not allow their preparation in a large amount for crystallization. Many of those cytotoxic proteins may play important biological roles, but their structural diversity may be left unexplored, especially in the context of the

Y. Endo et al. (eds.), *Cell-Free Protein Production: Methods and Protocols,* Methods in Molecular Biology, vol. 607
DOI 10.1007/978-1-60327-331-2_13,

structural genomics. In addition to novel structural folds that expand the universe of protein structure, elucidation of their structure may reveal novel modes of protein–DNA interaction and protein–protein interaction among others (1, 2).

Restriction endonucleases, essential tools for molecular biology, form a good example of such groups of cytotoxic proteins with important biological roles and structural diversity (3–5). A restriction endonuclease recognizes a specific DNA sequence and introduces a double-stranded break, while a paired modification enzyme can methylate a specific base of the same sequence. Their genes are usually tightly linked and form a restriction–modification gene complex. Various lines of evidence indicate that they behave as selfish mobile genetic elements just as viral genomes and transposons. Comparison of two closely-related genome sequences, such as those of two species within the same hyperthermophilic archaeon genus, *Pyrococcus abyssi* and *Pyrococcus horikoshii*, has indicated their mobility and linkage with various types of genome rearrangements (6). This property makes it possible to search restriction enzymes of a novel fold by bioinformatics methods. While the restriction endonucleases show little sequence conservation, the methyltransferases belong to a conserved protein family. Therefore, the bioinformatics strategy is to first identify methyltransferase genes and then search for their cognate restriction endonuclease genes in the neighboring open reading frames (ORFs) (7). If genome comparison reveals that the candidate gene has been moving between genomes together with the methyltransferase homolog, it is likely a restriction gene. This strategy allowed identification of PabI from *P. abyssi* as a unique 4-bp cutter restriction enzyme with novel properties, generation of TA3′ restriction terminus, and Mg^{2+} ion independence (7, 8).

PabI was predicted to adopt an entirely novel structure (7). However, PabI, like most of the restriction enzymes, is cytotoxic when expressed in vivo in the absence of appropriate and sufficient methylation on the genome. The strategies to bypass this toxicity include use of tightly repressible expression systems and expression of the cognate methyltransferase, but they did not work well. In contrast, in vitro translation systems can synthesize almost any protein, often with high accuracy and at a speed approaching in vivo rates. Among the cell-free systems, the wheat-germ-based system is of special interest: the translation machine has little codon preference and, especially, has little discrimination between methionine (Met) and selenomethionine (SeMet), a characteristic that is useful in crystal structure determination by SAD and MAD methods; translation is uncoupled from transcription, so that digestion of the template DNA by the produced restriction enzyme does not take place (8). The only problem with the system is the cost, especially in a large scale.

In this chapter, we describe successful preparation of PabI protein in this wheat-germ-based cell-free expression system in the native and SeMet-labeled forms (Fig. 13.1), their crystallization (Fig. 13.2), and structure determination. The crystal structure, mutant analysis and *in silico* modeling revealed that the protein adopts a novel fold and participates in a novel mode of protein–DNA interaction (8). The new nomenclature and abbreviations for restriction enzymes and their genes were used (9).

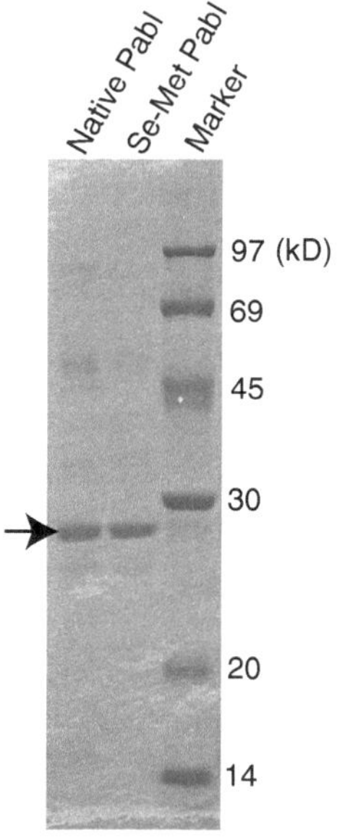

Fig 13.1 PabI restriction enzyme and its selenomethionine-substituted version expressed in a wheat-germ-based cell free expression system. SDS-PAGE.

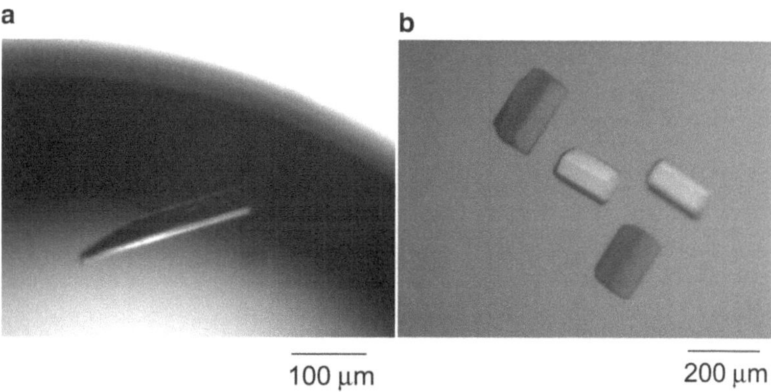

Fig 13.2. Crystals of PabI (**a**) and its seleomethionine-substituted version (**b**) prepared in a wheat-germ-based cell free expression system.

2. Materials

2.1. Production and Purification of the Native and Selenomethionine-Substituted Forms of the Protein

1. High-quality circular preparation of plasmid pEU-*pabIR*: the PabI gene (*pabIR*) is inserted into the multiple cloning site of pEU3b specialized for cell-free expression (10) prepared from *Escherichia coli* strain: JM109 {*recA1 endA1 gyrA96 thi hsdR17* (rK^- mK^+) e14 (*mcrA*) *supE44 relA1* Δ(*lac-proAB*)/[F′ *traD36 proAB+ lacIq lacZ* ΔM15]} (11) (see Note 1).
2. Milli-Q water, freshly prepared.
3. 5× Transcription buffer (TB): 400 mM HEPES-KOH, pH 7.8, 80 mM magnesium acetate, 10 mM spermidine, and 50 mM DTT.
4. Nucleotide tri-phosphates (NTPs) mix: a solution containing 25 mM each of ATP, GTP, CTP, and UTP.
5. SP6 RNA polymerase (80 units/μL, Promega).
6. RNasin (80 units/μL, Promega).
7. Translational substrate buffer (TSB): 30 mM HEPES-KOH, pH 7.8, 100 mM potassium acetate, 2.7 mM magnesium acetate, 0.4 mM spermidine, 2.5 mM DTT, 0.3 mM amino acid mix (20 standard amino acids), 1.2 mM ATP, 0.25 mM GTP, and 16 mM creatine phosphate.
8. SeMet buffer: 30 mM HEPES-KOH, pH 7.8, 100 mM potassium acetate, 2.7 mM magnesium acetate, 0.4 mM spermidine, 2.5 mM DTT, 0.3 mM each of 20 standard amino acids except for methionine, 1.2 mM ATP, 0.25 mM GTP, 16 mM creatine phosphate, and 250 μM selenomethionine.
9. 6-Well plate (3.2 cm in diameter, Whatman Inc., Clifton, NJ).
10. Wheat embryo extract (240 OD/mL) or WEPRO®1240 (CellFree Sciences Co., Ltd.).
11. 20 mg/mL creatine kinase (Roche Diagnostics K. K.).
12. Sephadex G-25 (fine) column (Amersham Biosciences).
13. Heparin–sepharose purification buffer: 10 mM Tris–HCl pH7.5, 1 mM DTT, 0–2 M NaCl gradient.
14. ÄKTA purifier chromatography system (GE healthcare).
15. Protein storage buffer: 10 mM Tris–HCl (pH 7.5), 200 mM NaCl, and 1 mM DTT.

2.2. Crystallization and Structure Determination

1. INTELLI-PLATEs and CrysChem plates for crystallization were purchased from Art Robbins Instruments and Hampton research, respectively.

2. Crystal screen HT, Index HT, Grid screening PEG6000 for the first screening of the crystallization conditions were purchased from Hampton research.
3. Wizard I and II purchased from Emerald BioStructure were used for the first screening of the crystallization condition.
4. Computer programs: HKL2000 (12) for integration and scaling of X-ray diffraction data; SnB (13) for determination of the selenium substructure of the Se-Met variant; SHARP (14) for refinement of the selenium sites and calculation of the initial phase; RESOLVE (15) for automated initial model building and refinement; MOLREP (16) for molecular replacement; Refmac5 (17) and CNS (18) for refinement of the model; XtalView (19) for manual model building.

3. Methods

Cell-free translation of proteins can be achieved mainly by three different modes, batch mode translation (20, 21), bilayer system (22), and CFCF protein synthesis method (10, 23). In the cell-free protein production, mRNA purification represents the most time-consuming step. Recently, based on the bilayer system, a sequential transcription–translation method was developed by directly mixing a transcribed mixture and the wheat germ extract (24). This method provides a very simple system to scale up translational mixture for large-scale protein production because it directly uses unpurified transcriptional mixture without mRNA purification. We used the method for mass production of PabI protein with Met (native) or SeMet (SeMet-substituted) (Fig. 13.1). After purification, native and SeMet-substituted proteins were used for crystallization (Fig. 13.2) and structure determination by X-ray crystallography (8).

3.1. Production and Purification of the Native and SeMet-Substituted Forms of the Protein

3.1.1. mRNA Synthesis

1. Prepare, in a 15-ml tube, 150 μg of high-quality circular plasmid molecules as templates, 1.2 mL of 5× TB, 0.6 mL of 25 mM NTP mix, 70 μL of SP6 RNA polymerase, 70 μL of RNasin, and Milli-Q water (to 6 mL).
2. Incubate the 15-mL tube at 37°C for 4 h (termed "mRNA solution").

3.1.2. Synthesis of Native Protein

1. Prepare four plates each with 6 wells (24 wells in all) (see Note 2).
2. Add 5.5 ml of TSB to each well.
3. Resuspend the white pellet present in the mRNA solution (Subheading 3.1.1) by gently mixing up and down without bubbling.

4. Prepare the translation mixture: 6 mL of the above mRNA solution, 6 mL of wheat-germ extract, and 24 μL of 20 mg/mL creatine kinase.
5. Mix carefully the translational mixture without bubbling.
6. Transfer carefully 500 μL of the translational mixture into the bottom of each well containing the substrate mixture.
7. Incubate the four plates at 26°C for 16 h (see Note 3).

3.1.3. Preparation of SeMet-Substituted Wheat Germ Extract

1. Gel-filtrate 6 mL of the wheat germ extract through 25-mL Sephadex G-25 (fine) column, equilibrated with 50 mL of SeMet buffer.
2. Collect the void fraction.
3. Store at –80°C until use.

3.1.4. Synthesis of SeMet-Substituted Protein

1. Prepare four plates (24 wells in all) (see Note 2).
2. Add 5.5 ml of SeMet buffer to each well.
3. Resuspend the white pellet present in the mRNA solution (Subheading 3.1.1) by gently mixing up and down without bubbling.
4. Prepare SeMet-substituted translation mixture: 6 mL of the above mRNA solution, 6 mL of the SeMet-substituted wheat-germ extract (Subheading 3.1.3), and 24 μL of 20 mg/mL creatine kinase.
5. Mix carefully the SeMet-substituted translational mixture without bubbling.
6. Transfer carefully 500 μL of the SeMet-substituted translational mixture into the bottom of each well containing the SeMet buffer.
7. Incubate the four plates at 26°C for 16 h (see Note 3).

3.1.5. Purification of the Synthesized PabI Protein (a Heat-Resistant Protein)

1. Heat the translational mixture at 90°C for 15 min.
2. Remove the denatured proteins and insoluble materials by centrifugation (10,000×*g*; 15 min; 4°C).
3. Purify PabI protein in the supernatant by a Heparin–Sepharose affinity column using ÄKTA purifier chromatography system.
4. Store the purified PabI protein in protein storage buffer at 4°C.

3.2. Crystallization and Structure Determination

In this section, we describe our experience with one protein in the indicative mood instead of the imperative mood.

3.2.1. Crystallization

1. The crystallization of PabI was performed by the sitting drop vapor diffusion method using the INTELI-PLATE (for initial screening of 96 conditions) and the CrysChem plate (for crystallization screening of 24 conditions).

2. Initial screening of crystallization condition was performed by the sparse-matrix screening method using commercially available screening kits of Crystal screen HT (96 conditions), Index HT (96 conditions), Grid screening PEG6000 (24 conditions), and Wizard I and II (96 conditions). Each crystallization drop was made by mixing 1 μL of the protein solution [0.5 mg/mL protein in 10 mM Tris-HCl (pH 7.5), 200 mM NaCl, and 1 mM dithiothreitol (DTT)] and an equal volume of the reservoir solution.
3. Because of the low solubility of PabI, we used a low concentration solution (0.5 mg/ml) for the first screening. Though the concentration of protein solution was quite low, crystals of PabI could be obtained in the reservoir solution containing PEG6000 as precipitant. The best crystal of PabI appeared under the reservoir conditions containing 100 mM MES (pH 6.0) and 5% PEG6000 after 2 weeks. Typical size of these crystals was 50 × 50 × 200 μm.
4. Although crystals of the Se-Met variant of PabI were also obtained under identical protein solution condition and reservoir solution condition, their size and quality were worse than those of the native crystals. For the improvement of crystal quality, the protein solution condition was modified to increase the solubility of the Se-Met variant. Se-Met variant of PabI was dialyzed against 10 mM MES (pH 6.0), 200 mM NaCl, 10 mM $MgCl_2$, and 10 mM DTT and concentrated to 1.9 mg/ml. The best crystal of the Se-Met variant of PabI was grown under the reservoir conditions containing 50 mM MES (pH 6.8) and 1% PEG6000 using the concentrated protein solution after 1 day.

3.2.2. Structure Determination

1. The X-ray diffraction datasets of the crystals of native PabI and its Se-Met variant were collected using high-brilliance X-ray generated by synchrotron radiation in Photon Factory (Tsukuba, Japan).
2. All the measurements were carried out under cryogenic conditions to reduce radiation damage. Each crystal was soaked in a corresponding reservoir solution containing 20% ethylene glycerol (final concentration) as the cryoprotectant, before being picked up and flash-cooled in a dry nitrogen stream at 95 K.
3. The diffraction data of the native crystal were collected at a wavelength of 1.000 Å using a Quantum 315 CCD detector (ADSC) at the BL-5A beamline in Photon factory. A native crystal of PabI diffracted X-rays to a resolution of 3.0 Å. The X-ray diffraction data were integrated and scaled with the program HKL2000 (23). Analysis of the diffraction date showed that the native PabI crystal belonged to the primitive monoclinic space group $P2_1$ with the unit cell parameters of a = 84.6 Å, b = 114.0 Å, c = 89.2 Å, and β = 116.3°. Data collection, phasing, and refinement statistics of PabI are summarized in Table 13.1.

Table 13.1
Data collection, phasing, and refinement statistics of R.PabI

	Native crystal	Se-Met labeled crystal
Data collection		
Beamline	Photon Factory BL-5A	Photon Factory NW12
Detector	ADSC Quantum 315	ADSC Quantum 210r
Data collection temperature (K)	95	95
Wave length (Å)	1.0000	0.9792
Unit Cell dimensions *a*, *b*, *c* (Å) β (degree)	84.6, 114.0, 89.2 116.3	84.6, 114.5, 89.4 116.3
Space Group	$P2_1$	$P2_1$
Resolution (Å)	20–3.0 (3.11–3.00)	20–2.9 (3.00–2.90)
Completeness	99.3 (96.7)	100.0 (100.0)
Unique reflection	30687	33650
Averaged redundancy	3.6 (3.4)	7.5 (7.6)
R_{merge} (%)	6.7 (28.3)	7.7 (28.0)
I/sigma	14.8 (4.5)	17.8 (5.1)
Phasing		
R_{cullis}		0.748
Phasing power		1.27
FOM before density modification		0.31
FOM after density modification		0.86
Refinement		
Number of nonhydrogen atoms		
Protein	10,599	
Water	0	
R/R_{free} (%)	24.9/31.8	
RMSD bond length (Å)	0.009	
RMSD bond angle (deg.)	1.5	
Ramachandran plot		
In most favored regions (%)	74.0	
In additional allowed regions (%)	22.3	
In Generously allowed regions (%)	2.3	
In disallowed regions (%)	1.4	

Values in parentheses are for the highest resolution shell

4. To determine the structure of PabI, which was predicted to possess novel protein fold (7), we applied SAD (single-wavelength anomalous dispersion) phasing method using the Se-Met variant crystal. The diffraction data of Se-Met crystal were collected at a wavelength of 0.9792 Å using a Quantum 210r CCD detector (ADSC) at the NW12 beam line in Photon Factory. Diffraction data of the Se-Met variant of PabI were processed in the same way. Analysis of the diffraction date showed that a Se-Met variant crystal diffracted X-rays to a resolution of 2.9 Å and belonged to the same space group as the native crystal, $P2_1$ with the unit cell parameters of $a = 84.6$ Å, $b = 114.2$ Å, $c = 89.4$ Å, and $\beta = 116.3°$.
5. Consideration of the Matthew's coefficient (25) (V_M) suggests that native and Se-Met variant crystals of PabI have six protein molecules per asymmetric unit ($V_M = 2.5$ Å^3/Da).
6. The crystal structure of PabI was determined by the SAD phasing method using the diffraction data set of the crystal of the Se-Met variant. The selenium substructure was determined by a direct method program, SnB (13). A total of 19 selenium sites was determined in the asymmetric unit. Because each PabI molecule possesses six methionine sites, this result indicated that approximately 80% of the selenium sites were detected by the SnB calculation.
7. Refinement of the coordinates of the selenium sites and calculation of the initial phase were performed using the program SHARP (14). Phase calculation resulted in an overall figure of merit (FOM) of 0.31 for the resolution range of 20–2.9 Å. After that, density modification and initial model building were performed with the program RESOLVE (15). Molecular models of 759 residues (56% of the total) were automatically built with this calculation.
8. The initial model of the Se-Met variant of PabI was refined and manually rebuilt with the programs CNS (18) and XtalView (19), using 10% (randomly chosen) of the reflections to calculate the R_{free}. The partially built protomer model was transformed into the other five subunits using the program MOLREP (16) in CCP4 (26) and refined with the program REFMAC5 (17) with noncrystallographic symmetry (NCS) restraints.
9. Crystal structure of the native PabI was determined by a molecular replacement method using the coordinates of the partially built structure of the Se-Met variant crystal with the program MOLREP (16). The final structure of PabI was refined and built using the diffraction data set (20–3.0 Å) of native crystal with the program CNS (without NCS restraints) and XtalView (see Note 4).
10. As a result, we were able to determine the structure of PabI at 3.0 Å resolution. The current model has been refined to an

R-factor and R_{free} values of 24.9% and 31.8%, respectively. The paper on the structure of PabI represents the first report of structure determination by X-ray crystallography using the protein overexpressed with the wheat-germ-based cell-free protein expression system (8).

4. Notes

1. RNase contamination is dramatically decreased by protein production in the cell-free system. Unfortunately, however, it is difficult to control the contamination even though commercially available kits are used. To prepare high-quality circular plasmid molecules, we use extraction with phenol/chloroform (phenol:chloroform:isoamyl alcohol = 24:24:1, pH 7.9) and with chloroform after plasmid purification by commercially available kits such as those by Qiagen.
2. To obtain a sufficient amount of PabI, we used four plates (24 wells in all). Twenty-four wells of the reaction gave approximately 3 mg of native and SeMet-substituted PabI.
3. For unstable proteins, incubation at 17°C for 18 h is better.
4. We did not use NCS refinement in the final step of refinement because R-factor and R_{free} values became worse when refined with NCS restraint. We could not build coordinates of water molecules and other nonprotein atoms in this structure.

Acknowledgments

The study of PabI was carried out in collaboration with Jan Kosinski, Ken Ishikawa, Masayuki Kamo, Koji Nagata, and Janusz M. Bujnicki. We are grateful to Kazuyuki Takai for encouragements and patience. IK was supported, during writing, by twenty-first century COE program "Genome and Language" from MEXT and "Grants-in-Aid for Scientific Research" (19657002, 21370001) from JSPS.

References

1. Zhang, Q. and Wang, Y. (2008) High mobility group proteins and their post-translational modifications. *Biochim. Biophys. Acta* **1784,** 1159–1166.
2. Hay, R.T. (2005) SUMO: a history of modification. *Mol. Cell* **18,** 1–12.
3. Kobayashi, I. (2001) Behavior of restriction-modification systems as selfish mobile elements and their impact on genome evolution. *Nucleic Acids Res.* **29,** 3742–3756.
4. Kobayashi, I. (2004) Restriction – modification systems as minimal forms of life. In

Restriction Endonucleases (Pingoud, A., ed.), Berlin, Springer. pp. 19–62.

5. Orlowski, J. and Bujnicki, J.M. (2008) Structural and evolutionary classification of Type II restriction enzymes based on theoretical and experimental analyses. *Nucleic Acids Res.* **36,** 3552–3569.
6. Chinen, A., Uchiyama, I. and Kobayashi, I. (2000) Comparison between Pyrococcus horikoshii and Pyrococcus abyssi genome sequences reveals linkage of restriction-modi fication genes with large genome polymorphisms. *Gene* **259,** 109–121.
7. Ishikawa, K., Watanabe, M., Kuroita, T., Uchiyama, I., Bujnicki, J.M., Kawakami, B., Tanokura, M. and Kobayashi, I. (2005) Discovery of a novel restriction endonuclease by genome comparison and application of a wheat-germ-based cell-free translation assay: PabI (5′-GTA/C) from the hyperthermophilic archaeon *Pyrococcus abyssi. Nucleic Acids Res.* **33,** e112.
8. Miyazono, K., Watanabe, M., Kosinski, J., Ishikawa, K., Kamo, M., Sawasaki, T., Nagata, K., Bujnicki, J.M., Endo, Y., Tanokura, M. and Kobayashi, I. (2007) Novel protein fold discovered in the PabI family of restriction enzymes. *Nucleic Acids Res.* **35,** 1908–1918.
9. Roberts, R.J., Belfort, M., Bestor, T., Bhagwat, A.S., Bickle, T.A., Bitinaite, J., Blumenthal, R.M., Degtyarev, S.Kh., Dryden, D.T., Dybvig, K., Firman, K., Gromova, E.S., Gumport, R.I., Halford, S.E., Hattman, S., Heitman, J., Hornby, D.P., Janulaitis, A., Jeltsch, A., Josephsen, J., Kiss, A., Klaenhammer, T.R., Kobayashi, I., Kong, H., Krüger, D.H., Lacks, S., Marinus, M.G., Miyahara, M., Morgan, R.D., Murray, N.E., Nagaraja, V., Piekarowicz, A., Pingoud, A., Raleigh, E., Rao, D.N., Reich, N., Repin, V.E., Selker, E.U., Shaw, P.C., Stein, D.C., Stoddard, B.L., Szybalski, W., Trautner, T.A., Van Etten, J.L., Vitor, J.M., Wilson, G.G. and Xu, S.Y. (2003) A nomenclature for restriction enzymes, DNA methyltransferases, homing endonucleases and their genes. *Nucleic Acids Res.* **31,** 1805–1812.
10. Sawasaki, T., Ogasawara, T, Morishita, R. and Endo, Y. (2002) A cell-free protein synthesis system for high-throughput proteomics. *Proc. Natl. Acad. Sci. U.S.A.* **99,** 14652–14657.
11. Yanisch-Perron, C., Vieira, J. and Messing, J. (1985) Improved M13 phage cloning vectors and host strains: nucleotide sequences of the M13mp18 and pUC19 vectors. *Gene* **33,** 103–119.
12. Otwinowski, Z. and Minor, W. (1997) Processing of X-ray diffraction data collected in oscillation mode. *Methods Enzymol.* **276,** 307-326.
13. Weeks, C. and Miller, R. (1999) The design and implementation of SnB version 2.0. *J. Appl. Cryst.* **32,** 120–124.
14. Bricogne, G., Vonrhein, C., Flensburg, C., Schiltz, M. and Paciorek, W. (2003) Generation, representation and flow of phase information in structure determination: recent developments in and around SHARP 2.0. *Acta Crystallogr. D Biol. Crystallogr.* **59,** 2023–2030.
15. Terwilliger, T. (2003) Automated main-chain model building by template matching and iterative fragment extension. *Acta Crystallogr. D Biol. Crystallogr.* **59,** 38–44.
16. Vagin, A. and Teplyakov, A. (1997) MOLREP: an automated program for molecular replacement. *J. Appl. Cryst.* **30,** 1022-1025.
17. Murshudov, G. N., Vagin, A. A. and Dodson, E. J. (1997) Refinement of macromolecular structures by the maximum-likelihood method. *Acta Crystallogr. D Biol. Crystallogr.* **53,** 240–255.
18. Brunger, A. T., Adams, P. D., Clore, G. M., DeLano, W. L., Gros, P., Grosse-Kunstleve, R. W., Jiang, J. S., Kuszewski, J., Nilges, M., Pannu, N. S., Read, R. J., Rice, L. M., Simonson, T. and Warren, G. L. (1998) Crystallography & NMR system: A new software suite for macromolecular structure determination. *Acta Crystallogr. D Biol. Crystallogr.* **54,** 905–921.
19. McRee, D. E. (1999) XtalView/Xfit – a versatile program for manipulating atomic coordinates and electron density. *J. Struct. Biol.* **125,** 156–165.
20. Madin, K., Sawasaki, T., Ogasawara, T. and Endo, Y. (2000) A highly efficient and robust cell-free protein synthesis system prepared from wheat embryos: plants apparently contain a suicide system directed at ribosomes. *Proc. Natl. Acad. Sci. U.S.A.* **97,** 559-564
21. Hanes, J. and Pluckthun, A. (1997) *In vitro* selection and evolution of functional proteins by using ribosome display. *Proc. Natl. Acad. Sci. U.S.A.* **94,** 4937–4942.
22. Sawasaki, T., Hasegawa, Y., Tsuchimochi, M., Kamura, N., Ogasawara, T. and Endo Y. (2002) A bilayer cell-free protein synthesis system for high-throughput screening of gene products. *FEBS Lett.* **514,** 102–105.
23. Spirin, A. S., Baranov, V. I., Ryabova, L. A., Ovodov, S. Y. and Alakhov, Y. B. (1988) A continuous cell-free translation system capable of producing polypeptides in high yield. *Science* **242,** 1162–1164.

24. Sawasaki, T., Gouda, M.D., Kawasaki, T., Tsuboi, T., Tozawa, Y., Takai, K. and Endo, Y. (2005) The wheat germ cell-free expression system: methods for high-throughput materialization of genetic information. *Methods Mol. Biol.* **310,** 131–144.

25. Matthews, B. W. (1968) Solvent content of protein crystals. *J. Mol. Biol.* **33,** 491–497.

26. Collaborative, Computational, Project, Number, 4. (1994) The CCP4 suite: programs for protein crystallography. *Acta Crystallogr. D Biol. Crystallogr.* **50,** 760–763.

Chapter 14

Production of Multi-Subunit Complexes on Liposome Through an *E. coli* Cell-Free Expression System

Yutetsu Kuruma, Toshiharu Suzuki, and Takuya Ueda

Abstract

Recently cell-free translation systems became a laboratory research tool to obtain an objective protein. However, a general method for in vitro protein synthesis and multi-subunit complex formation on lipid membrane has not been established. Here, we describe the procedure for the production of subcomplexes of F_oF_1-ATP synthase using a reconstructed cell-free translation system. As for the membrane part F_o ($a_1b_2c_{(10-15)}$), cosynthesis of *c*-subunit and UncI proteins, which are integral membrane proteins, and further c_{11}-ring formation in lipid bilayer are performed by use of a liposomes-containing cell-free system. Moreover, as for the cytoplasm part F_1 ($\alpha_3\beta_3\gamma\delta\varepsilon$), subcomplex formations are successfully achieved by mixing the translation products.

Key words: Cell-free translation system, Liposomes, ATP synthase, Membrane protein, Multi-subunit complex

1. Introduction

Although cell-free translation system has been highly developed and applied in various research fields (1), production of a multi-subunit complex in a cell-free system is still a challenging task. Especially, super molecular complexes consisting of several membrane proteins have been regarded as out of the target of production by a cell-free system, so far. In this chapter, we introduce a practical example of in vitro synthesis of F_oF_1-ATP synthase (F_oF_1) (2) by use of a reconstructed cell-free expression system (PURE system) that consists of translation factors, RNA polymerase, ribosomes, and low molecular weight compounds (3). F_oF_1 is composed of two large complexes, i.e., the membrane integrated F_o part and the water-soluble F_1 part. The F_1 complex has $\alpha_3\beta_3\gamma\delta\varepsilon$ subunits and catalytic sites for ATP-synthesis/hydrolysis. The F_o complex consists of $a_1b_2c_{(10-15)}$

Y. Endo et al. (eds.), *Cell-Free Protein Production: Methods and Protocols,* Methods in Molecular Biology, vol. 607
DOI 10.1007/978-1-60327-331-2_14,

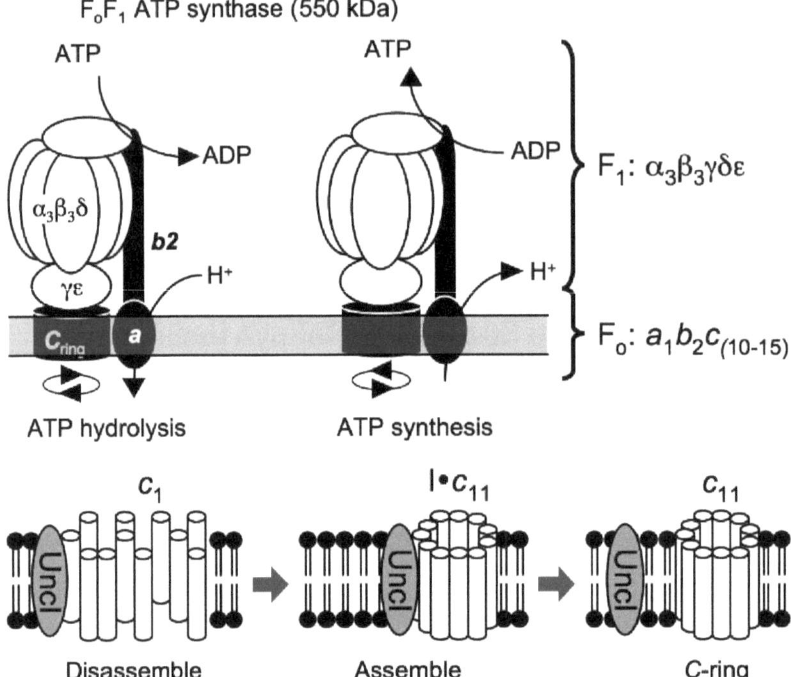

Fig. 14.1. Schematics of F_oF_1-ATP synthase. (*Upper*) F_oF_1 catalyzes ATP-production by the energy of proton motive force that is formed across membranes, and in reverse, has an ability to pump H^+ by the energy of ATP-hydrolysis. F_oF_1 is composed of two domains, water-soluble F_1 ($\alpha_3\beta_3\gamma\delta\varepsilon$) and membrane integral F_o ($a_1b_2c_{(10-15)}$). The catalytic reaction is achieved by rotating its rotor moiety ($\gamma\varepsilon + c_{11}$-ring). (*Bottom*) In addition to the eight kinds of subunits forming F_oF_1, a molecular chaperon, UncI, is necessary to form the c_{11}-ring architecture from 11 *c*-subunits [2, 4]. The figure shows the schematic model of the c_{11}-ring formation assisted by UncI protein.

and bears a function of H^+-transport via membrane that generates membrane potential in cells (Fig. 14.1).

The in vitro formation of various subcomplexes of F_1 can be demonstrated by synthesizing F_1-subunit proteins in the PURE system and mixing the reacted mixtures in the given combinations. Additionally, in the F_o part, c_{11}-ring formation in lipid bilayer is practicable by simultaneous in vitro synthesis of *c*-subunit and a membrane chaperon protein (UncI) (4) in the presence of liposomes, which are beforehand constructed and added in the PURE system. The final goal of this study is to synthesize all responsible proteins of F_oF_1 by the cell-free system and to assemble them on the lipid bilayer to obtain the functional F_oF_1-ATP synthase.

2. Materials

2.1. Common Materials

1. The PURE system (see Notes 1 and 2) is obtainable from the dealing companies (New England BioLabs or Wako, etc.).

2. A slab gel electrophoresis instrument: Mini-PROTEAN® Tetra Cell (Bio-Rad).
3. A power supply.
4. 30% Acrylamide/bis: 30% (w/v) acrylamide, 0.8% (w/v) *N,N'*-methylenebisacrylamide. Store at 4°C.
5. 10% (w/v) Ammonium peroxodisulfate (APS). Store at 4°C.
6. *N,N,N,N,'*-tetramethyl-ethylenediamide (TEMED). Store at 4°C.

2.2. F_1 Complex Formation

2.2.1. Cell-Free Translation

1. Plasmids pET17b-alpha, -beta, -gamma, and -epsilon encoding the genes of α, β, γ, and ε-subunit proteins of thermophilic *Bacillus* PS3 F_oF_1, respectively (see Note 3).
2. L-(^{35}S)methionine (1,000 Ci/mmol, in vitro translation quality).

2.2.2. Native-Page Analysis

1. 10× Native-gel Buffer: 200 mM Hepes–KOH (pH 7.6), 1 M KCl, 100 mM $MgCl_2$.
2. 80% glycerol.
3. Bromophenol blue.
4. Stain Buffer: 0.5% CBB R-250, 50% methanol, 10% acetic acid. Store at room temperature (r.t.).
5. Destain Buffer: 30% methanol, 10% acetic acid. Store at r.t.

2.2.3. RI Imaging

1. BAS-5000 (FUJIFILM).
2. IMAGING PLATE BAS-IP SR 2025 (FUJIFILM).
3. BAS CASSETTE 2025 (FUJIFILM).
4. IMAGE READER software (FUJIFILM).
5. Gel Dryer Model 583 (Bio-Rad).
6. Filter paper 3MM (Whatman).

2.3. F_o Complex Formation

2.3.1. Liposome Preparation

1. Soybean L-α-phosphatidylcholine (Type II-S) (Sigma).
2. 10 mM Hepes–KOH (pH 7.5) containing 5 mM $MgCl_2$ and 10% glycerol.
3. Astrason ultrasonic processor XL2020 (Misonic incorporated) equipped with a tip (tip diameter is 12 mm).

2.3.2. Cell-Free Translation

1. Plasmids pET23c-*c* and pET23c-*I* encoding genes of *c*-subunit and UncI protein of *Propionigenium modestum* F_oF_1, respectively (see Note 4).

2.3.3. Liposome Isolation and c_{11}-Ring Purification

1. Transsonic water-bath type sonicator 460/H (Elma).
2. 10 mM Hepes–KOH (pH. 7.5).
3. 1,2-cyclohexanediaminetetraacetic acid (CDTA) (nacalai tesque).

4. 4 μg/mL ProteinaseK.
5. 4 μg/mL RNaseA (ribonuclease A).
6. 0.5 M $MgCl_2$ prepared by dissolving the powder in MIlliQ water.
7. 10 mM Hepes–KOH (pH 7.5) containing 50 mM $MgCl_2$

2.3.4. Disassembly of c_{11}-Ring by TCA

1. 10% (w/v) trichloroacetic acid (TCA) prepared by diluting a 100% solution with MIlliQ water.
2. 10 mM Hepes–KOH (pH 7.5)

2.3.5. SDS-PAGE Analysis

1. 4× Separating-gel Buffer: 1.5 M Tris–HCl (pH 8.8), 0.4% (w/v) sodium dodecyl sulfate (SDS). Store at room temperature (r.t.).
2. 4× Stacking-gel buffer: 0.5 M Tris–HCl (pH 6.8), 0.4% SDS. Store at r.t.
3. 10× SDS-running buffer: 250 mM Tris, 1.92 M glycine, 1% SDS. Store at r.t.
4. 2× SDS loading buffer: 125 mM Tris–HCl (pH 9.0), 20% (v/v) glycerol, 4.6% SDS, 0.006% (w/v) blomophenol blue (BPB), 10% (v/v) β-mercaptoethanol (add when you use). Store at r.t.
5. Isopropanol.
6. Silver Staining Kit, Protein (GE Healthcare).

3. Methods

In F_oF_1, F_o is composed of three kinds of membrane integral subunits (the stoichiometry is $a_1b_2c_{(10-15)}$) and functions inside lipid bilayer. During synthesizing the subunits, therefore, the addition of liposomes into the cell-free reaction mixture is inevitable for their regular folding and complex formation. On the other hand, five kinds of subunits constituting the F_1 complex ($\alpha_3\beta_3\gamma\delta\varepsilon$) are water-soluble. Thus, the lipid is not necessary. In this section, first, the methods for the F_1 $\alpha_3\beta_3\gamma\varepsilon$ subcomplex formation by use of the PURE system are described. Second, the methods for demonstrating in vitro formation of the F_o *c*-subunits ring with coupling of the protein synthesis reaction are described.

3.1. F_1 Formations

3.1.1. Cell-Free Translation

1. Mix 25 μL of solution A (buffer mix) and 10 μL of solution B (enzyme mix) in a sterile Eppendorf tube (for 50 μL reaction mixture) according to the manufacturer's manual.

2. Add the plasmids (pET17b-alpha, pET17b-beta, pET17b-gamma, or pET17b-epsilon) individually, at the concentration of 20 ng/μL.
3. Add 1 μL of L-(^{35}S)methionine.
4. Fill them up with MIlliQ water up to 50 μL.
5. Incubate the prepared cell-free mixtures at 37°C for >1 h (see Note 5).
6. Mix equal volumes of the respective cell-free mixtures in given combinations (e.g., αβ, αγ, αβγ, etc., *see* Fig. 14.2).
7. Incubate the mixtures at 37°C for 1 h.

3.1.2. Native-PAGE Analysis

The subcomplex formations of F_1 are detectable by Native-PAGE analysis with a control of the purified F_1 sample. In the following methods, a radioisotope-labeled amino acid is used for visualization of the products; therefore, all steps must be carried out in a radiation control area.

1. Clean up the two glass plates of the electrophoresis instrument with MIlliQ water and pure ethanol and assemble them in a proper way.
2. Mix the Native-PAGE gel solution as follows. For 10 mL of 7.5% gel, mix 2.5 mL of 30% acrylamide/bis, 1 mL of

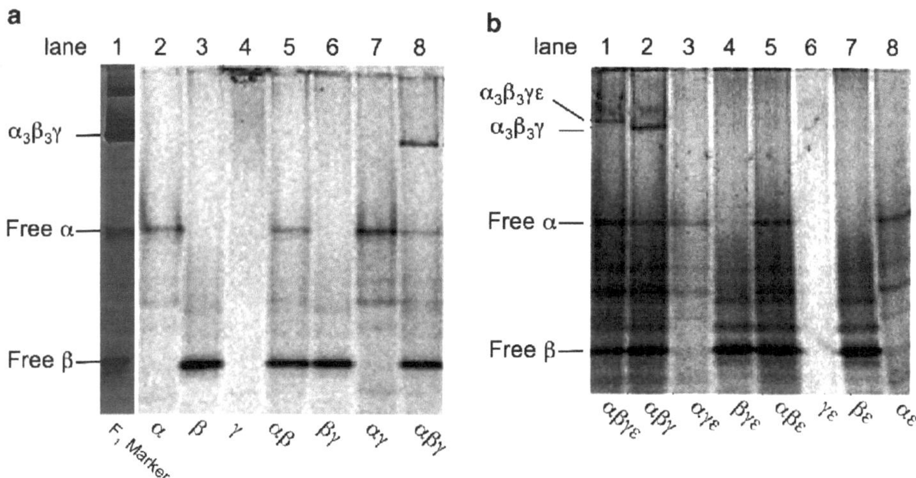

Fig. 14.2. Subunit synthesis and complex formation of *Bacillus* PS3 F_1 by the PURE system. Subunits of F_1 were individually synthesized by the PURE system, and then mixed in the given combinations. (**a**) Combinations among α, β, and γ. The synthesized subunits were mixed as the indicated sets (lanes 5–8) and subjected to Native-PAGE analysis. Purified F_1 was used as a molecular weight marker (lane 1). In addition to free α and β subunits, which did not participate in the complex formation, a band of F_1 sub-complex ($\alpha_3\beta_3\gamma$) is observed at upper region of the gel as indicated. (**b**) Combinations among four kinds of subunits, α, β, γ, and ε. Above the $\alpha_3\beta_3\gamma$ band (lane 2), a band for an $\alpha_3\beta_3\gamma\varepsilon$ sub-complex appeared on the gel (lane 1). Of the four subunits, γ and ε show no band on the gels, because the peptides migrate in an opposite direction during the electrophoresis, owing to their high alkaline pI values. All experiments, except F_1 marker, were performed by using radioisotope labeled amino acid.

10× Native-gel buffer, and 100 μL of 10% APS, and add MIlliQ water up to 10 mL.

3. Add 10 μL of TEMED and mix gently the gel solution (this step should be just before the next step).
4. Pour the gel solution into the slit of the assembled glass plates up to top.
5. Insert a comb on the top of the gel solution.
6. Incubate for 30 min in a cold room.
7. Remove the comb and set the gel on the instrument.
8. Supply the 1× Native-gel Buffer on the upper and lower chambers of the instrument.
9. Connect the instrument to a power supply with cables (see Note 6).
10. Pre-run for 30 min at 50 mA.
11. Prepare samples as follows: 12.5 μL of the cell-free mixture, 7.5 μL of 80% glycerol, and 0.006% (w/v) bromophenol blue (see Note 7).
12. Terminate the prerunning.
13. Carefully load the samples into wells of the gel.
14. Run the gel at 30 mA in cold room (see Note 8).
15. Terminate the electrophoresis when the blue line reach to the bottom of the gel.
16. Remove the glass plates from the instrument, and carefully peel the gel off from the glass plates.
17. Soak the gel in Stain Buffer for 30 min with gently shaking at room temperature.
18. Discard the Stain Buffer.
19. Soak the gel in Desteining Buffer until the protein bands appear at room temperature.
20. Take a photograph of the gel.

3.1.3. RI Imaging

After the gel running and staining, the gel can be further analyzed by autoradiography.

1. Put the gel on Whatman filter paper in the Gel Dryer.
2. Cover the gel with a plastic wrap and the transparent sealing gasket.
3. Dry the gel for 30 min at 80°C.
4. Put the dried gel into the BAS CASSETTE with the IMAGING PLATE.
5. Leave the cassette over night.
6. Visualize the imaging plate by the IMAGE READER. An example of the results produced is shown in Fig. 14.2.

3.2. F_o Formation

3.2.1. Liposome Preparation

1. Suspend the phospholipid in 10 mM Hepes–KOH (pH 7.5) containing 5 mM $MgCl_2$ and 10% glycerol, at the concentration of 44 mg/mL, by calmly stirring at least for 30 min.
2. Sonicate the lipid solution for 3 min on ice water (see Note 9).
3. Aliquot the sonicated solution in Eppendorf tubes (<1ml), dip them in liquid nitrogen for 3 min, and store at –80°C until use.

3.2.2. Liposome-Coupled Cell-Free Translation

1. Mix the PURE system mixture as in **step 1** of Subheading 3.1.1.
2. Add the plasmids pET23c-*c* and/or pET23c-*I*, at the final concentration of 35 ng/μL.
3. Mix with the liposome solution at the concentration of 0.6 μg/μL with adjusting the volume with MIlliQ water.
4. Incubate the liposomes-containing mixtures for 90 min at 30°C.

3.2.3. Liposome Isolation and c_{11}-Ring Purification

1. Dilute the cell-free reaction mixtures fivefold with 10 mM Hepes–KOH (pH. 7.5), and sonicate for 3 min by the water-bath sonicator.
2. Add CDTA at the concentration of 10 mM in order to terminate the translation reaction.
3. Add proteinaseK and RNaseA at the concentration of 0.4 μg/mL for each, and chill for 30 min on ice (see Note 10).
4. Add 0.5 M $MgCl_2$ at the concentration of 50 mM (see Note 11).
5. Precipitate the liposomes by centrifugation, 20,000×*g* for 15 min at 4°C.
6. Remove the supernatant and wash the resulting precipitates enriched liposomes with 10 mM Hepes–KOH (pH 7.5) containing 50 mM $MgCl_2$.
7. Precipitate the liposomes by centrifugation, 20,000×*g* for 15 min at 4°C.
8. 2. Suspend the precipitates in appropriate volume of 10 mM Hepes–KOH (pH 7.5) for further SDS-PAGE analysis.
9. Subject the suspensions to additional treatment with 5% TCA as in Subheading 3.2.4, if c_{11}-ring is necessary to be disassembled to monomeric *c*-subunits.

3.2.4. Disassembly of c_{11}-Ring by TCA

1. Add equal volume of 10% TCA into the suspension sample.
2. Keep it on ice for 30 min, and then, centrifuge at 20,000×*g* for 10 min at 4°C.
3. Remove the supernatants by pipetting.
4. Suspend the precipitates with appropriate volume of 10 mM Hepes–KOH (pH 7.5) (see Note 12).

3.2.5. SDS-PAGE Analysis

1. Clean up the two glass plates as in step 1 of Subheading 3.1.2.
2. Mix the separating gel (lower gel) solution as follows. For 10 mL of 12% gel, mix 4 mL of 30% acrylamide/bis, 2.5 mL of 4× Separating-gel Buffer, and 100 μL of 10% APS, and add MIlliQ water up to 10 mL.
3. Add 10 μL of TEMED and mix gently the gel solution (this step should be just before the next step) (see Note 13).
4. Pour the gel solution into the slit of the assembled glass plates with leaving space for a stacking gel.
5. Overlay isopropanol on the poured solution to avoid drying of the solution surface and to allow formation of a horizontal surface (see Note 14).
6. Incubate for 30 min at room temperature for the gel polymerization.
7. Pour off the isopropanol layer completely.
8. Mix the stacking gel solution (upper gel) as follows. For 3 mL of 5% gel, mix 0.5 mL of 30% acrylamide/bis, 0.8 mL of 4× Stacking-gel Buffer, and 30 μL of 10% APS, and add MilliQ water up to 3 mL.
9. Add 3 μL of TEMED and mix gently the gel solution (this step should be just before the next step).
10. Pour the upper gel solution over the separating gel, and insert the comb (see Note 14).
11. Prepare the SDS-running Buffer by diluting 50 mL of 10× SDS-running Buffer with 450 mL of MIlliQ water.
12. Set the polymerized gel unit to the instrument and supply the SDS-running Buffer on the upper and lower chambers.
13. Carefully remove the comb within the SDS-running Buffer and wash the wells by gentle pipetting (see Note 15).
14. Prepare samples by mixing them with the equal volume of 2× SDS Loading Buffer, e.g., 5 μL of protein sample and 5 μL of Loading Buffer.
15. Boil them at 95°C for 30 s.
16. Carefully load the samples into wells of the upper gel.
17. Connect the instrument to a power supply. The voltage of power supply should be set less than 100 V until the dye of the sample enter the separating gel. After the samples enter into the separating gel, the voltage can be increased up to 200 V.
18. Stop the run when the dye front arrive at the bottom of the gel and carefully remove the gel from the glasses.

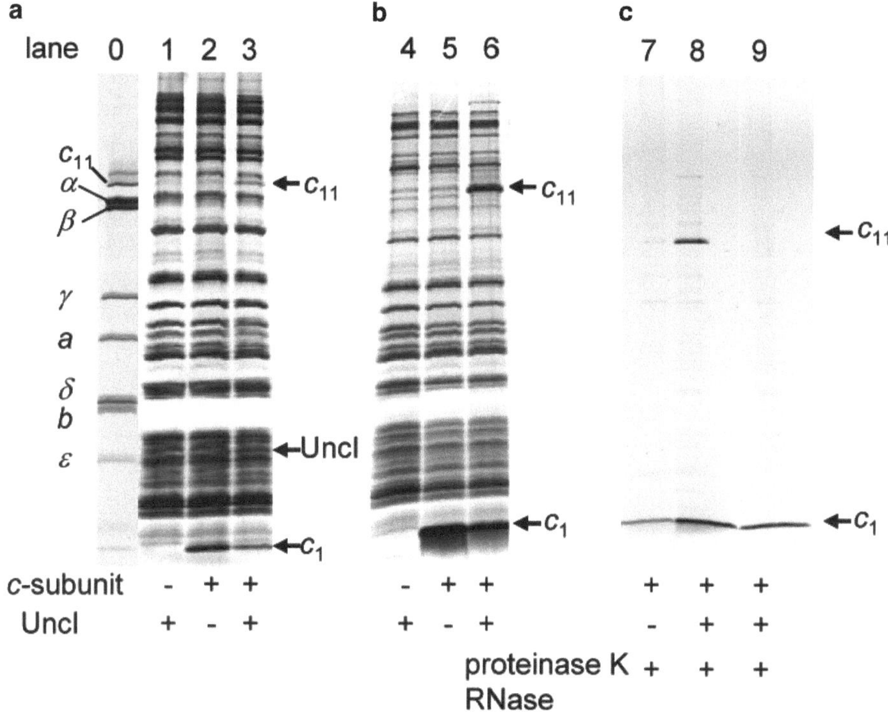

Fig. 14.3. Uncl-dependent c_{11}-ring assembly in vitro. The *c*-subunit and Uncl from *P. modestum* were synthesized in the PURE system. The *c*-subunit gene in pET23c-*c* and/or the Uncl gene in pET23c-I were synthesized in the PURE system mixtures in the presence of liposomes. After the reaction, (**a**) the protein synthesis mixtures were directly analyzed with SDS-PAGE, (**b**) liposomes in the mixture were enriched by precipitation in 50 mM $MgCl_2$ and analyzed with SDS-PAGE, or (**c**) ribosomes and other free proteins were digested by proteinase-K and RNase, and liposomes were isolated by subsequent precipitation in 50 mM $MgCl_2$, which were analyzed with SDS-PAGE. The indicated position of Uncl was known from immunoblotting. The positions of *c*-subunit monomer and c_{11}-ring are indicated as c_1 and c_{11}. Lane 0, the chimeric F_0F_1 preparation made of thermophilic F_1 and *P. modestum* F_0. Lane 9, treated with 10% trichloroacetic acid prior to SDS-PAGE. (Reproduced from ref. 2 with permission of Elsevier Publisher).

19. Stain the gel with the Silver Staining Kit to visualize the product bands.
20. Take a photograph of the stained gel. An example of the results produced is shown in Fig. 14.3.

4. Notes

1. The PURE system contains 10 translation factors (IF1, IF2, IF3, EF-G, EF-Tu, EF-Ts, RF1, RF2, RF3, RRF), 20 aminoacyl-tRNA synthetases, T7 RNA polymerase, 70S ribosomes, methionyl-tRNA formyltransferase, creatine kinase, myokinase, nucleoside diphosphate kinase, pyrophosphatase, and a

mix of tRNAs. In addition, low molecular weight compounds are included (ATP, GTP, CTP, UTP, DTT, spermidine, creatine phosphate, 20 amino acids, 10-formyl-5,6,7,8-tetrahydrofolic acid, potassium glutamate and magnesium acetate) (3).

2. Although *Escherichia coli* cell extract-based cell-free translation systems have been generally used for in vitro protein synthesis, possibility of contamination of membrane constituent cannot be completely eliminated when the extract is prepared from cells. This fact may bring unexpected high background at the stage of the activity assay for the synthesized membrane protein. On the contrary, the PURE system is completely free of any membrane constituent and also of proteases and ribonucleases.
3. All plasmids contain an *Nde*I and an *Eco*RI cleavage site at the upstream and downstream of the carrying open reading frame, respectively. The genes have been cloned from plasmid pTR19-ASDS (5) and introduced into pET17b (Novagen) by PCR-amplification.
4. Both plasmids contain an *Nde*I and a *Hin*dIII cleavage site at the upstream and downstream of the carrying open reading frame, respectively. The genes have been cloned from pTRN-IPbT (4) and introduced into pET23c (Novagen) by PCR-amplification. The pET23c-*I* contains 6 histidine residues at the C-terminus of the UncI.
5. The PURE system is deactivated within 4 h.
6. Check the correct direction (+ or –) of cables.
7. For the F_1 marker samples, Native-gel Buffer with 30% of glycerol is used.
8. The current should not exceed over 50 mA.
9. At this step, the probe sonicator is used for the continuous 3-min sonication.
10. Through this treatment, ribosomes and all the soluble proteins are digested.
11. This step is needed to stimulate clustering of the liposomes for the following precipitation.
12. Be sure that the color of resuspended sample solution is blue, which shows basic or neutral pH. If the color is yellow (acidic pH), the sample must be neutralized by adding a basic buffer, for example, a small volume (e.g. 1 or several μL) of 100 mM Tris–KOH (pH 9.0), until the color changes to blue.
13. The volumes of acrylamide/bis and water can be modified in order to obtain a desired acrylamide concentration.
14. About 30 min are needed to polymerize the gel.
15. Be sure that there is no air bobble at the bottom of the gel. If there, remove them by syringe and needle.

Acknowledgments

The authors thank Dr. Yoko Ozaki for the works on the *c*-ring formation and Mr. Kurumada for the works on in vitro production of F_1 subunits. The authors also thank Prof. Masasuke Yoshida (Tokyo Institute of Technology, Japan) who kindly provided us the F_oF_1 samples.

References

1. Shimizu, Y., Kuruma, Y., Ying, B. W., Umekage, S., and Ueda T. (2006) Cell-free translation systems for protein engineering. *FEBS J.* **273,** 4133–4140.
2. Ozaki, Y., Suzuki, T., Kuruma, Y., Ueda, T., and Yoshida, M. (2008) UncI protein can mediate ring-assembly of *c*-subunits of F_oF_1-ATP synthase *in vitro. Biochem. Biophys. Res. Commun.* **367,** 663–666.
3. Shimizu, Y., Inoue, A., Tomari, Y., Suzuki, T., Yokogawa, T., Nishikawa, K., and Ueda, T. (2001) Cell-free translation reconstituted with purified components. *Nat. Biotechnol.* **19,** 751–755.
4. Suzuki, T., Ozaki, Y., Sone, N., Feniouk, B. A., and Yoshida, M. (2007) The product of *uncI* gene in F_1F_o-ATP synthase operon plays a chaperone-like role to assist *c*-ring assembly. *Proc. Natl. Acad. Sci. U.S.A.* **104,** 20776–20781.
5. Suzuki, T., Ueno, H., Mitome, N., Suzuki, J., Yoshida, M. (2002) F_o of ATP synthase is a rotary proton channel: Obligatory coupling of proton translocation with rotation of *c*-subunit ring. *J. Biol. Chem.* **277**, 13281–13285.

Chapter 15

Synthesis of a Hetero Subunit RNA Modification Enzyme by the Wheat Germ Cell-Free Translation System

Hiroyuki Hori

Abstract

Cell-free translation systems are a powerful tool for the production of many kinds of proteins. However, there are some barriers to improve the system in order to make it a more convenient approach. These include the fact that the production of proteins made up of hetero subunits is difficult. In this chapter, we describe the synthesis of yeast tRNA (m^7G46) methyltransferase as a model protein. This enzyme catalyzes transfer of a methyl group from S-adenosyl-L-methionine to guanine at position 46 in tRNA and generates N^7-methylguanine. Yeast tRNA (m^7G46) methyltransferase is composed of two protein subunits, Trm8 and Trm82. To obtain the active Trm8-Trm82 complex, co-translation of both subunits is necessary. Preparation of mRNAs, in vitro synthesis and purification of the complex are explained in this chapter.

Key words: Hetero subunit, RNA modification enzyme, tRNA methyltransferase, Trm8–Trm82 complex, N^7-methylguanine

1. Introduction

Cell-free translation systems are a powerful tool for the production of many kinds of proteins as described in this book (1–5). The systems allow the production of proteins otherwise toxic to cell viability and can introduce unnatural amino acids into the polypeptide chain (6, 7). Indeed, several cell-free translation systems are now used in proteomics studies (8, 9). Furthermore, the systems are also utilized for protein structural studies such as selenomethionine incorporation in X-ray crystal structure analyses (10) and stable isotope labeling for NMR studies (11, 12). Moreover, recently, we have reported the production of an apo-form of the cofactor binding protein by a wheat germ cell-free translation system (6).

Y. Endo et al. (eds.), *Cell-Free Protein Production: Methods and Protocols,* Methods in Molecular Biology, vol. 607
DOI 10.1007/978-1-60327-331-2_15,

One of the features of the wheat germ cell-free translation system is absence of the translational inhibitors such as ribosome inactivation proteins (13, 14), nucleases and proteases. Therefore, typical protein synthesis continues more than 120 h (2). However, there are some barriers in improving the system in order to make it a more convenient approach. These include the fact that the production of proteins made up of hetero subunits is difficult. In a living cell, many hetero subunit proteins exist and their translation is likely to be strictly regulated. Therefore, development of an in vitro production system of hetero subunit proteins will contribute to the understanding of the translational control.

To generate hetero subunit proteins, several methods have been devised. Although in vivo expression systems in bacteria or yeast cells are convenient, detailed analysis of protein–protein interactions is not so easy because multisubunit proteins are generally expressed in a holo-form in living cells and deletion mutant proteins are often digested by intrinsic proteases. Of the various in vivo expression systems, the baculovirus system has merit with multiple virus infections allowing the expression of hetero subunit proteins in insect cells (15). However, the system requires specialized techniques to maintain high infectivity of virus and degradation of the expressed protein by intrinsic proteases needs to be prevented because baculovirus infection induces cell death.

Compared to in vivo expression systems, the wheat germ cell-free translation system has several advantages. For example, proteases are removed from the extract and the in vitro system enables the analysis of detailed protein–protein interactions. Further, unnatural amino acid and/or isotope labeling can be used. However, synthesis of different proteins (or subunits) in the same reaction vessel is not so easy because translational initiation factors and ribosomes are disperse to hetero mRNA molecules. Thus, in vitro production of hetero subunit proteins is a challenging task.

In this chapter, we describe the synthesis of yeast tRNA (m^7G46) methyltransferase (tRNA (guanine-N^7-)-methyltransferase, EC 2. 1. 1. 33] (16) as a model protein. This enzyme catalyzes transfer of a methyl group from S-adenosyl-L-methionine (AdoMet) to guanine at position 46 in tRNA (Fig. 15.1a) and generates N^7-methylguanine at position 46 (m^7G46). The modified m^7G46 base forms a tertiary base pair with a C13-G22 base pair in the L-shaped tRNA structure (Fig. 15.1b and c). For a long time, the gene(s) responsible for the m^7G46 modification in yeast tRNA remained unidentified. However, it has been reported that yeast tRNA (m^7G46) methyltransferase is composed of two protein subunits, Trm8 and Trm82 (16). The Trm8 protein has amino acid sequence motifs (for example, GXGXG motif) found in AdoMet dependent methyltransferases, strongly suggesting that this subunit has an AdoMet binding site and contains a catalytic

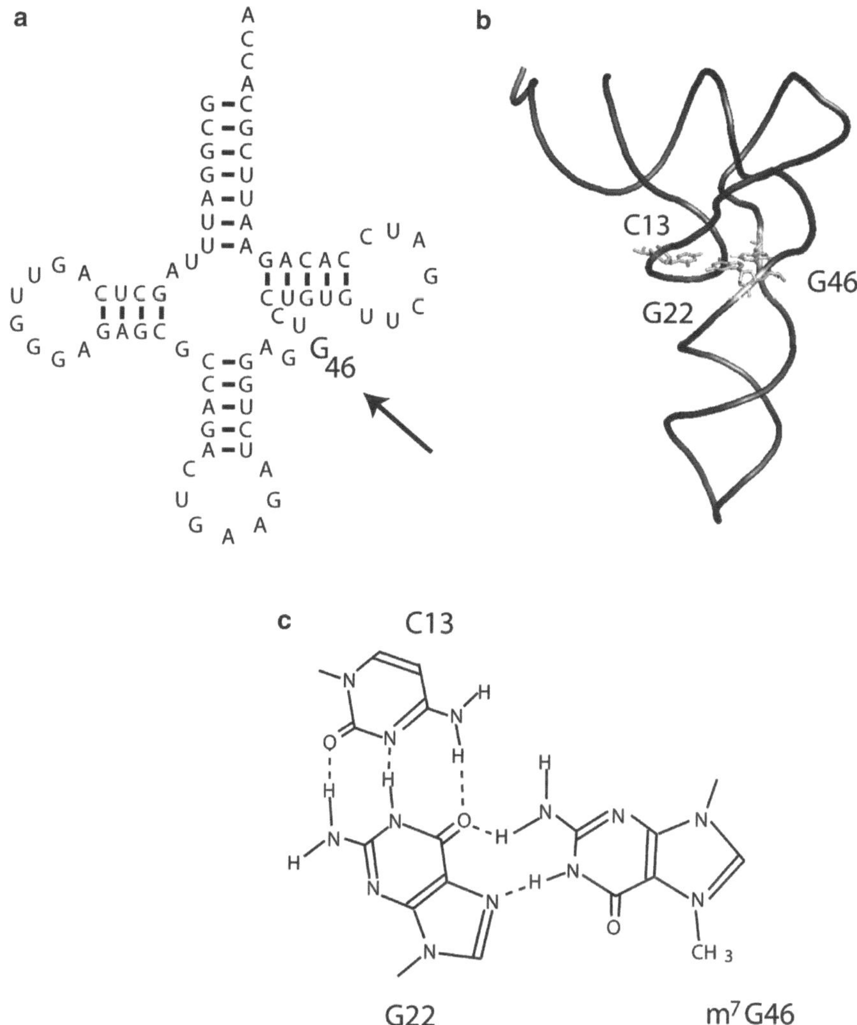

Fig. 15.1. Modification site of tRNA (m^7G46) methyltransferase and the C13-G22-m^7G46 tertiary base pair. (**a**) Yeast $tRNA^{Phe}$ is depicted in a cloverleaf structure. An arrow shows the modification site, G46. (**b**) The G46 base forms a tertiary base pair with the C13-G22 base pair in the L-shaped tRNA structure. (**c**) Hydrogen bonds in the C13-G22-m^7G46 base pair are shown by *broken lines*. These hydrogen bonds stabilizes the L-shaped tRNA structure.

center. In contrast, the Trm82 protein has five WD40 repeats not seen in the other RNA methyltransferases but in the macromolecule binding proteins. Furthermore, during the preparation of this manuscript, the crystal structure of the Trm8–Trm82 complex was reported (17).

In this chapter, I describe the synthesis of the Trm8/Trm82 complex as one example of the synthesis of hetero subunit proteins (18, 19).

2. Materials

All chemical reagents are analytical grade if without special descriptions (see Note 1).

2.1. Transcription of mRNAs

1. KOD-plus ver. 2.0 DNA polymerase from Toyobo.
2. Transcription buffer (5×): 400 mM Hepes-KOH (pH 7.6), 80 mM $Mg(OAc)_2$, 16 mM spermidine, and 80 mM ditiothreitol. Store at −30°C.
3. NTP mixture (10×): 25 mM ATP, 25 mM GTP, 25 mM CTP, and 25 mM UTP. Each NTP (100 mM) solution is adjusted pH 7.0 with 2 M Tris base and then mixed. Store at −30°C.
4. SP6 RNA polymerase from Promega.
5. RNasin (recombinant human placenta RNase inhibitor) from Takara.
6. An agarose gel electrophoresis system. A Mupid submarine gel electrophoresis system (Mupid) is used in our laboratory, however, the other agarose gel electrophoresis systems can be used.

2.2. Synthesis of Each Subunit by Batch Method

1. Reaction mixture (2×): 60 mM Hepes-KOH (pH 7.6), 4 mM $Mg(OAc)_2$, 200 mM KOAc, 32 mM creatine phosphate, 2.4 mM ATP, 0.5 mM GTP, 0.8 mM spermidine, and 0.6 mM amino acid mixture. Twenty amino acids are passed through a Steradisc 13 (0.2 μm) filter device (Kurabo) before mixing to prepare the amino acid mixture. Store at −30°C.
2. Creatine kinase (Roche) is dissolved in 30 mM Hepes-KOH (pH 7.6) at 40 mg/ml, divided to small aliquots (e. g. 10 μl), frozen by liquid nitrogen and stored at −80°C. Avoid freeze and thaw cycles.
3. L-[U-^{14}C]-Leucine from ICN.
4. Wheat germ extract (200 A260 units/ml) from Cell-Free Sciences.
5. 12.5% SDS-polyacrylamide gel electrophoresis system. An ATTO AE-7800 system is used in our laboratory however the other systems can be used.
6. Fuji Photo Film BAS2000 imaging analyzer system. If not available, the other ^{14}C-autoradiography system is required for monitoring the protein synthesis.

2.3. Preparation of Yeast tRNA^{Phe} Transcript for Enzyme Assay

1. DNA oligomers (Invitrogen) for template DNA preparation are as follows: YFF, 5′-AAA TTC CTC GAG TAA TAC GAC TCA CTA TAG CGG ATT TAG CTC AGT TGG GAG AGC GCC AGA CTG AAG A-3′; YFR 5′-TGG TGC GAA TTC

TGT GGA TCG AAC ACA GGA CCT CCA GAT CTT CAG TCT GGC GCT CTC CCA ACT-3′.

2. T7 RNA polymerase from Takara.
3. Ex Taq DNA polymease from Takara.
4. 1 M Hepes-KOH (pH 7.6).
5. 1 M ditiothreitol.
6. 1 M $MgCl_2$.
7. 100 mM spermidine.
8. 500 μg/ml bovine serum albumin.
9. NTP mixture (10×) (see Subheading 2. 1, item 3). Store at −30°C.
10. Pyrophosphatase (2.5 U/μl) from Sigma. Store at −30°C.
11. RNasin (see Subheading 2. 1, item 4).
12. 10% polyacrylamide (7 M urea) gel electrophoresis system.
13. A thermal cycler system. A PCR Thermal Cycler MP system (Takara) is used in our laboratory however, another PCR systems can be used.
14. Gel elution buffer: 500 mM ammonium acetate, 10 mM $MgCl_2$, 1 mM EDTA, 0.1% SDS.
15. Thin layer plates from Merck (code No. 1.05565. cellulose F: 10×10 cm).
16. An UV 254 nm hand monitor. A Mineralight lamp (model UVGL-25) from UVP is used in our laboratory however, another UV hand monitors can be used.
17. Glycogen solution for molecular biology (code 10 901 393 001) from Roche.

2.4. Enzyme Assay of the Trm8–Trm82 Complex

1. Methylation buffer (5×): 250 mM Tris-HCl (pH 7.5), 25 mM $MgCl_2$, 30 mM 2-mercaptoethanol, and 250 mM KCl. Store at −30°C.
2. [Methyl-^{14}C]-AdoMet (1.95 GBq/mmol) from ICN.
3. Yeast $tRNA^{Phe}$ transcript (see Subheading 2. 3).
4. A liquid scintillation counter system. If not available, another ^{14}C-quatification system such as a Fuji Photo Film BAS2000 imaging analyzer system is needed.
5. 5% trichloroacetic acid. Store at 4°C.

2.5. Synthesis of Hetero Subunit Protein by Dialysis Method

1. Reaction mixture (2×) (see Subheading 2. 2, item 1).
2. Dialysis external solution is 1× Reaction mixture.
3. 40 mg/ml creatine kinase (see Subheading 2. 2, item 2).
4. Wheat germ extract (200 A260 units/ml) from Cell-Free Sciences.

5. Dialysis internal solution (translation mixture) is the mixture of 1,200 μl wheat germ extract (200 A260 unit/ml), 1,400 μl 2× Reaction mixture, 40 μl creatine kinase and 160 μl water. Prepare at the start of the experiment and use immediately.

2.6. Purification of Hetero Subunit Complex

1. HiTrap Chelating HP (code number, 17-0408-01) column from GE Healthcare.
2. His wash buffer: 50 mM Tris-HCl (pH 7.6), 5 mM $MgCl_2$, 6 mM 2-mercaptoethanol, 20 mM imidazole, and 200 mM KCl.
3. His elution buffer: 50 mM Tris-HCl (pH 7.6), 5 mM $MgCl_2$, 6 mM 2-mercaptoethanol, 350 mM imidazole, and 200 mM KCl.
4. Dialysis buffer: 50 mM Tris-HCl (pH 7.6), 5 mM $MgCl_2$, 6 mM 2-mercaptoethanol, and 50 mM KCl.
5. Development an imidazole linear gradient (20-350 mM) by His wash and elution buffers, and collect the fractions.

3. Methods

In order to synthesize an equal amount of hetero subunits (i.e. Trm8 and Trm82 in this case) in the same reaction vessel, pilot experiments to monitor the synthesis of each protein are important. To maintain the same translational efficiency, 5′-leader sequences between the SP6 RNA polymerase promoter (or T7 RNA polymerase promoter) and the first methionine codon (AUG) of the mRNAs should be unified. If mRNAs, which contain different 5′-leader sequences, are mixed in the same vessel, dispersion of the translational factors and ribosomes is difficult to be controlled. In fact, in the case of the Trm8 and Trm82, the multi-cloning linker of pEU3b vector was customized: the linker region between Xho I and Nco I sites was deleted and made a new Nco I site at the position of Xho I site. After the vector construction, quantification of transcribed mRNA is necessary. Although mRNA can be directly transcribed from the plasmid, use of amplified DNA by PCR is recommended for the preparation of template DNA because mRNA transcribed from the amplified DNA is homogeneous and easily quantified by agarose gel electrophoresis (Fig. 15.2).

Before the hetero dimer synthesis, the translation efficiency of each subunit should be checked (Fig. 15.3). In the absence of the partner subunit, the synthesized protein may be precipitated. However, the translation efficiency is able to be monitored (Fig. 15.3a and b). If the translation efficiencies of the subunits

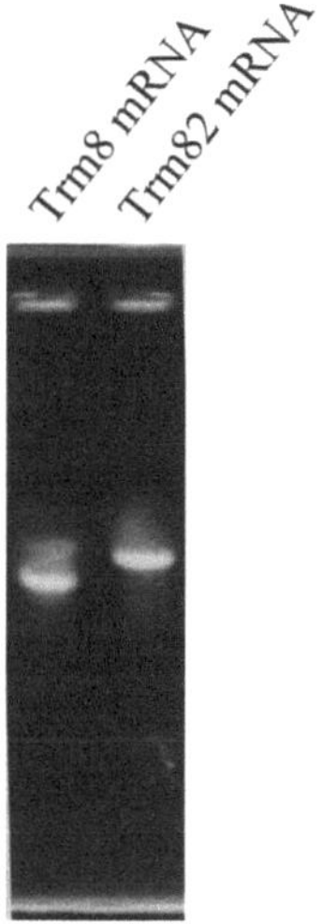

Fig. 15.2. Monitoring of mRNA molecules. Trm8 and Trm82 mRNA molecules were analyzed by 1.8% agarose gel electrophoresis. These mRNAs were transcribed from PCR amplified DNA templates. The gel was stained with ethidium bromide.

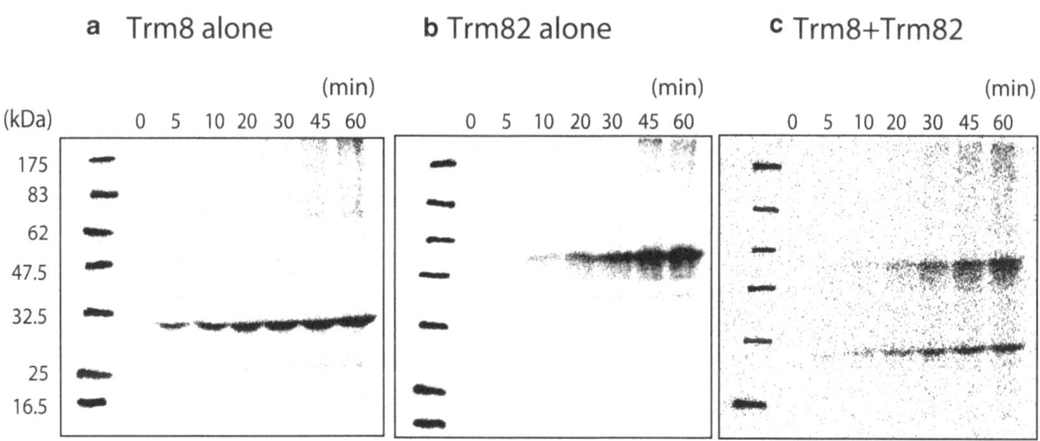

Fig. 15.3. Batch synthesis of the Trm8 and/or Trm82 proteins. Time-dependent 14C-leucine incorporation into the polypeptides was analyzed by 12.5% SDS-polyacrylamide gel electrophoresis. The gels were scanned with a Fuji Photo Film BAS 2000 imaging analyzer. The Trm8 mRNA alone (**a**), Trm82 mRNA alone (**b**), and both mRNAs (**c**) were used for the synthesis. This figure is modified from the figure in reference (19).

are comparable, the hetero dimer synthesis can be carried out under the same condition. In some cases, the synthesis yield may be decreased as compared to the single subunit synthesis. In fact, in the case of Trm8–Trm82 complex preparation, mRNA concentration was determined to be a half of the single subunit synthesis based on the pilot experiments (Fig. 15.3c) (see Note 3).

3.1. Transcription of mRNAs

1. Amplify template DNA by PCR using following primers: SPU, 5 -GCG TAG CAT TTA GGT GAC ACT-3 ; AODA2306, 5 -AGC GTC AGA CCC CGT AGA AA-3 . The PCR mixture is comprised of 10 µl 10× KOD-plus buffer (Toyobo) 1 µl SPU primer (100 pmol/µl), 1 µl AODA2306 primer (100 pmol/µl), 10 µl 2 mM dNTPs (Toyobo), 2 µl 25 mM $MgSO_4$, 1 µl plasmid (pEU3b-trm8 or pEU3b-trm82), 1 µl KOD-plus ver. 2.0 DNA polymerase and 66.5 µl water. One PCR cycle is as follows; denature step at 98°C for 10 s, annealing step at 55°C for 30s, and elongation step at 68°C for 180s. This cycle is repeated for 25 times.
2. Recover the amplified DNA by phenol-chloroform treatment and ethanol precipitation.
3. Dissolve the DNA pellet in 10 µl water.
4. Prepare the transcription mixture comprised of 20 µl transcription buffer (5×), 10 µl 25 mM NTP mixture, 10 µl amplified DNA, 1 µl RNasin, 1 µl SP6 RNA polymerase and 58 µl water.
5. Incubate the transcription mixture at 37°C for 2 h.
6. Recover the mRNA by ethanol precipitation.
7. Dissolve the mRNA in 26 µl water.
8. Analyze 1 µl of the RNA by 1% agarose gel electrophoresis.
9. Usually, mRNA concentration is around 0.3 µg/µl. If necessary, absorbance at 260 nm should be measured. 1 A260 unit corresponds to about 50 µg.

3.2. Synthesis of Each Subunit by Batch Method

1. Dilute the mRNA to 0.3 µg/µl. In the case of hetero subunit synthesis, total mRNA concentration is adjusted to 0.3 µg/µl: equal volumes of 0.3 µg/µl mRNAs are mixed before use.
2. Mix 25 µl diluted mRNA, 35 µl 2× Reaction mixture, 30 µl wheat germ extract, 1.0 µl creatine kinase, and 8.0 µl L-[U-^{14}C]-Leucine in an eppendorf tube.
3. Incubate above translation mixture at 26°C and take 10 µl of the sample at appropriate time points (e.g. 0, 5, 10, 20, 30, 45, and 60 min).
4. 5 µl of each sample is loaded onto a 12.5% SDS-polyacrylamide gel.
5. After the electrophoresis, the gel is stained with Coomassie brilliant blue.
6. Dry the gel in vacuo.
7. Take the autoradiogram of the gel and visualize the image by a Fuji Photo Film BAS2000 imaging analyzer system.

3.3. Preparation of Yeast tRNA^{Phe} Transcript for Enzyme Assay

1. Mix 1 µl YFF primer (100 pmol/µl), 1 µl YFR primer (100 pmol), 12 µl dNTPs (Takara), 10 µl Ex Taq buffer (Takara) and 75.5 µl water in a PCR tube.
2. Add 0.5 µl Taq DNA polymerase to the mixture.
3. Elongation of the DNA primers is carried out using a thermal cycler. One cycle contains the following steps: denature step at 94°C for 30 s, annealing step at 55°C for 30 s, and elongation step at 72°C for 1 min. This cycle is repeated for 25 times.
4. Analyze 1 µl of the sample by 1.8% agarose gel electrophoresis.
5. Recover the template DNA by phenol–chloroform treatment and ethanol precipitation.
6. Prepare the tRNA transcription mixture comprised of 8 µl 1 M Hepes-KOH (pH 7.6), 1 µl 1 M DTT, 4 µl 1 M MgCl2, 2 µl 100 mM spermidine, 20 µl bovine serum albumin, 20 µl NTP mixture, 1 µl RNasin, 1 µl pyrophosphatase, template DNA, 2.5 µl T7 RNA polymerase, and 140.5 µl water.
7. Incubate the tRNA transcription mixture at 37°C for 3 h.
8. Recover the tRNA transcript by phenol–chloroform treatment and ethanol precipitation.
9. Dissolve the RNA pellet in 50 µl water and add 50 µl gel loading buffer.
10. Load the sample onto a 10% polyacrylamide gel (7 M urea).
11. After the electrophoresis, visualize the RNA on a thin layer plate containing fluorescence dye by UV 254 nm irradiation.
12. Excise the band of the tRNA transcript.
13. Extract the gel in 400 µl gel elution buffer at room temperature overnight.
14. Transfer the buffer to a new eppendorf tube.
15. Add 1 µl glycogen solution to the sample.
16. Add 1 ml ethanol and chill at –80°C for 15 min.
17. Recover the RNA by centrifugation at 8,000 × g for 10 min.
18. Dry the RNA pellet and dissolve it in 50 µl water.
19. Measure absorbance at 260 nm and calculate the quantity of the RNA. In the case of yeast tRNA^{Phe} transcript, 1.0 A260nm unit corresponds to 1.7 nmol.

3.4. Enzyme Assay of the Trm8–Trm82 Complex

One of the easy ways to monitor the synthesis is detection of the enzyme activity if possible. In the case of the Trm8–Trm82 complex, assay of methyl-transfer activity to yeast tRNA^{Phe} transcript is a sensitive method.

1. Methylation mixture contains 10 μl methylation buffer, 0.1 A260 unit yeast $tRNA^{Phe}$ transcript, 5 μl translation mixture, and 2 μl [Methyl-^{14}C]-AdoMet in a total volume of 50 μl.
2. Incubate the above mixture at 30°C for 30 min.
3. Spot 40 μl of the mixture onto a Whatman 3 mm filter.
4. Wash the filter for 15 min in 50 ml ice chilled 5% trichloroacetic acid. This treatment is repeated three times.
5. Wash the filter in 50 ml ice chilled ethanol and dry.
6. ^{14}C-incorporation into the transcript is monitored by a liquid scintillation counter.

3.5. Synthesis of Hetero Subunit Protein by Dialysis Method

The dialysis method is used for a large-scale synthesis of the complex. This method is basically the same as the batch method. (see Note 2)

1. Wash the dialysis bug by double deionized water.
2. Pour the dialysis internal solution into the bug.
3. Put the bug into a 50 ml Falcon centrifugation tube.
4. Pour the dialysis external solution into the tube.
5. Incubate the tube at 26°C for 24 h.

3.6. Purification of the Hetero Subunit Complex

We designed a His x 6 tag at the N-termini of the Trm8 protein. To prepare the hetero subunit complex, we recommend that the tag sequence is designed in only one subunit. When the tag was attached at each subunit, the purification became to be difficult because all monomer subunits were bound to the affinity column (Fig. 15.4).

1. Transfer the dialysis internal solution into a new 50 ml Falcon centrifugation tube.
2. Centrifuge the tube at 8,000 × g for 10 min to remove the precipitate. Repeat this step twice.
3. Load the sample into a HiTrap Chelating HP column equilibrated with His wash buffer.
4. Wash the column by 8 ml His wash buffer.
5. Develop an imidazole linear gradient (20–350 mM) and collect the fractions.
6. The Trm8–Trm82 fractions were accessed by 12.5% SDS-PAGE and combined.
7. Dialyze the sample against the dialysis buffer overnight.
8. Check the enzyme activity.
9. Concentrate the hetero subunit fraction by a Centricon YM-10 filter device (Amicon).

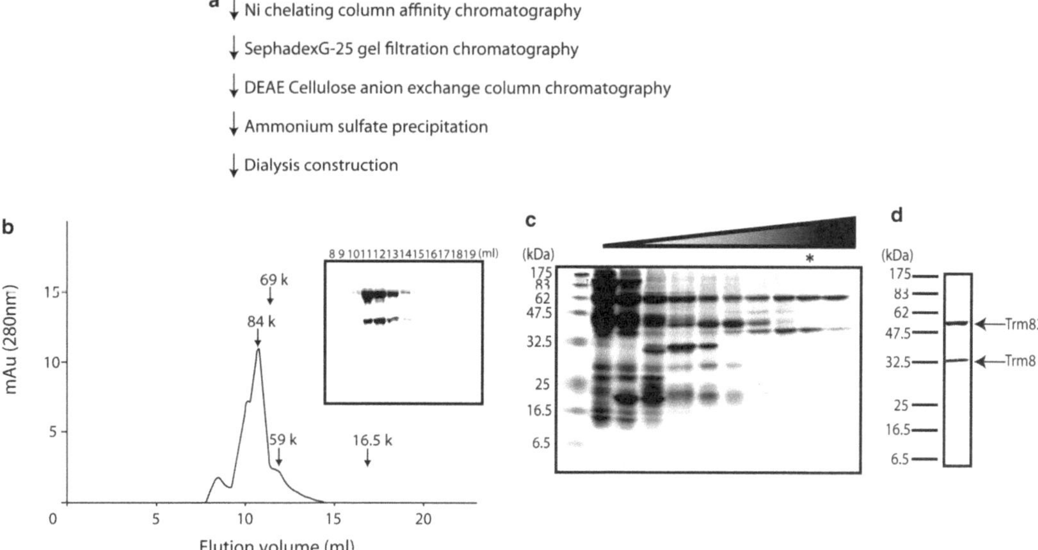

Fig. 15.4. Purification of the Trm8–Trm82 complex. (**a**) Scheme of the purification procedures. The Trm8–Trm82 complex shown in panel (**d**) was purified by this method. (**b**) Elution profile of the imidazole stepwise eulted sample analyzed by Superdex 75 gel-filtration chromatography. The elution positions of standard proteins are shown by *arrows*. *Inset*: 12.5% SDS-polyacrylamide gel electrophoresis of the eluted samples. Lane numbers correspond to the fractions from the Superdex 75 gel-filtration chromatography. (**c**) Elution profiles from Ni-NTA affinity chromatography using a 20–350 mM imidazole linear gradient. The Trm8–Trm82 complex fractions at around 300 mM imidazole (marked by an *asterisk*) were used for further purification. (**d**) The purified Trm8–Trm82 complex. The gel was stained with Coomassie brilliant blue. This figure is modified from the figure in reference (19).

4. Notes

1. Unless stated otherwise, all solutions should be prepared in water that has a resistivity of 18.2 MΩ-cm. This standard is referred to as "water" in this text.
2. Dialysis synthesis continues more than 120 h. In the case of a long time synthesis, sodium azide should be added into both internal and external solutions (final concentration is 0.02%).
3. The results concerning with synthesis of the Trm8–Trm82 complex were summarized in Fig. 15.5. Until now, we have tried synthesis of several multisubunit RNA modification enzymes as well as the Trm8–Trm82 complex. In any case, controls of mRNA concentrations were necessary. Further, in almost cases, cotranslation was required for the active enzyme formation.

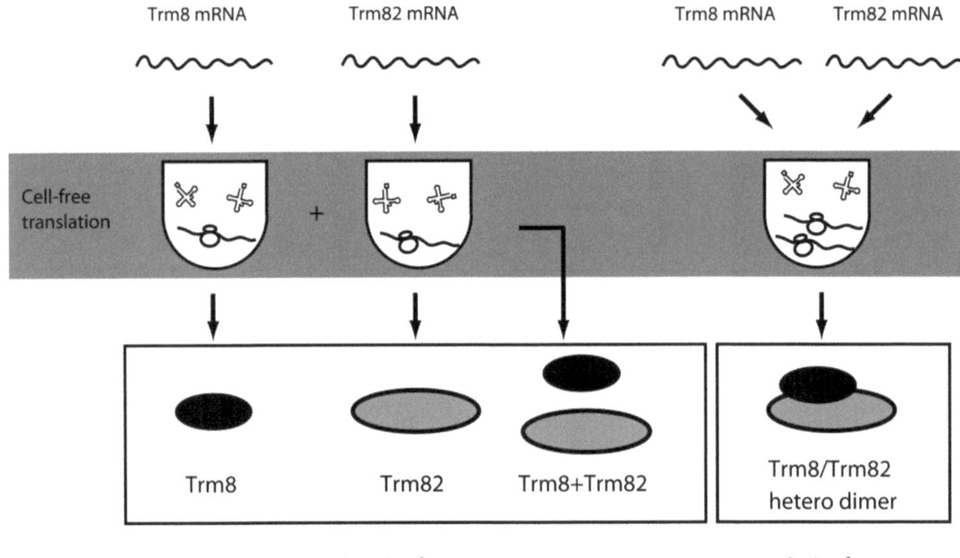

Fig. 15.5. Schematic drawing showing summary of the results. When the Trm8 mRNA alone or Trm82 mRNA alone was used for cell-free translation, Trm8 or Trm82 protein was synthesized. However, the synthesized proteins did not form the complex by mixing. The active Trm8–Trm82 complex was synthesized only by the co-translation of Trm8 and Trm82 mRNAs. (Reproduced from reference (19)

Acknowledgments

I thank Mr. Yuki Muneyoshi (Ehime University) for helping the preparation of this manuscript. This work was partly supported by Grant-in-aid (18048032, 20034041) for Science Research on Priority Areas, and Grant-in-aid (17613003, 19350087) for Science Research from the Ministry of Education, Science, Sports, and Culture of Japan.

References

1. Spirin, A. S., Baranov, V. I., Ryabova, L. A., Ovodov, S. Y., and Alakhov, Y. B. (1988) A continuous cell-free translation system capable of producing polypeptides in high yield. *Science* **242**, 1162–1164.
2. Madin, K., Sawasaki, T., Ogasawara, T., and Endo, Y. (2000) A highly efficient and robust cell-free protein synthesis system prepared from wheat embryos: plants apparently contain a suicide system directed at ribosomes. *Proc. Natl. Acad. Sci. USA* **97**, 559–564.
3. Shimizu, Y., Inoue, A., Tomari, Y., Suzuki, T., Yokogawa, T., Nishikawa, K., and Ueda, T. (2001) Cell-free translation reconstituted with purified components. *Nat. Biotechnol.* **19**, 751–755.

4. Kigawa, T., Yabuki, T., Yoshida, Y., Tsutsui, M., Ito, Y., Shibata, T., and Yokoyama, S. (1999) Cell-free production and stable-isotope labeling of milligram quantities of proteins. *FEBS Lett.* **442**, 15–19.

5. Endoh, T., Kanai, T., Sato, Y. T., Liu, D. V., Yoshikawa, K., Atomi, H., and Imanaka, T. (2006) Cell-free protein synthesis at high temperatures using the lysate of a hyperthermophile. *J. Biotechnol.* **126**, 186–195.

6. Abe, M., Ohno, S., Yokogawa, T., Nakanishi, T., Arisaka, F., Hosoya, T., Hiromatsu, T., Suzuki, M., Ogasawara, T., Sawasaki, T., Nishikawa, K., Kitamura, M., Hori, H., and Endo, Y. (2007) Detection of structural changes in a cofactor binding protein by using a wheat germ cell-free protein synthesis system coupled with unnatural amino acid probing. *Proteins* **67**, 643–652.

7. Kiga, D., Sakamoto, K., Kodama, K., Kigawa, T., Matsuda, T., Yabuki, T., Shirouzu, M., Harada, Y., Nakayama, H., Takio, K., Hasegawa, Y., Endo, Y., Hirao, I., and Yokoyama, S. (2003) An engineered *Escherichia coli* tyrosyl-tRNA synthetase for site-specific incorporation of an unnatural amino acid into proteins in eukaryotic translation and its application in a wheat germ cell-free system. *Proc. Natl. Acad. Sci. USA* **99**, 9715–9720.

8. Sawasaki, T., Ogasawara, T., Morishita, R., and Endo, Y. (2002) A cell-free protein synthesis system for high-throughput proteomics. *Proc. Natl. Acad. Sci. USA* **99**, 14652–14657.

9. Yokoyama, S. (2003) Protein expression systems for structural genomics and proteomics. *Curr. Opin. Chem. Biol.* **7**, 39–43.

10. Kigawa, T., Yamaguchi-Nunokawa, E., Kodama, K., Matsuda, T., Yabuki, T., Matsuda, N., Ishitani, R., Nureki, O., and Yokoyama, S. (2001) Selenomethionine incorporation into a protein by cell-free synthesis. *J. Struct. Funct. Genomics* **2**, 29–35.

11. Morita, E. H., Shimizu, M., Ogasawara, T., Endo, Y., Tanaka, R., and Kohno, T. (2004) A novel way of amino acid-specific assignment in (1)H-(15)N HSQC spectra with a wheat germ cell-free protein synthesis system. *J. Biomol. NMR* **30**, 37–45.

12. Kigawa, T., Muto, Y., and Yokoyama, S. (1995) Cell-free synthesis and amino acid-selective stable isotope labeling of proteins for NMR analysis. *J. Biomol. NMR* **6**, 129–134.

13. Ogasawara, T., Sawasaki, T., Morishita, R., Ozawa, A., Madin, K., and Endo, Y. (1999) A new class of enzyme acting on damaged ribo somes: ribosomal RNA apurinic site specific lyase found in wheat germ. *EMBO J.* **18**, 6522–6531.

14. Ozawa, A., Sawasaki, T., Takai, K., Uchiumi, T., Hori, H., and Endo, Y. (2003) RALyase; a terminator of elongation function of depurinated ribosomes. *FEBS Lett.* **555**, 455–458.

15. Mikhailov, M. V., and Ashcroft, S. J. (2000) Interactions of the sulfonylurea receptor 1 subunit in the molecular assembly of beta-cell K(ATP) channels. *J. Biol. Chem.* **275**, 3360–3364.

16. Alexandrov, A., Martzen, M. R., and Phizicky, E. M. (2002) Two proteins that form a complex are required for 7-methylguanosine modification of yeast tRNA. *RNA* **8**, 1253–1266.

17. Leulliot, N., Chaillet, M., Durand, D., Ulryck, N., Blondeau, K., and Tilbeurgh, van H. (2008) Structure of the yeast tRNA m7G methylation complex. *Structure* **16**, 52–61.

18. Matsumoto, K., Toyooka, T., Tomikawa, C., Ochi, A., Takano, Y., Takayanagi, N., Endo, Y., and Hori, H. (2007) RNA recognition mechanism of eukaryote tRNA (m^7G46) methyltransferase (Trm8–Trm82 complex). *FEBS Lett.* **581**, 1599–1604.

19. Matsumoto, K., Tomikawa, C., Toyooka, T., Ochi, A., Takano, Y., Takayanagi, N., Abe, M., Endo, Y., and Hori, H. (2008) Production of yeast tRNA (m^7G46) methyltransferase (Trm8–Trm82 complex) in a wheat germ cell-free translation system. *J. Biotechnol.* **133**, 453–460.

Chapter 16

Strategies for the Cell-Free Expression of Membrane Proteins

Sina Reckel, Solmaz Sobhanifar, Florian Durst, Frank Löhr, Vladimir A. Shirokov, Volker Dötsch, and Frank Bernhard

Abstract

Cell-free expression offers an interesting alternative method to produce membrane proteins in high amounts. Elimination of toxicity problems, reduced proteolytic degradation and a nearly unrestricted option to supply potentially beneficial compounds like cofactors, ligands or chaperones into the reaction are general advantages of cell-free expression systems. Furthermore, the membrane proteins may be translated directly into appropriate hydrophobic and membrane-mimetic surrogates, which might offer significant benefits for the functional folding of the synthesized proteins. Cell-free expression is a rapidly developing and highly versatile technique and several systems of both, prokaryotic and eukaryotic origins, have been established. We provide protocols for the cell-free expression of membrane proteins in different modes including their expression as precipitate as well as their direct synthesis into detergent micelles or lipid bilayers.

Key words: Continuous exchange cell-free expression, S30 extract, P-CF, D-CF, L-CF, Detergents, Lipids, NMR, Transmembrane segments, Stable isotope labelling

1. Introduction

Integral membrane proteins (MPs) constitute nearly one-third of the typical mammalian genome and serve as more than half of the current available drug targets. However, the challenges, in particular with regard to MP expression, remain significant. Often, the over-expression of MPs in vivo leads to toxicity problems due to overloading of the translocon machinery or to excessive aggregation in the form of inclusion bodies. Also, mis-targeting followed by protein degradation poses a significant issue. Even if MPs are expressed in high amounts, their extraction from the membrane and their transfer into artificial hydrophobic

Y. Endo et al. (eds.), *Cell-Free Protein Production: Methods and Protocols,* Methods in Molecular Biology, vol. 607
DOI 10.1007/978-1-60327-331-2_16, © Humana Press, a part of Springer Science+Business Media, LLC 2010

environments is often achieved with relatively harsh conditions, which may affect protein structure and function. Thus, the cell-free (CF) expression approach provides a powerful alternative as the absence of compartmentalisation circumvents the targeting process and in addition, the open nature of CF systems offers the unique opportunity to create an appropriate hydrophobic environment, allowing for the soluble expression of MPs. The inclusion of co-factors and chaperones can be further alternatives to stabilise the structure of synthesized MPs.

While the CF system is reported to work in batch (1) as well as in continuous exchange configurations (2), the focus of this chapter is set on the continuous-exchange cell-free expression (CECF) system which, in our hands, has proven to be most efficient for the expression of MPs. The principle of CECF systems relies on the existence of two different compartments. One of these holds the reaction mixture (RM) that contains all necessary components for the transcription-translation process. This RM is dialysed against a feeding mixture (FM) that holds all lower molecular weight components, supplying the RM with essential precursors, meanwhile removing undesirable side products that may inhibit protein expression. The CF system in general offers different modes of expression resulting in either solubilised MP or formation of an insoluble MP precipitate which, however, differs considerably from the well known inclusion bodies produced in cellular expression systems (*see* Fig. 16.1). The MP precipitate, which forms upon expression in the absence of any hydrophobic environment (precipitate-forming CF expression: P-CF), readily resolubilizes in relatively mild detergents and has been shown to give functionally folded samples (3–5). The expression of MPs directly in the presence of detergent allows the formation of proteo-micelles already during the translation process (*see* Fig. 16.1). This detergent-based CF expression (D-CF) mode is already well established and *E. coli* extracts are tolerant to a considerable variety of different detergents (6). Amongst those, the Brij detergents show an outstanding performance also with regard to larger α-helical MPs such as G-protein coupled receptors (GPCRs). The lipid-based expression mode (L-CF) is a currently emerging approach, and the in vitro incorporation mechanisms of MPs into supplied lipid bilayers are still not understood. While some MPs incorporate into empty liposomes directly (7), others may require additional co-factors to assist in the insertion process. Expression in the presence of preformed translocons could be accomplished by addition of inner membrane vesicles into the CF reaction (8). Very recently, the CF expression in the presence of disc-shaped nano-lipoprotein particles of defined size has been shown to work for a variety of different MP targets, and it may offer yet another interesting alternative expression mode (9).

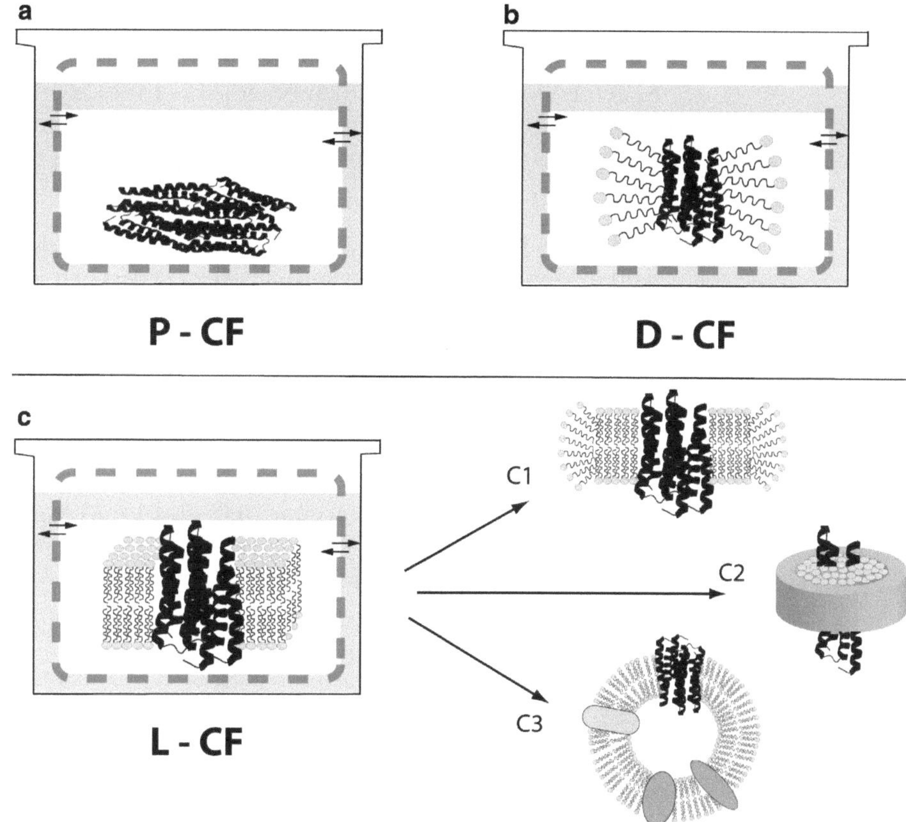

Fig. 16.1. CF expression modes for MPs. (**a**) The absence of any hydrophobic compounds cause precipitation of the synthesized MPs (P-CF). (**b**) Formation of proteo-micelles upon MP expression in the presence of detergents (D-CF). (**c**) Formation of lipid-inserted MPs upon expression in the presence of defined liposomes (L-CF) bicelles (C1), nanodiscs (C2) or membrane vesicles (C3).

We give detailed protocols for the expression of MPs in the P-CF and D-CF modes, and we provide preliminary recommendations for the synthesis of MPs in presence of lipids. We further emphasize new strategies for the production of labelled MPs by CF expression as a prerequisite for structural studies.

2. Materials

2.1. Preparation of E. coli S30 Extract

1. Fermenter (5–15 L)
2. Spectral photometer for the measurement of the optical density at 600 nm.
3. Centrifuge and rotors which can fit various volumes.

4. High pressure cell disrupter (IUL Instruments GmbH, Königswinter, Germany) (see Note 1).
5. Spectra/Por® 4 dialysis membrane, 12–14 kDa molecular weight cut off (MWCO) (Roth, Karlsruhe, Germany).
6. 100 ml of sterilized LB medium, 10 g/l tryptone, 5 g/l yeast extract, 10 g/l sodium chloride.
7. 10 L of 2× YTPG medium, 22 mM KH_2PO_4, 40 mM K_2HPO_4, 100 mM glucose, 16 g/l tryptone, 10 g/l yeast extract, 5 g/l NaCl.
8. *Escherichia coli* strain A19 (see Note 2).
9. Liquid nitrogen.
10. S30-A buffer, 10 mM Tris-acetate pH 8.2, 14 mM $Mg(OAc)_2$, 0.6 mM KCl, 6 mM 2-mercaptoethanol.
11. S30-B buffer, 10 mM Tris-acetate pH 8.2, 14 mM $Mg(OAc)_2$, 0.6 mM KCl, 1 mM 1,4-dithiotreitol (DTT), 0.1 mM phenylmethansulfonyl fluoride (PMSF).
12. S30-C buffer, 10 mM Tris-acetate pH 8.2, 14 mM $Mg(OAc)_2$, 0.6 mM KCl, 0.5 mM DTT.
13. 4 M NaCl

2.2. CECF Expression

Stock solutions should be prepared with Milli Q water and kept at –20°C until use.

1. Reaction containers, e.g. Mini- and Maxi-CECF-Reactors (see Fig. 16.2) (see Note 3).
2. Spectra/Por® 4 dialysis membrane (12–14 kDa MWCO, width 25 mm, diameter 15.9 mm) for analytical scale expression and 0.5–3 ml Slide-A-Lyzer dialysis cassettes with a 10 kDa MWCO (Pierce, Rockford, IL, USA) for preparative scale expression (see Note 4).
3. Thermo shaker for incubation of the reaction containers.
4. 500 mM 1,4-Dithiotreitol (DTT).
5. 1 M acetyl phosphate lithium potassium salt (AcP), adjusted to pH 7.0 with potassium hydroxide (see Note 5).
6. 1 M phospho(enol)pyruvic acid (PEP) monopotassium salt, adjusted to pH 7.0 with potassium hydroxide (see Note 5).
7. Amino acid mixture containing 4 mM of each of 20 natural amino acids (see Note 6).
8. RCWMDE mixture containing 16.7 mM of each of the amino acids R, C, W, M, D, E (see Note 7).
9. Complete® protease inhibitor cocktail (Roche Diagnostics, Penzberg, Germany).
10. 10 mg/ml folinic acid calcium salt.

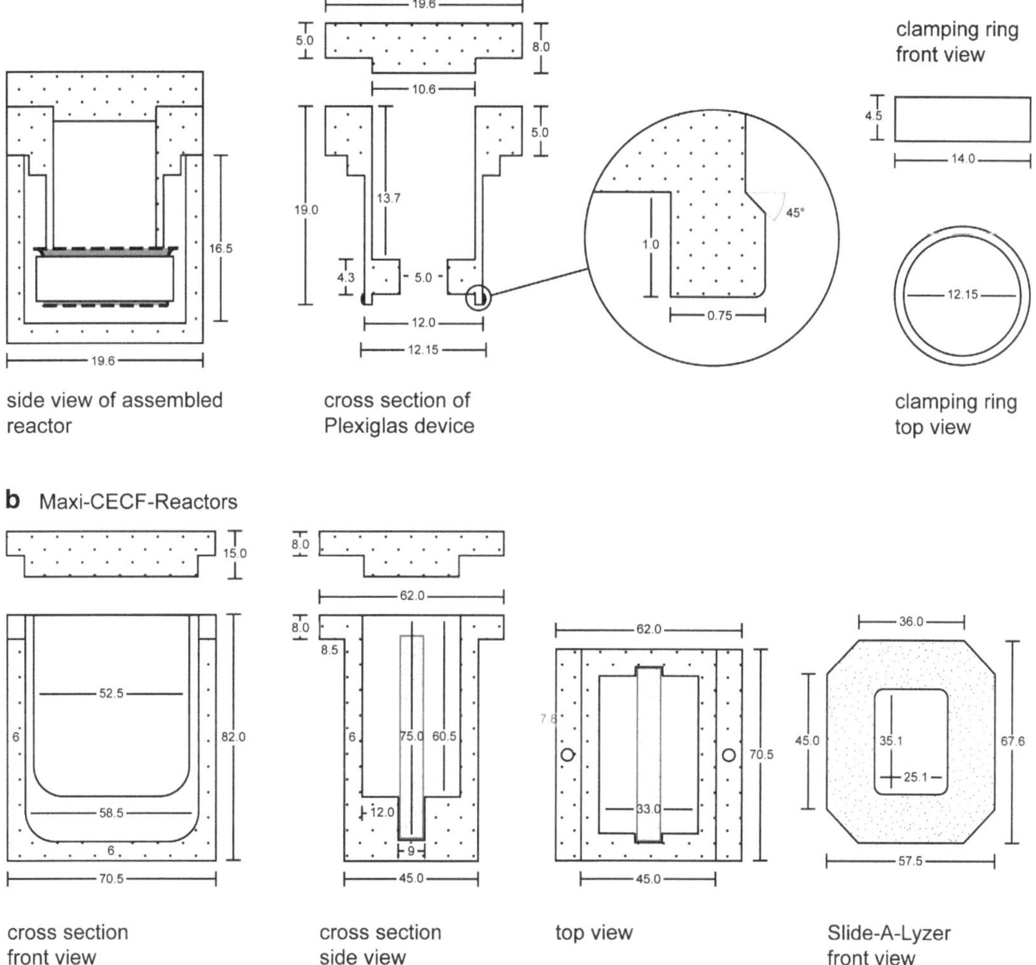

Fig. 16.2. Design of CECF reaction containers. (**a**) Mini-CECF-Reactors for 50–70 µl analytical scale expressions. These containers are suitable for the incubation in 24-well plates. The RM is placed in the inner cavity of the reactor. (**b**) Maxi-CECF-Reactors for 1–3 ml preparative scale production of MPs. These containers are used in combination with commercially available Slide-A-Lyzer dialysis cassettes (Pierce, Rockford, IL, USA).

11. HEPES buffer, 2.4 mM HEPES adjusted to pH 8.0 with potassium hydroxide.
12. 1 M $Mg(OAc)_2$.
13. NTP mixture containing 90 mM ATP, 60 mM CTP, 60 mM GTP and 60 mM UTP, adjusted to pH 7.0 with sodium hydroxide.
14. Plasmid DNA containing the gene of interest with T7 promotor regulatory elements (preferably at concentrations above 250 ng/µl) (see Note 8).

15. 40% (w/v) polyethyleneglycol 8,000 (PEG 8,000).
16. 4 M KOAc.
17. 10 mg/ml pyruvate kinase (Roche Diagnostics, Penzberg, Germany).
18. 40 U/µl RiboLock® RNase inhibitor (Fermentas, St. Leon-Rot, Germany).
19. S30-C buffer, 10 mM Tris-acetate pH 8.2, 14 mM $Mg(OAc)_2$, 0.6 mM KCl.
20. *E. coli* S30 extract.
21. 10% (w/v) sodium azide (NaN_3).
22. 400 U/µl T7-RNA polymerase (see Note 9).
23. 40 mg/ml total *E. coli* tRNA (Roche Diagnostics, Penzberg, Germany).

2.3. Solubilisation of MPs and Expression in the D- and L-CF Mode

1. Solubilization buffer A, 20 mM Tris-HCl, pH 7.5, 150 mM NaCl, onefold Complete® protease inhibitor cocktail (Roche Diagnostics, Penzberg, Germany)
2. Solubilization buffer B, same as Solubilization buffer A, but supplemented with an appropriate detergent at its working concentration (see Table 16.1).
3. Detergents: Stock solutions are made up in Milli Q water at the corresponding concentrations (see Note 10).

   Sodium dodecylsulfate (SDS) ultrapure (Roth, Karlsruhe Germany);

   1-myristoyl-2-hydroxy-sn-glycero-3-(phopho-rac(1-glycerol)) (LMPG), 1-palmitoyl-2-hydroxy-sn-glycero-3-(phopho-rac (1-glycerol)) (LPPG), 1-myristoyl-2-hydroxy-sn-glycero-3-phosphocholine (LMPC) and 1,2-dihexanoyl-sn-glycero-3-phosphocholine (DHPC) (Avanti Polar Lipids, AL, USA);

   n-dodecylphosphocholine (Fos-choline-12) and n-decyl-β-D-maltoside (DM) (Anatrace, OH, USA);

   Digitonin, Triton X-100, polyethylene-(23)-laurylether (Brij®35), polyoxyethylene-(20)-cetylether (Brij®58), polyoxyethylene-(20)-stearylether (Brij®78) and polyoxyethylene-(20)-oleylether (Brij®98) (Sigma-Aldrich, Taufkirchen, Germany);

   n-dodecyl-β-D-maltoside (DDM) (AppliChem, Darmstadt, Germany);

   n-octyl-β-D-glucopyranoside (β-OG) (Glycon Biochemicals, Luckenwalde, Germany).
4. Lipids (Avanti Polar Lipids, AL, USA): *E. coli* total lipid extract, 1,2-dimyristoyl-sn-glycero-3-phosphocholine (DMPC), 1,2-dioleyl-sn-glycero-3-phosphocholine (DOPC),

Table 16.1
Detergents for MP expression and resolubilisation[a]

Detergent	Monomer Ref	CMC		Work conc.[b]	Removal[c]	NMR/Ref.[d]
	(Da)	mM	(%)	(%)		
D-CF						
Brij-35	1,200	0.08	(0.001)	1.5	(B)	–
Brij-58	1,123	0.075	(0.008)	1.5	(B)	–
Brij-78	1,152	0.046	(0.005)	1.0	(B)	–
Brij-98	1,150	0.025	(0.003)	0.2	(B)	–
Digitonin	1,229	0.73	(0.08)	0.4	(B)	–
DDM	511	0.17	(0.009)	0.1	B	+ (15)
Triton X100	650	0.3	(0.021)	0.1	B	–
P-CF						
Anionic						
SDS	288	2.6	(0.075)	1	D, B	+ (16)
LMPG	479	0.2	(0.01)	2	B	+ (15)
LPPG	507	0.02	(0.001)	2	B	+ (15)
Zwitterionic						
Fos-12	352	1.5	(0.047)	2	B	+ (17)
LMPC	468	0.04	(0.002)	2	B	+ (16)
DHPC	453	1.4	(0.063)	2–5	D	+ (18)
LDAO	229	2	(0.046)	2	D	+ (19, 20)
Nonionic						
Triton X100	650	0.3	(0.021)	2	B	–
DM	483	1.8	(0.09)	2–5	D, B	+ (21)
DDM	511	0.17	(0.009)	2–5	B	+ (22)
β-OG	292	18	(0.53)	2–5	D	–
L-CF						
E. coli lipids	930			0.1–0.4	–	–
DMPC	678			0.1–0.4	–	–
DOPC	786			0.1–0.4	–	–
POPC	760			0.1–0.4	–	–

[a]Selection of detergents and lipids typically used for MP production
[b]Suggested range of initial working concentrations
[c]Recommended methods of detergent removal: *D* dialysis, *B* adsorption to Biobeads
[d]Recommended for NMR measurements according to indicated references

1-palmitoyl-2-oleyl-sn-glycero-3-phosphocholine (POPC) (see Note 10).

5. Extrusion buffer, 100 mM HEPES pH 8.0.
6. Ultra sonication water bath (Bandelin, Berlin, Germany).
7. Mini extruder (Avanti Polar Lipids, AL, USA).

2.4. Quality Improvement of CF Produced MPs

1. Solubilization buffer A, 20 mM Tris-HCl, pH 7.5, 150 mM NaCl, onefold Complete® protease inhibitor cocktail (Roche Diagnostics, Penzberg, Germany).
2. BT Bio-beads SM-2 (Bio-Rad, München, Germany).
3. Detergent which preferentially can be adsorbed with Bio-beads SM-2 (*see* Table 16.1).
4. *E. coli* polar lipid extract (Avanti Polar Lipids, AL, USA), dissolved in chloroform at 20 mg/ml

2.5. Production of Labelled MPs for Evaluation by NMR Spectroscopy

1. 96–98% U-^{15}N cell-free amino acid mix (20 aa), 98% ^{15}N, ^{13}C labelled individual amino acids, 95–99% (Cambridge Isotope Laboratories Inc., MA, USA).
2. NMR buffer; 20 mM Bis-Tris propane pH 6.8 or 20 mM MES/Bis-Tris pH 6.0.

3. Methods

3.1. Preparation of S30 CF Extract

In general, it is highly recommended to use a fermenter rather than shaking flasks, as higher cell densities can be achieved and aeration is much more efficient. The quality of each batch of extract can vary and might require optimisation at least with regard to the final concentrations of Mg^{2+} and K^+ ions in the CF reaction. We describe the protocol for S30 extract preparation from 10 L of culture, although other volumes may also be considered.

Day 1:

1. Inoculate 10 litres of sterilized 2× YTPG media in an appropriate fermenter with 100 ml of a fresh overnight culture from *E. coli* strain A19.
2. Grow the cells at 37°C with vigorous stirring and aeration until they reach the mid-log phase corresponding to an OD_{600} of approximately 5.0–8.0 (see Note 11).
3. Chill down the broth to below 12°C as quickly as possible.
4. Harvest the cells by 15 min centrifugation at 7,000×*g* and 4°C. Cells should always be kept at 4°C hereafter.
5. Carefully resuspend the cell pellet in 300 ml pre-chilled S30-A buffer and centrifuge again for 10 min at 8,000×*g* and 4°C.

Repeat this step twice and extend the final centrifugation step to 30 min (see Note 12).

6. Resuspend the cell pellet in 110% (v/w) pre-chilled S30-B buffer.
7. Lyse the cells by passing through a high pressure cell disrupter (1.6 kbar).
8. Centrifuge the lysate for 30 min at 30,000×*g* and 4°C. Fill the upper 2/3 of the supernatant in a fresh vial and repeat the centrifugation step once.
9. Adjust the supernatant to a final concentration of 400 mM NaCl and subsequently incubate at 42°C for 45 min. Significant precipitation is usually observed (see Note 13).
10. Dialyse the turbid extract at least twice against 100-fold excess of S30-C buffer using a dialysis membrane with a 12-14 kDa MWCO.

Day 2:

1. Clear the extract by 30 min centrifugation at 30,000×*g* and 4°C.
2. Aliquot the final extract and immediately freeze it in liquid nitrogen. The frozen extract can be stored at −80°C for several months (see Note 14).

3.2. P-CF Expression of MPs

CF reactions can be set up in analytical scale for optimization and screening as well as in preparative scale for the production of milligram quantities of recombinant MPs. Analytical scale reactions with a volume of 50–70 µl and a RM: FM ratio of 1: 14 (v/v) can be carried out in Mini-CECF-Reactors (see Note 15). For preparative scale expression with a RM of 1-3 ml and a RM: FM ratio of 1: 17 (v/v) the Maxi-CECF-Reactors can be used (see Fig. 16.2). Final concentrations of the individual reaction components (see Table 16.2) and a sample pipetting protocol (see Table 16.3) are given. The P-CF expression mode is often the most productive approach, giving the highest MP yields. Expression of the same target MP from identical templates in the D-CF mode requires additional optimization steps, and is sometimes associated with a significant reduction in MP yields. More laborious techniques for expression analysis such as western blotting might then become necessary, while Coomassie Blue staining after SDS-PAGE could already be sufficient for the detection of P-CF expressed MPs (see Fig. 16.3). The P-CF mode is therefore recommended as the first choice for the initial screening of new MP targets.

1. Thaw stock solutions and mix carefully. Enzymes, tRNA and S30 extract should be kept on ice after thawing.
2. First prepare the master mix FRM by pipetting the components common to FM and RM (see Table 16.3).

Table 16.2
Cell-free reaction components

Component	Stock concentration	Concentration in RM and FM[a]	
		Start	Optimization
Acetyl phosphate	1 M	20 mM	
ATP	90 mM	1.2 mM	
Complete® protease inhibitor	50×	1×	
CTP, GTP, UTP	60 mM each	0.8 mM each	
1,4-dithiothreitol (DTT)	500 mM	2 mM	
Folinic acid	10 mg/ml	0.1 mg/ml	
HEPES	2.4 M	100 mM	
Magnesium acetate[b]	1 M	15 mM	13–17 mM
Phospho(enol)pyruvic acid	1 M	20 mM	
Polyethylenglycol 8,000	40%	2%	
Potassium acetate[b]	4 M	290 mM	270–310 mM
RCWMDE mixture	16.7 mM each	1 mM	0–1 mM
Sodium azide	10%	0.05%	
Only in FM		Concentration in FM	
S30-C buffer	100%	35%	
Amino acid mixture	4 mM each	1 mM each	
Only in RM		Concentration in RM	
Amino acid mixture	4 mM each	0.5 mM	
E. coli tRNA	40 mg/ml	0.5 mg/ml	
Plasmid DNA	300 ng/µl	15 ng/µl	15–30 ng/µl
Pyruvat kinase	10 mg/ml	0.04 mg/ml	
RiboLock RNase inhibitor	40 U/µl	0.3 U/µl	
S30 extract	100%	35%	
T7 RNA-polymerase	400 U/µl	6 U/µl	

[a]Concentrations for initial start reactions as well as recommended ranges for optimization are given

[b]The concentration of potassium and magnesium ions is subject to optimization. Other sources of potassium and magnesium within the reaction are the HEPES buffer, the PEP solution, the AcP solution and the S30 extract

3. Take the appropriate aliquots from the master mix FRM and complete the FM first (*see* Table 16.3).
4. Incubate the completed FM at 30°C.

Table 16.3
Pipetting protocol for a CECF reaction with 1 ml RM

Stock solution	Add to FRM[a]	Add to FM	Add to RM
10% Sodium azide	90 µl		
40% Polyethylenglycol 8,000	900 µl		
4 M Potassium acetate[b]	679 µl		
1 M Magnesium acetate[b]	182 µl		
2.4 M HEPES buffer	660 µl		
50× Complete protease inhibitor	360 µl		
10 mg/ml Folinic acid	180 µl		
500 mM 1,4-dithiotreitol (DTT)	72 µl		
75× NTP mixture	240 µl		
1 M Phospho(enol)pyruvic acid	360 µl		
1 M Acetyl phosphate	360 µl		
4 mM Amino acid mixture	2,856 µl	2,697 µl	
16.7 mM RCWMDE mixture	1,080 µl		
Master mix FRM		7,573 µl	445 µl
S30-C buffer		5,950 µl	
10 mg/ml Pyruvat kinase			4 µl
40 mg/ml *E. coli* tRNA			13 µl
400 U/µl T7 RNA-polymerase			15 µl
40 U/µl RiboLock			8 µl
E. coli S30 extract			350 µl
300 ng/µl Plasmid DNA			50 µl
Bi-destilled water		fill up to 17,000 µl	fill up to 1,000 µl
Total volume	8.019 ml	17 ml	1 ml

[a]Master Mix FRM for FM and RM

[b]Volumes of potassium acetate and magnesium acetate are calculated under consideration of additional ion sources (HEPES buffer, PEP, AcP, S30 extract)

5. Complete the RM by pipetting the enzymes, the DNA template and the S30 extract (see Table 16.3) and mix well. Do not vortex.
6. Assemble the chosen CECF-Reactor and fill the FM into the FM-compartment (see Note 16).

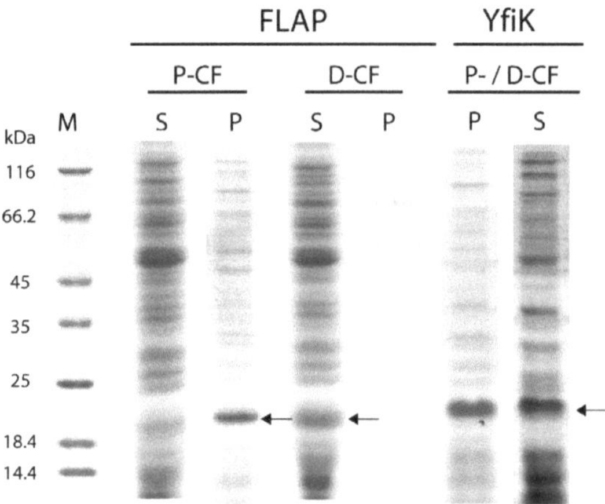

Fig. 16.3. P-CF and D-CF expression of MPs. Expression of 5-lipoxygenase activating protein (FLAP) and of the *E. coli* amino acid transporter YfiK. Almost complete soluble expression of the two MPs is achieved by D-CF expression. 1 µl samples are loaded. Samples were separated on 16.5% SDS-gels and stained with Coomassie Blue. *Arrows* indicate the expressed MPs. *S* supernatant, *P* resolubilized pellet fraction, *M* marker.

7. Fill the RM into the RM-compartment (see Note 17). Try to avoid air bubbles which might interfere with an efficient exchange between the two compartments.
8. Assemble the CECF-Reactor and incubate the reaction on an appropriate shaker at 30°C with moderate agitation for 15–20 h (see Note 18). During this time formation of MP precipitates will appear resulting in turbidity of the RM.
9. Harvest the MP by 10 min centrifugation at 18,000×*g*. Analyze supernatant and pellet fractions separately by SDS PAGE and/or western blotting if necessary (see Fig.16.3).

3.2.1. Solubilisation of P-CF Expressed MPs

Unlike proteins deposited into inclusion bodies during conventional in vivo expression, CF generated MP precipitates do not seem to be completely unfolded and therefore often do not require extensive unfolding and refolding procedures, such as treatments with urea or guanidinium hydrochloride. In contrast, many MP precipitates derived from P-CF reactions readily solubilize according to the following protocol by addition of appropriate detergents (see Table 16.1).

1. Centrifuge the RM containing the desired MP as precipitate at 18,000×*g* for 10 min.
2. Wash the precipitated MP in Solubilization buffer A by resuspending it in a volume similar to that of the initial RM. Centrifuge as mentioned above.
3. Discard the supernatant and repeat this washing step (see Note 19).

4. Resuspend the pellet with Solubilization buffer B and incubate the mixture at 30–37°C for up to 4 h to allow folding of the MP (see Note 20).
5. Centrifuge at 18,000×*g* for 10 min and analyse the supernatant for the solubilized MP by SDS-PAGE.

3.3. D-CF Expression of MPs

As a first step, the optimal detergent for efficient solubilisation of the target protein should be identified since protein yield and solubility vary for each target. Apart from detergent properties, the concentration used for the reaction plays an important role for the general expression efficiency and can undergo optimisation. While some detergents such as DPC and β-OG have an inhibitory effect already at low concentrations close to their critical micellar concentration (CMC), other detergents such as the polyoxyethylene alkyl ethers covering the Brij detergent family are tolerated up to 100-fold CMC.

1. Design detergent screen in analytical scale reactions. For initial trials, use the working concentrations indicated in Table 16.1 (see Note 21).
2. Prepare 10–15-fold detergent stock solutions. The amount of added detergent has to be subtracted from the amount of H_2O in the pipetting scheme given in Table 16.3 and is thus limited by this volume.
3. The detergent must be added into both reaction compartments. Mix well by pipetting, but avoid air bubbles especially in the RM.
4. Incubate the reaction as described in Subheading 3.2.
5. Separate the RM by centrifugation at 18,000×*g* for 10 min.
6. The soluble part of the RM is then transferred into a fresh tube while the remaining precipitate should be suspended in an equal amount of H_2O.
7. Prepare samples for initial analysis by SDS PAGE. Use 1–3 µl of the reaction per gel lane (see Fig. 16.3) (see Note 22).

3.4. L-CF Expression of MPs

Lipids are the natural environment of MPs and insertion into membranes is often indispensable for functional studies. In addition, lipids appear sometimes to be essential for the stability of MPs and they may play additional roles in folding or catalytic activity in particular cases. Some residual lipids in amounts of approx. 100 µg/ml are already present in the S30 extract after preparation. CF expression in the presence of additionally supplied lipids may soon become an important option for functional studies or in cases where the protein quality obtained with the other CF expression modes is unsatisfactory. This L-CF mode, however, more closely resembles expression approaches in cellular environments and thus problems associated with the efficient targeting and integration of MPs into lipid bilayers may become an issue again. The L-CF mode expression of MPs is an emerging technique, currently with only few examples, and no general protocols yet described. However, it is already evident that CF systems tolerate a wide range of lipids in high concentrations. Furthermore, lipids can be offered in a variety of different formulations which might affect the efficiencies of MP insertion (see Fig. 16.1).

3.4.1. Expression in Presence of Detergent-Solubilized Lipids

Expression in the presence of detergent-solubilised lipids is a mixture between the D-CF and the L-CF modes. The chosen ratio of detergent: lipid will determine the preferential formation of mixed micelles or bicelles. Mixed detergent micelles might accommodate few lipid molecules which could then be easier to interact with the inserted MPs.

1. Choose a lipid or lipid mixture appropriate for the target protein and a detergent that is capable of solubilising the lipid (see Note 23). Well known bicelle forming agents are DMPC and DHPC (a long and short-chain lipid) which may serve as a starting point.
2. Make up at least tenfold stocks of detergent and lipid in H_2O or a buffer such as 100 mM HEPES, pH 8.0. The volume of the detergent/lipid mixture is limited, and at maximum can only be equal to the volume of water in the pipetting scheme (see Table 16.3).
3. Vortex the detergent until complete solubilisation. To homogenize the lipid suspension, apply repeated cycles of 10 min sonication and vortexing.
4. Then combine the detergent and the lipid in the desired molar ratios and mix thoroughly by vortexing before adding to the RM.
5. While lipids remain in the RM, the detergent should be added to both compartments in order to prevent dilution.
6. Proceed as described in Subheading 3.2.

3.4.2. Expression in Presence of Defined Liposomes

Preformed liposomes can be added directly to the RM prior to expression at concentrations between approx. 1–4 mg/ml. Unilamellar vesicles might be most effective for the insertion of synthesized MPs. Upon hydration of lyophilized lipids, large multilamellar vesicles are usually formed, which can then be converted to more desirable unilamellar vesicles of defined size by extrusion. MP insertion efficiencies might depend on the acyl chain length and on the type of the polar head group of the lipid. The phosphocholine head groups, as found in DMPC and POPC, might be favourable for some targets, but a variety of lipids should be tested as indicated in Table 16.1.

1. Prepare ≥10-fold concentrated lipid solutions.
2. If lipids are dissolved in chloroform, dry first thoroughly using a SpeedVac or rotary evaporator. Chloroform must be completely removed as it inhibits the CF reaction.
3. Dissolve the appropriate amount of lipids in 100 mM HEPES, pH 8.0. A homogenous suspension should be obtained by several cycles of vortexing and sonication.
4. Take a pre-soaked membrane of 0.2–0.4 μm pore size and assemble the extruder parts according to the manufacturer's instructions.
5. Fill one of the two 1 ml Hamilton syringes of the extruder with 100 mM HEPES buffer and push through the extruder twice to equilibrate the assembly.
6. Fill the lipid suspension into one of the syringes.
7. Pass the lipid suspension through the extruder. A minimum of 15 cycles are recommended in order to obtain a homogenous suspension of even vesicle sizes.
8. Liposomes can be stored at 4°C after extrusion for few days. For a longer storage of hydrated lipids, store at −20°C and repeat extrusion prior to use.
9. Add preformed liposomes into the RM compartment of the CECF reaction and proceed as described in Subheading 3.2.

3.5. Quality Improvement of CF Produced MPs

Integrity of the synthesized MPs, the sample homogeneity, and in particular functional folding are important parameters which can be modulated by the selected CF expression conditions. Optimization of reaction conditions therefore requires relatively fast evaluation approaches that provide the necessary feedback information on the quality of the synthesized MPs. Important are the full-length expression of MPs and their efficient reconstitution into liposomes, which are the prerequisites for many activity assays.

3.5.1. Improving Integrity of CF Produced MPs

Evaluation of full-length MP production is important as in particular with targets of heterologous origin premature termination of the translation may occur (see Note 24). MPs often denature incompletely even in SDS-containing sample buffers, and the detected molecular size after SDS-PAGE can therefore differ from the calculated size. Immunodetection of the expressed MPs with antibodies directed against C-terminal tags like a poly(His)$_6$-tag will therefore help to identify full-length translated proteins. Additional prematurely terminated products could be monitored by immunodetection of N-terminal attached tags. Optimising the reaction conditions according to the following modifications can help at least to reduce the fraction of prematurely terminated products.

1. Increase the amount of bulk *E. coli* tRNA to 1 mg/ml (see Note 25).
2. Generally increase the amount of amino acids from the 4 mM stock or adjust the individual amino acid concentrations according to the protein sequence.
3. Analyse DNA sequence for clusters of rare codons and make possible changes by selected mutations. If defined fragments occur on SDS-PAGE, bands should be analysed using mass spectrometry to identify problematic DNA sequences.
4. A C-terminal affinity tag would allow purification of only the full-length protein.

3.5.2. Reconstitution of MPs for Functional Assays

The success of a reconstitution procedure largely depends on the right choice of lipids and detergents. Proteo-liposomes could then be further analyzed by freeze-fracture electron microscopy for validation of the density and homogeneity of reconstituted MPs. As an example, we describe the reconstitution of MPs after P-CF expression into *E. coli* total lipids.

1. Solubilize MPs in a detergent which can be removed with BT Bio-beads SM-2 (see Note 26) and which has shown good solubilisation behaviour with the target protein. Suitable detergents for reconstitution include Triton X100, DDM and LMPG.
2. Hydrate *E. coli* lipids in Milli Q water for a stock solution of 50 mg/ml. Use repeated cycles of vortexing and sonication in order to receive a homogenous suspension.
3. Add the lipid suspension to the solubilized MP in a molar ratio of about 1,000:1 (lipid:MP) and mix carefully by pipetting (see Note 27).
4. Incubate at 30°C with moderate agitation for at least 1 h.
5. Prepare Bio-beads by washing in 10 ml 100 % methanol per gram Bio-beads. Let the beads settle down and discard the supernatant.
6. Wash the Bio-beads in 10 ml of Milli Q water. Repeat this step twice.
7. Equilibrate the Bio-beads by washing twice with Resolubilization buffer A.
8. Pre-saturate the Bio-beads with lipids by 30 min incubation with 10 ml of Resolubilization buffer A supplemented with 5 mg lipid mixture at 30°C.
9. Add lipid pre-saturated Bio-beads to the reconstitution reaction. Choose a ratio of about 100:1 (w/w) (Bio-beads: detergent).
10. Incubate over night at 30°C with moderate agitation.
11. Replace Bio-beads with a fresh sample and incubate for another 6 h at 30°C with moderate agitation.

12. Let the Bio-beads settle down and harvest the supernatant containing the proteo-liposomes. The proteo-liposomes can further be evaluated by sucrose gradient ultracentrifugation or by freeze-fracture electron microscopy.

3.6. CF Production of Labelled MPs for Structural Studies

The structural analysis of MPs by Nuclear Magnetic Resonance (NMR) technologies is primarily limited by their difficult isolation and their shear size, while the surrounding membrane-like environments often further contribute to line broadening and signal overlap. The number of NMR-suitable detergents is thus limited and case-dependent and only few detergents are of somewhat general use (see Table 16.1). This owes to the fact that certain detergents such as SDS and LMPG do not appear to contribute significantly to the tumbling rate of the inserted proteins, possibly by allowing free rotation within the micelle. However, no universal detergent exists for MP analysis and appropriate detergents must rather be selected by way of numerous trials (see Note 28). It is therefore recommended to first evaluate NMR spectra of the target MP embedded in different detergent micelles in order to identify the most suitable conditions (see Fig.16.4). The relatively easy to identify signals of glycine residues and of tryptophan side chains in corresponding NMR spectra might serve as preliminary indicators for good dispersion and sample homogeneity. P-CF or D-CF expression protocols could then be considered for the production of highly concentrated and labelled MP samples in the desired detergents.

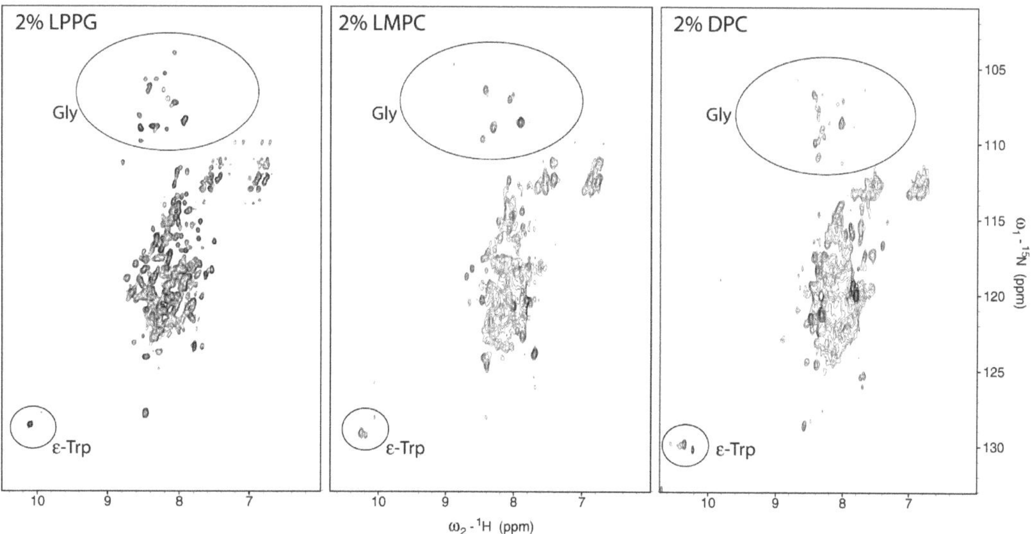

Fig. 16.4. $^{15}N,^{1}H$-HSQC spectra of the 5-lipoxygenase activating protein FLAP in three different detergents upon expression in the P-CF mode. All detergent members are lipid-like with either a phosphocholine (LMPC, DPC) or a phosphoglycerol (LPPG) head group. Differences in sample heterogeneity as well as stability become clear by number of peaks, line width and chemical shift dispersion in the proton dimension. In particular, the side-chain signals of tryptophan as well as a region in which mostly glycines occur (*both circled*) indicate a beneficial effect of LPPG for the spectral quality of the protein. In addition, the spectrum in LPPG displays the best dispersion and the least extent of line-broadening.

3.6.1. Preparation of Uniformly Labelled MPs

The only modification to the CECF protocol is the replacement of unlabelled amino acids by their labelled counterparts.

1. Depending on the given experiment, use either a U-^{15}N or a U-$^{15}N/^{13}C$ amino acid mix.
2. Prepare a 4 mM stock solution of labelled amino acids (see Note 29).
3. Proceed as described in Subheading 3.2 or 3.3

3.6.2. Selective Labelling Strategies

Selective labelling techniques reduce the severe resonance overlap observed for many MPs and simplify spectra. CECF offers full labelling flexibility with limited scrambling (some degree of scrambling has been observed in cases of aspartate, asparagine, glutamate, and glutamine). Apart from labelling single amino acids (see Note 30), sophisticated selective and combinatorial labelling schemes have been developed.

Transmembrane Segment (TMS) Enhanced Labelling for Backbone Assignment of Hydrophobic Regions

This technique exploits the unique nature of the transmembrane α-helical amino acid composition by targeting primarily these for labelling (10) (see Fig. 16.5). Statistical studies have revealed that the amino acids alanine, leucine, valine, phenylalanine, isoleucine,

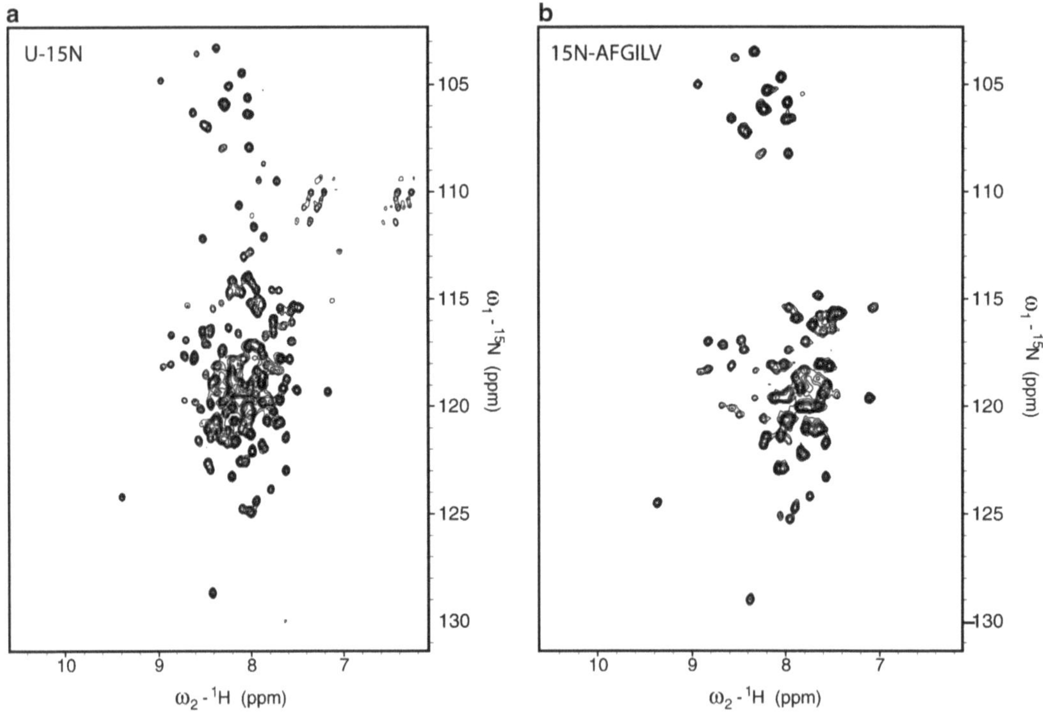

Fig. 16.5. Spectral improvement by TMS labelling of a 19 kDa C-terminal fragment of human presenillin-1. TROSY-HSQC spectra of (**a**) Uniformly ^{15}N labelled and (**b**) A,F,G,I,L,V-^{15}N labelled samples are shown. While the uniformly labelled sample shows a very high signal density, the selectively labelled sample has less peaks with reduced overlap, thus facilitating analysis of predominant TMSs of the protein.

Table 16.4
Chemical shift data for TMS-labelled amino acids[a]

Residue	Atom	Chemical shift range (ppm)	Chemical shift average (ppm)
GLY	Cα	35.50–55.80	45.34 ± 1.32
ILE	Cα	51.15–71.70	61.58 ± 2.72
VAL	Cα	44.98–70.02	62.44 ± 2.91
ALA	Cα	43.00 - 65.52	53.17 ± 2.01
LEU	Cα	44.60–67.11	55.65 ± 2.16
PHE	Cα	47.31–69.82	58.14 ± 2.63
ILE	Cβ	20.94–51.88	38.59 ± 2.08
VAL	Cβ	20.55–42.84	32.70 ± 1.84
ALA	Cβ	9.79–32.99	18.95 ± 1.84
LEU	Cβ	20.55–42.84	42.25 ± 1.90
PHE	Cβ	25.52–50.40	39.89 ± 2.10

[a]Residues with overlapping chemical shift ranges are grouped together

and glycine constitute nearly 60% of TMSs. Specific labelling of this amino acid subset will therefore simplify NMR spectra while predominantly providing information from residues located in the hydrophobic TMSs (see Fig. 16.5). This allows the unambiguous assignment of unique stretches within these regions via typical 3D NMR measurements, such as HNCA and HNCACB, which detect the amino acid Cα and Cβ resonances, respectively. Assignment can then be achieved according to the characteristic chemical shifts of these residues (see Table 16.4).

1. Produce a 20 amino acid mixture composed of the six ^{15}N, ^{13}C labelled TMS selected amino acids and the remaining 14 non-labelled amino acids. The average concentration of the stock solution should be 100 mM.
2. Add amino acids to the RM at a final concentration of 0.5–1 mM and proceed as described in Subheading 3.2 or 3.3.
3. Obtain HNCA spectrum and follow with assignment
4. If required, obtain HNCACB spectrum to resolve ambiguities in assignment

Selective Combinatorial Labelling for Backbone Assignment of Sequential Amino acids Pairs

In this scheme, certain selected amino acids are ^{15}N labelled and others ^{13}C labelled (11). Three samples are then prepared, each having a different labelling combination (see Fig. 16.6). Selective two-dimensional NMR experiments, namely HSQC and HNCO,

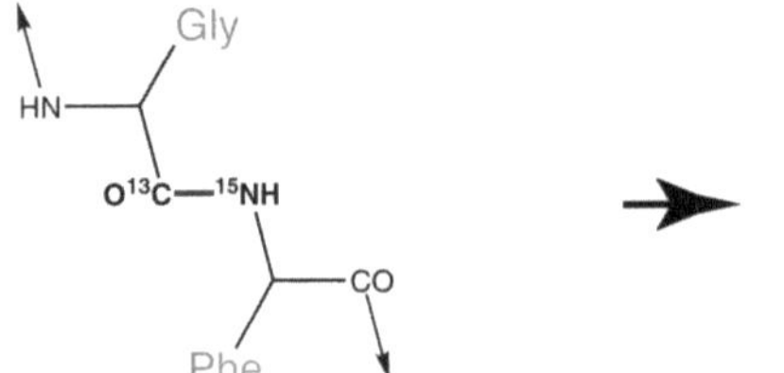

amino acid	sample 1	sample 2	sample 3
Ala	^{15}N	^{15}N	^{15}N
Phe	^{15}N	^{15}N	Δ
Ile		^{15}N	^{15}N
Ser	^{13}C	^{13}C	
Leu		^{13}C	^{13}C
Gly	^{13}C	Δ	Δ
Val			^{13}C

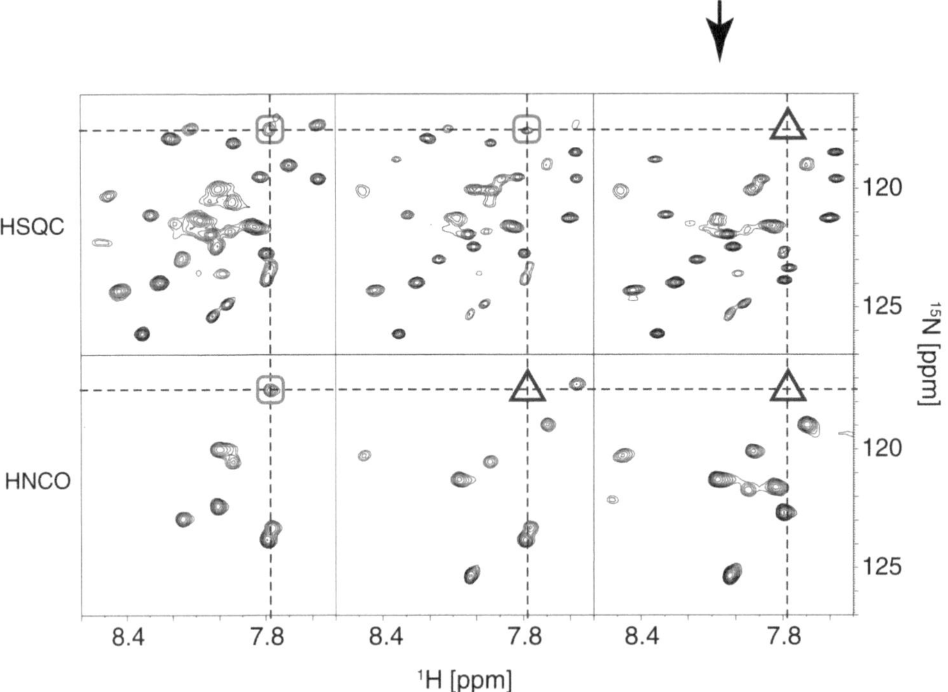

Fig. 16.6. Example of a combinatorial labelling scheme with three differentially ^{15}N and ^{13}C labelled samples of the *E. coli* tellurite transporter Δ-TehA. The labelling scheme is shown in the table on top. In the ^{15}N, ^{1}H-TROSY spectra, backbone amide protons of the ^{15}N-labelled amino acids are visible. The dashed crossed lines indicate the peak to be identified. The indicated peak at position 117.5/7.78 is identified as phenylalanine as it is present in samples 1 and 2, but not in sample 3. The corresponding HNCO spectra show the amide crosspeaks after carbonyl transfer and indicate that the preceding residue of this phenylalanine must be a glycine, as no crosspeaks are visible in sample 2 and 3 without ^{13}C-glycine. The analysed phenylalanine residue can thus be localized as a Gly-Phe pair in the primary sequence which in this example corresponds to Phe97 of Δ-TehA.

are then measured. The HSQC experiment maps the position of the ^{15}N amide resonance of each amino acid within the protein, whereas the HNCO experiment relies on the transfer of magnetism from the amide to the preceding carbonyl and back along the peptide bond in order to provide a signal. Therefore, one should observe all ^{15}N amide resonances in the HSQC spectra, and only the resonances of those ^{15}N amides which are preceded by a ^{13}C carbonyl moiety in the HNCO spectra. These spectra together would then allow identification of amino acid pairs thus permitting specific assignments assuming that the pairs are unique.

1. Assign backbone as far as possible using standard NMR methods.
2. Choose regions where assignment is sparse and look for unique combinations of amino acid pairs. Group amino acids in three different samples in such a way that the chemical shift pattern in the six recorded spectra will be unique for each pair of amino acids.
3. Prepare a labelling scheme where the second amino acid in each pair is ^{15}N labelled and the first, preceding amino acid is ^{13}C labelled (see Fig. 16.6).
4. Prepare 3 amino acid mixtures, where the selected amino acids are ^{15}N and ^{13}C labelled accordingly while all others remain unlabelled. The amino acid composition should reflect the percent composition of each amino acid in the given protein and the average stock concentration should be 100 mM. Add amino acids to RM and FM to a final concentration of 0.5-1 mM and proceed as described in Subheading 3.2 or 3.3.
5. Proceed with the measurement of 2D versions of the TROSY-HSQC and HNCO experiments.

4. Notes

1. A French press can also be used for cell lysis, but sonication should be avoided due to the potential disintegration of ribosomes.
2. For the CF S30 extract preparation, various other *E. coli* strains like BL21 (DE3) or D10 have also been used.
3. Alternatively, commercially available Micro DispoDialysers (Roth, Karlsruhe, Germany) or D-tube dialysers (Merck, Darmstadt, Germany) can be used. Generally, MWCOs in the range of 10–20 kDa and regenerated cellulose as membrane material are preferable.
4. In general, Slide-A-Lyzer cassettes can be reused several times. Used cassettes should be stored in 0.01% sodium azide or 20% ethanol at 4°C for no more than a couple of weeks. Before reusing, the Slide-A-Lyzer should be rinsed with Milli Q water. Membranes of the Mini-CECF-Reactors should be used only once.

5. Since expression rates strongly depend on the concentration of potassium ions, we recommend always using the same concentration of potassium hydroxide for pH adjustment. In keeping the amount of potassium added per gram of AcP or PEP constant, the pH may vary slightly with each preparation. However, the buffer system in the RM will compensate for this.
6. A 20 mM stock solution of tyrosine and 100 mM stock solutions of each of the 19 remaining amino acids should be made first. Some amino acids might not completely dissolve at these concentrations and can be handled as suspensions. Aliquots of the individual amino acid stocks are then combined to give the 4 mM mixture. The final concentration of amino acids in the standard reaction is 1 mM. However, the expression of larger proteins or of MPs having a significant bias in amino acid composition may benefit from increased final concentrations of amino acid subsets predominant in the sequence.
7. Increased concentrations of RCWMDE compensate for losses caused by instability and metabolic degradation of these amino acids.
8. The DNA quality is critical for expression rates. No RNA contamination should be present and residual RNase I should be removed by phenol/$CHCL_3$ extraction. Commercial plasmid preparation kits usually yield sufficiently pure DNA for the CECF reaction. For DNA resolubilisation, we suggest dissolving the DNA in Milli Q water rather than in buffers. With regard to vector construction, the pET series as well as the higher copy pIVEX plasmids are recommended. Increasing the amount of DNA in the CECF reaction may also help to increase protein yield and can thus undergo optimisation.
9. T7-RNA polymerase is one of the most expensive components and is required in high concentrations. We therefore recommend recombinant expression of this enzyme in *E. coli* BL21 (DE3) from which yields of approximately $0.5–1 \times 10^6$ units of T7-RNA polymerase can be obtained per one litre of culture (12, 13). Alternatively, commercial sources (e.g. Roche Diagnostics, Penzberg, Germany) could be used.
10. The lists of detergents and lipids give only a selection of some of the current most commonly used compounds and are given to provide a starting point for optimization. Some detergents such as Brij derivatives might require heating up to 50°C and excessive vortexing in order to get dissolved completely.
11. It is crucial for the activity of the S30 extract to harvest the cells in the mid-log phase. We therefore recommend an initial pilot experiment to record the growth kinetics.

12. While we suggest continuing with extract preparation according to the following steps, the protocol may as well be interrupted at this point and the pellets could be stored at -80°C.
13. This step is necessary to remove endogenous mRNA from the ribosomes in order to limit background expression.
14. Extract aliquots should be made according to the commonly used volumes in order to avoid repeated freezing and thawing steps.
15. The RM to FM ratio largely influences the expression rates and larger FM volumes generally correlate with higher protein yields. However, the gain of yield increasingly becomes limited by the exchange rate between RM and FM. Therefore we recommend the indicated RM: FM ratios as good economic compromises.
16. CF reactors should be completely dry prior use. Residual water drops will dilute the reaction so that ion concentrations are not accurate. For assembly of the Mini-CECF-reactors, the dialysis membrane first has to be cut in 2.5 × 2.5 cm squares and stored in Milli Q water. For the RM compartment, turn the Plexiglas device (see Fig. 16.2) upside down and place the wet membrane on its top. Avoid excessive water. Carefully pose the plastic ring over the membrane to fix it and adjust it by pushing the device onto a piece of parafilm to ensure an even surface for the membrane. Then, place the reactor into a 24-well plate containing the feeding mix. Air bubbles between FM and dialysis membrane must be avoided. For assembly of the Maxi-CECF-reactor, the FM is first filled into the Plexiglas container. It is absolutely required to remove residual water from this container and from the dialysis cassette which is then filled with the RM. The dialysis cassette is inserted into the Plexiglas container with the filling hole on the top.
17. To fill the RM into the appropriate cavity, approach the membrane with the pipet tip along the wall of the inner opening of the Plexiglas device. The pipette tip has to almost touch the dialysis membrane. Capillary forces will then direct the RM into the cavity without generating air bubbles. As this requires a bit of practice, first trials should be made using coloured solutions, which in addition will be a good way to check whether the reactor is correctly assembled without leakage. For the Maxi-CECF-reactors, fill the RM into the dialysis cassette with a syringe by penetrating the gasket as recommended by the supplier. Before removing the syringe from the dialysis cassette, adjust the levels of RM and FM in the reactor by removing some air from the cassette. This ensures optimal substance exchange through the dialysis membrane.

18. The standard temperature for the CECF reaction is 30°C. However, for the optimization of e.g. folding kinetics, lower incubation temperatures may be evaluated.
19. Wash-buffer volumes may be varied. Nonspecific co-precipitated proteins might be more efficiently removed by addition of low concentrations of detergents that do not solubilise the target MP such as a Brij-derivatives to Solubilization buffer A.
20. The temperature for MP solubilisation depends on the target protein and on the detergent used. We recommend the evaluation of different detergent concentrations as well as different temperatures.
21. Detergent concentrations must be adjusted at the best compromises of yield and soluble MPs. Mixtures of different detergents might also be considered for evaluation.
22. It is recommended to employ immunoblotting for verification and analysis of MP expression. Terminal antigen epitopes such as the T7-tag or a poly(His)$_6$-tag are helpful for the specific detection of the target MP, and for the verification of full-length expression, in addition to their benefits in purification.
23. Depending on the ratio between the detergent and lipid, the solution may remain turbid as solubilising properties of detergents vary with the types of detergent and lipid used. The solubilisation process can be followed photometrically by measuring the light scattering at OD 540 nm, which typically decreases with the increasing lipid solubility.
24. Strong differences in codon usage of genes from the *E. coli* repertoire, secondary structure formation of the mRNA, or amino acid shortage can be reasons for premature termination.
25. Using S30 extract sources from *E. coli* strains overproducing rare tRNAs such as *E. coli* Rosetta® or BL21 (DE3) Codon Plus® (Merck, Darmstadt, Germany) may as well address problems caused by codon bias (14).
26. Alternatively, detergent removal could be achieved by dialysis or by using specific detergent removal kits (Pierce, Rockford, IL, USA).
27. For some detergents such as Triton X100 and DDM, the lipids will be solubilised so that the turbidity of the lipid suspension changes. Upon incubation with biobeads, the solution will become turbid again indicating detergent removal.
28. An extensive list of detergent and additive conditions for solved structures of MPs is available online: (www.mpibp-frankfurt.mpg.de/michel/ public/memprotstruct.html) MPs may have to be first soluble expressed in the D-CF mode and then exchanged into detergents that are better suitable for structural

analysis. The mode of CF expression might also have a significant impact on the quality of NMR spectra.

29. Commercial amino acid mixtures differ significantly in their composition of individual amino acids. In particular, histidine and tryptophan frequently tend to be rather low and concentrations of these may need to be increased as required. In addition, non-hydrolysed peptides might still be present in substantial amounts so that the effective concentrations of free amino acids may be considerably lower than indicated.

30. Labelling the amino acids glycine and tryptophan of MPs is particularly useful for rapid optimization of sample preparation conditions. Signals from glycine are well-dispersed, and the side-chain of tryptophan occupies a unique region in corresponding NMR spectra. Dispersion and number of signals from these two amino acids can therefore give rapid information on the integrity, conformational exchange and folding of the analysed MP.

References

1. Kim, T. W., Keum, J. W., Oh, I. S., Choi, C. Y., Kim, H. C., and Kim, D. M. (2007) An economical and highly productive cell-free protein synthesis system utilizing fructose-1, 6-bisphosphate as an energy source. *J. Biotechnol.* **130,** 389–393.
2. Shirokov, V. A., Kommer, A., Kolb, V. A., and Spirin, A. S. (2007) Continuous-exchange protein-synthesizing systems. *Methods Mol. Biol.* **375,** 19–55.
3. Gourdon, P., Alfredsson, A., Pedersen, A., Malmerberg, E., Nyblom, M., Widell, M., Berntsson, R., Pinhassi, J., Braiman, M., Hansson, O., Bonander, N., Karlsson, G., and Neutze, R. (2008) Optimized in vitro and in vivo expression of proteorhodopsin: a seven-transmembrane proton pump. *Protein Expr. Purif.* **58,** 103–113.
4. Keller, T., Schwarz, D., Bernhard, F., Dotsch, V., Hunte, C., Gorboulev, V., and Koepsell, H. (2008) Cell free expression and functional reconstitution of eukaryotic drug transporters. *Biochemistry* **47,** 4552–4564.
5. Elbaz, Y., Steiner-Mordoch, S., Danieli, T., and Schuldiner, S. (2004) In vitro synthesis of fully functional EmrE, a multidrug transporter, and study of its oligomeric state. *Proc. Natl. Acad. Sci. USA* **101,** 1519–1524.
6. Klammt, C., Schwarz, D., Fendler, K., Haase, W., Dotsch, V., and Bernhard, F. (2005) Evaluation of detergents for the soluble expression of alpha-helical and beta-barrel-type integral membrane proteins by a preparative scale individual cell-free expression system. *FEBS J.* **272,** 6024–6038.
7. Kalmbach, R., Chizhov, I., Schumacher, M. C., Friedrich, T., Bamberg, E., and Engelhard, M. (2007) Functional cell-free synthesis of a seven helix membrane protein: in situ insertion of bacteriorhodopsin into liposomes. *J. Mol. Biol.* **371,** 639–648.
8. Wuu, J. J., and Swartz, J. R. (2008) High yield cell-free production of integral membrane proteins without refolding or detergents. *Biochim. Biophys. Acta* **1778,** 1237–1250.
9. Katzen, F. (2008) Cell-free protein expression of membrane proteins using nanolipoprotein particles. *Biotechniques* **45,** 190.
10. Reckel, S., Sobhanifar, S., Schneider, B., Junge, F., Schwarz, D., Durst, F., Lohr, F., Guntert, P., Bernhard, F., and Dotsch, V. (2008) Transmembrane segment enhanced labeling as a tool for the backbone assignment of alpha-helical membrane proteins. *Proc. Natl. Acad. Sci. USA* **105,** 8262–8267.
11. Koglin, A., Klammt, C., Trbovic, N., Schwarz, D., Schneider, B., Schafer, B., Lohr, F., Bernhard, F., and Dotsch, V. (2006) Combination of cell-free expression and NMR spectroscopy as a new approach for structural investigation of membrane proteins. *Magn. Reson. Chem.* **44 Spec No,** S17–S23.
12. Li, Y., Wang, E., and Wang, Y. (1999) A modified procedure for fast purification of T7 RNA polymerase. *Protein Expr. Purif.* **16,** 355–358.

13. Schwarz, D., Junge, F., Durst, F., Frolich, N., Schneider, B., Reckel, S., Sobhanifar, S., Dotsch, V., and Bernhard, F. (2007) Preparative scale expression of membrane proteins in Escherichia coli-based continuous exchange cell-free systems. *Nat. Protoc.* **2,** 2945–2957.
14. Chumpolkulwong, N., Sakamoto, K., Hayashi, A., Iraha, F., Shinya, N., Matsuda, N., Kiga, D., Urushibata, A., Shirouzu, M., Oki, K., Kigawa, T., and Yokoyama, S. (2006) Translation of 'rare' codons in a cell-free protein synthesis system from Escherichia coli. *J. Struct. Funct. Genomics* **7,** 31–36.
15. Krueger-Koplin, R. D., Sorgen, P. L., Krueger-Koplin, S. T., Rivera-Torres, I. O., Cahill, S. M., Hicks, D. B., Grinius, L., Krulwich, T. A., and Girvin, M. E. (2004) An evaluation of detergents for NMR structural studies of membrane proteins. *J. Biomol. NMR* **28,** 43–57.
16. Chill, J. H., Louis, J. M., Miller, C., and Bax, A. (2006) NMR study of the tetrameric KcsA potassium channel in detergent micelles. *Protein Sci.* **15,** 684–698.
17. Lee, S., Howell, S. B., and Opella, S. J. (2007) NMR and mutagenesis of human copper transporter 1 (hCtr1) show that Cys-189 is required for correct folding and dimerization. *Biochim. Biophys. Acta* **1768,** 3127–3134.
18. Schnell, J. R., and Chou, J. J. (2008) Structure and mechanism of the M2 proton channel of influenza A virus. *Nature* **451,** 591–595.
19. Malia, T. J., and Wagner, G. (2007) NMR structural investigation of the mitochondrial outer membrane protein VDAC and its interaction with antiapoptotic Bcl-xL. *Biochemistry* **46,** 514–525.
20. Roosild, T. P., Greenwald, J., Vega, M., Castronovo, S., Riek, R., and Choe, S. (2005) NMR structure of Mistic, a membrane-integrating protein for membrane protein expression. *Science* **307,** 1317–1321.
21. Schubert, M., Kolbe, M., Kessler, B., Oesterhelt, D., and Schmieder, P. (2002) Heteronuclear multidimensional NMR spectroscopy of solubilized membrane proteins: resonance assignment of native bacteriorhodopsin. *Chembiochem* **3,** 1019–1023.
22. Takeuchi, K., Takahashi, H., Kawano, S., and Shimada, I. (2007) Identification and characterization of the slowly exchanging pH-dependent conformational rearrangement in KcsA. *J. Biol. Chem.* **282,** 15179–15186.

Chapter 17

Production of Membrane Proteins Through the Wheat-Germ Cell-Free Technology

Akira Nozawa, Hideaki Nanamiya, and Yuzuru Tozawa

Abstract

Membrane proteins play crucial roles in various processes. However, biochemical characterization of the membrane proteins remains challenging due to the difficulty in producing membrane proteins in a functional state. Here, we describe a novel method for the production of functional membrane proteins based on a wheat germ cell-free translation system. Using this method, functional membrane proteins are successfully synthesized in the presence of liposomes and a detergent. In addition, the synthesized membrane proteins are easily purified from the cell-free translation mixture as proteoliposomes by sucrose density gradient ultracentrifugation. These advantages over conventional approaches are very helpful for the clarification of the function of membrane proteins.

Key words: Liposome, Membrane protein, Proteoliposome, Wheat-germ cell-free protein synthesis system

1. Introduction

Completion of entire genome sequences of various organisms revealed the existence of a large number of genes encoding membrane proteins in their genomes (1, 2). Membrane proteins are associated with a variety of essential processes such as perception and transduction of external signals, import or export of substances through the membranes, and the production of energy. However, only a limited number of membrane proteins have been functionally characterized so far, largely due to the difficulty in obtaining sufficient amounts of membrane proteins in a functional state. For example, preparation of membrane proteins from living cells is only applicable to particular proteins which are fully expressed in the cells. Furthermore, overexpression of recombinant membrane proteins often causes growth arrest of the host

Y. Endo et al. (eds.), *Cell-Free Protein Production: Methods and Protocols,* Methods in Molecular Biology, vol. 607
DOI 10.1007/978-1-60327-331-2_17, © Humana Press, a part of Springer Science+Business Media, LLC 2010

cells , and thus sufficient amounts of proteins for function analysis are hardly obtained.

In contrast, a cell-free translation system is able to circumvent these difficulties associated with the expression of recombinant membrane proteins in living cells. Any type of proteins is able to be potentially synthesized by a cell-free translation system, except for proteins which negatively regulate translation activity (3, 4). Recently, the production of membrane proteins by cell-free systems has been extensively attempted by several groups (5–9). In addition to these, we have recently demonstrated a successful production of functional plant transporter proteins by a modified wheat cell-free system in the presence of detergent and liposomes (10). In this report, we describe the details of a novel cell-free synthesis method for the production of membrane proteins, and a method for the partial purification of proteoliposomes obtained by the integration of synthesized proteins into liposomes during cell-free synthesis. This method is a promising way for the production of sufficient amount of membrane protein in a functional state.

2. Materials

2.1. Preparation of Liposomes

1. Asolectin from soybean from Sigma-Aldrich, Tokyo, Japan.
2. Chloroform and acetone from Nacalai tesque, Kyoto, Japan.
3. Sonicator (Digital Sonifier model 250D, 200 W, 20 kHz) from Branson, Danbury, CT.

2.2. Wheat Germ Cell-Free Protein Synthesis System

1. Wheat germ extract (WEPRO1240, $OD_{260}=240$) from CellFree Science Co. Ltd., Matsuyama, Japan.
2. Creatine kinase (Roche Diagnostics K.K., Tokyo, Japan) is dissolved in water at 40 mg/mL and stored in aliquots at −80°C.
3. Dialysis buffer (4×): 120 mM HEPES-KOH (pH 7.8), 400 mM potassium acetate, 10.8 mM magnesium acetate, 1.6 mM spermidine, 10 mM dithiothreitol, 4.8 mM ATP, 1 mM GTP, 64 mM creatine phosphate, 0.02% NaN_3, and 1.2 mM of each amino acid. Store at −80°C (*see* Note 1).
4. Poly(oxyethylene)(23) lauryl ether (Brij35; Sigma-Aldrich) is dissolved in Milli-Q water at 0.8% (w/v). Store at −20°C.

2.3. Sucrose Density Gradient Ultracentrifugation

1. Sucrose (Nuclease and Protease tested) from Nacalai tesque.
2. Sephadex G-25 from GE healthcare UK Ltd., Buckinghamshire, England.
3. Ultracentrifugation buffer: 140 mM NaCl, 5.4 mM KCl, and 10 mM Tris–HCl (pH 8.0). Store at room temperature.

4. Ultracentrifugation tube (2.2 PA) and rotor (S55S) from Hitachi Koki, Tokyo, Japan.

3. Methods

3.1. Preparation of Phospholipid

1. Dissolve 10 g of asolectin in 30 ml of chloroform.
2. Add 180 ml of ice-cold acetone to the solution and stir the suspension on a magnetic stirrer for 2 h at room temperature.
3. Turn off the stirrer and allow the solution to stand overnight at 4°C for precipitating phospholipids.
4. Aspirate as much as possible of the supernatant fluid using a glass pipette and dry the pellet completely under a flow of nitrogen gas.
5. Store the dried phospholipid mixture at –20°C until use.

3.2. Preparation of Liposomes

1. Fifty milligrams of the acetone-washed phospholipid mixture is rehydrated in 500 μL of Milli-Q water.
2. Sonicate the suspension (10% amplitude and 30% duty cycle) for 5 min on ice until the suspension changes from milky to nearly clear in appearance (*see* Note 2).

3.3. Synthesis of Proteins by Wheat Germ Cell-Free System

Several methods are so far established for cell-free protein synthesis system. Among them, the bilayer-based cell-free protein synthesis system is a simple, powerful and less time-consuming method with low cost (11). In this method, continuous supply of amino acids and energy components and removal of small byproducts is progressed through a phase between the translation mixture and substrate mixture. Therefore, the reaction is able to be prolonged for a long time and the target protein is successfully obtained in good yield. In this section, we introduce the cell-free protein synthesis of membrane proteins by bilayer method. Liposomes and a nonionic detergent, Brij35, are added to both mixtures, providing a successful integration of the synthesized membrane proteins into liposomes during cell-free synthesis. This method is optimized for the synthesis of *Arabidopsis thaliana* phosphoenolpyruvate/phosphate translocator 1 (AtPPT1) (10) (*see* Note 3). The amount of membrane proteins obtained by this method is usually 1–10 μg in 150 μL scale reaction. These amounts are sufficient for functional analysis.

1. Thaw wheat germ extract, creatine kinase, and dialysis buffer (4×) on ice. Place and keep all reagents on ice during handling. Prepare 25 μL of translation mixture on ice according to the mixing formula as follows: 6.25 μL of wheat germ

extract, 0.25 μL of 40 mg/mL creatine kinase, 4.6875 μL of dialysis buffer (4×), 1.25 μL of 0.8% (w/v) Brij35, 2.5 μL of 100 mg/mL liposome, X μL of mRNA (*see* Note 4), and (10.3 – X) μL of Milli-Q water.

2. Prepare 125 μL of substrate mixture on ice according to the mixing formula as follows: 31.25 μL of dialysis buffer (4×), 6.25 μL of 0.8% (w/v) Brij35, 12.5 μL of 100 mg/mL liposome, and 75 μL of Milli-Q water.
3. Add 125 μL of the substrate mixture to a well of U-shaped microtiter plate.
4. Carefully transfer 25 μL of the translation mixture into the bottom of the well containing substrate mixture to form bilayer.
5. Seal the plate to avoid evaporation.
6. Incubate the plate at 26°C for 16 h.
7. After the reaction, the reaction mixture (150 μL) is gently mixed, and then transferred to the new microfuge tube for further analysis.
8. Five microliters aliquot of the reaction mixture is subjected to SDS–PAGE analysis to check the proper synthesis of the target protein (*see* Note 5).

3.4. Sucrose Density Gradient Ultracentrifugation

The membrane proteins synthesized by liposome-containing wheat germ cell-free system are expected to be incorporated into liposomes during translation, if the synthesized proteins are functional. Therefore, the reaction mixture is subjected to functional analysis after exchanging the buffer by gel filtration using Sephadex G-25. Alternatively, the proteoliposome obtained by liposome-containing cell-free translation reaction is able to be partially purified by discontinuous sucrose density gradient ultracentrifugation. This procedure is also useful for confirming the integration of the synthesized membrane protein into liposome.

1. Prepare 10% (w/v) and 30% (w/v) sucrose solution in ultracentrifugation buffer.
2. Add 600 μL of 30% sucrose solution to ultracentrifugation tube (2.2 PA), then 1,300 μL of 10% sucrose solution is carefully overlaid on the 30% solution.
3. Overlay the reaction mixture (100 μL) carefully onto discontinuous sucrose gradients.
4. Centrifuge at 105,000 × *g* for 4 h at 4°C (S55S rotor).
5. Collect 200-μL fractions from the top of the tubes. Proteoliposomes are found at the interface between 10% and 30% sucrose.

4. Notes

1. All solutions are prepared in Milli-Q water synthesized by Autopure WT101UV (Yamato Scientific Co., Ltd., Tokyo, Japan) with BioPak filter (Nihon Millipore K. K., Tokyo, Japan), which is able to eliminate DNases, RNases and Pyrogens.
2. We newly prepare the suspension for each experiment and store it on ice until use.
3. Goren and Fox (2008) have recently reported that human stearoyl-CoA desaturase was successfully synthesized by a wheat germ cell-free protein synthesis system in the presence of liposomes, but without adding any detergents (12).
4. Quality of templates for transcription is critical either for quantity and quality of protein production. We, therefore, use a plasmid preparation kit, JETSTAR 2.0 Plasmid Midi Kit (Genomed, Bad Oeynhausen, Germany) for the isolation of pure plasmids. Using the plasmids as templates, mRNAs are synthesized by in vitro transcription as described previously (13). The volume of the mRNA (X μL) added to the reaction is determined in order to contain 25 μg of the synthesized mRNA.
5. If the target protein is synthesized in the presence of [^{14}C] leucine, the amount of synthesized protein is calculated by incorporation of [^{14}C]leucine into the synthesized protein, which is measured by liquid scintillation spectroscopy.

Acknowledgments

This study was supported by a Grant-in-Aid for Scientific Research on Priority Areas (Plant Membrane Transport) from the Ministry of Education, Culture, Sports, Science, and Technology of Japan (no. 18056016 to Y.T.)

References

1. The Arabidopsis Genome Initiative (2000) Analysis of the genome sequence of the flowering plant *Arabidopsis thaliana*. *Nature* **408,** 796–815.
2. Venter, J.C., Adams, M.D., Myers, E.W., Li, P.W., Mural, R.J., Sutton, G.G., et al. (2001) The sequence of the human genome. *Science* **291,** 1304–1351.
3. Katzen, F., Chang, G., and Kudlicki, W. (2005) The past, present and future of cell-free protein synthesis. *Trends Biotechnol.* **23,** 150–156.

4. Endo, Y., and Sawasaki, T. (2006) Cell-free expression systems for eukaryotic protein production. *Curr. Opin. Biotechnol.* **17,** 373–380.
5. Klammt, C., Löhr, F., Schäfer, B., Haase, W., Dötsch, V., Rüterjans, H., Glaubitz, C., and Bernhard, F. (2004) High level cell-free expression and specific labeling of integral membrane proteins. *Eur. J. Biochem.* **271,** 568–580.
6. Ishihara, G., Goto, M., Saeki, M., Ito, K., Hori, T., Kigawa, T., Shirouzu, M., and Yokoyama, S. (2005) Expression of G protein coupled receptors in a cell-free translational system using detergents and thioredoxin-fusion vectors. *Protein Expr. Purif.* **41,** 27–37.
7. Kalmbach, R., Chizhov, I., Schumacher, M.C., Friedrich, T., Bamberg, E., and Engelhard, M. (2007) Functional cell-free synthesis of a seven helix membrane protein: *in situ* insertion of bacteriorhodopsin into liposomes. *J. Mol. Biol.* **371,** 639–648.
8. Liguori, L., Marques, B., Villegas-Méndez, A., Rothe, R., and Lenormand, J.-L. (2007) Production of membrane proteins using cell-free expressing systems. *Expert Rev. Proteomics* **4,** 79–90.
9. Schwarz, D., Junge, F., Durst, F., Frölich, N., Schneider, B., Reckel, S., Sobhanifar, S., Dötsch, V., and Bernhard, F. (2007) Preparative scale expression of membrane proteins in *Escherichia coli*-based continuous exchange cell-free systems. *Nat. Protoc.* **2,** 2945–2957.
10. Nozawa, A., Nanamiya, H., Miyata, T., Linka, N., Endo, Y., Weber, A.P.M., and Tozawa, Y. (2007) A cell-free translation and proteoliposome reconstitution system for functional analysis of plant solute transporters. *Plant Cell Physiol.* **48,** 1815–1820.
11. Sawasaki, T., Hasegawa, Y., Tsuchimochi, M., Kamura, N., Ogasawara, T., Kuroita, T., and Endo, Y. (2002) A bilayer cell-free protein synthesis system for high-throughput screening of gene products. *FEBS Lett.* **514,** 102–105.
12. Goren, M.A. and Fox, B.G. (2008) Wheat germ cell-free translation, purification, and assembly of a functional human stearoyl-CoA desaturase complex. *Protein Expr. Purif.* **62,** 171–178.
13. Sawasaki, T., Gouda, M.D., Kawasaki, T., Tsuboi, T., Tozawa, Y., Takai, K., and Endo, Y. (2005) The wheat germ cell-free expression system: methods for high-throughput materialization of genetic information. *Methods Mol. Biol.* **310,** 131–144.

Chapter 18

Ribosome Display with the PURE Technology

Takuya Ueda, Takashi Kanamori, and Hiroyuki Ohashi

Abstract

The ribosome display utilizes the formation of the mRNA–ribosome–polypeptide ternary complex in the cell-free protein synthesis system as the linking of the genotype (mRNA) to the phenotype (polypeptide). However, the presence of intrinsic components such as nucleases in the cell-extract based cell-free protein synthesis systems inevitably reduces the stability of the ternary complex, which would prevent attainment of reliable results. We have developed an efficient and highly controllable ribosome display system using the Protein synthesis Using Recombinant Elements (PURE) system. The mRNA–ribosome–polypeptide ternary complex is highly stable in the PURE system and then the selected mRNA can be easily recovered, because activities of nucleases and other inhibitory factors are very low in the PURE system. Furthermore, omission of the release factors within the original PURE system can aid stalling of the ribosome at the termination codon to form the mRNA–ribosome–polypeptide ternary complex. We believe that these advantages assure the usability of the modified PURE system for ribosome display.

Key words: PURE system, Ribosome display, Cell-free protein synthesis system, In vitro selection, Single chain Fv, Protein engineering

1. Introduction

Recently, the ribosome display technology has become one of the most widely used tools for in vitro selection of functional proteins and peptides. This technology is based on the formation of a messenger RNA (mRNA)–ribosome–nascent polypeptide ternary complex in a cell-free protein synthesis system (1, 2). The formation of such complex enables physical linkage between phenotype (polypeptide) and genotype (mRNA). The sequence information for polypeptide of interest can be selected by affinity purification of this complex. Several applications of ribosome display technology have been reported such as screening of single chain Fv (scFv) with high affinity to the antigens (3, 4). Ribosome display was

Y. Endo et al. (eds.), *Cell-Free Protein Production: Methods and Protocols,* Methods in Molecular Biology, vol. 607
DOI 10.1007/978-1-60327-331-2_18, © Humana Press, a part of Springer Science+Business Media, LLC 2010

originally developed based on a cell-free protein synthesis system using *Escherichia coli* (*E. coli*) S30 extract (1, 5). Because *E. coli* S30 extract contains endogenous components such as proteases and nucleases, degradation of mRNA and nascent polypeptide reduces a stability of the mRNA–ribosome–polypeptide ternary complex. Furthermore, incomplete information of components within cell extract makes it difficult to optimize the system for ribosome display.

We have established a highly controllable cell-free translation system called the protein synthesizing using recombinant elements (PURE) system (6, 7). Because the PURE system is reconstructed by using only the essential purified factors and enzymes responsible for gene expression in *E. coli*, it contains no nucleases and proteases. We have developed a new strategy that allows us to prepare the ternary complex that is more stable than that obtained by the cell-extract-based ribosome display, and termed "Pure Ribosome Display (PRD)" (8). We have refined the original PURE system for efficient selection of a functional scFv gene. Developed system enabled specific selection from a 1:10^{11} dilution of scFv mRNA into the competitor mRNA by several selection rounds (8).

2. Materials

2.1. Preparation of mRNA for Translation

1. Template DNA; an example of a template DNA construct for PRD is showed in Fig. 18.1 (see Note 1).
2. mRNA synthesis kit; for example, *CUGA7 in vitro* Transcription Kit (Wako, Japan).
3. RNA purification kit; for example, RNeasy MinElute Cleanup Kit (Qiagen).

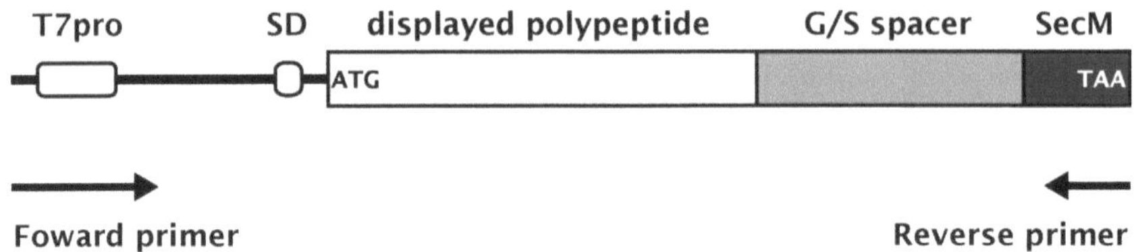

Fig. 18.1. The DNA construct used for the PRD. A T7 promoter (T7pro) and a Shine-Dalgarno sequence (SD) are located upstream the gene encoding for the displayed polypeptide followed by a Gly/Ser rich spacer sequence from gene III (G/S spacer) and SecM elongation arrest sequence (SecM). The initiation and the termination codon are inserted at the 5′- and 3′- terminus of the gene, respectively. The primers used for RT-PCR after in vitro selection are indicated at the bottom.

2.2. In Vitro Translation

1. PURE system reagent (6, 7) (see Note 2):

   0.33 μM Ribosome, 2.70 μM Initiation Factor (IF) 1, 0.40 μM IF2, 1.50 μM IF3, 0.26 μM Elongation Factor (EF)-G, 0.92 μM EF-Tu, 0.66 μM EF-Ts, 0.25 μM Release Factor (RF) 1, 0.24 μM RF2, 0.17 μM RF3, 0.50 μM Ribosome Recycling Factor, 1,900 U/ml Alanyl-tRNA synthetase (AlaRS), 2,500 U/ml ArgRS, 20 μg/ml AsnRS, 2,500 U/ml AspRS, 630 U/ml CysRS, 1,300 U/ml GlnRS, 1,900 U/ml GluRS, 5,000 U/ml GlyRS, 630 U/ml HisRS, 2,500 U/ml IleRS, 3,800 U/ml LeuRS, 3,800 U/ml LysRS, 6,300 U/ml MetRS, 1,300 U/ml PheRS, 1,300 U/ml ProRS, 1,900 U/ml SerRS, 1,300 U/ml ThrRS, 630 U/ml TrpRS, 630 U/ml TyrRS, 3,100 U/ml ValRS, 10 μg/ml T7 RNA polymerase, 4,500 U/ml Methionyl-tRNA transformylase, 4.0 μg/ml Creatine kinase, 3.0 μg/ml Myokinase, 1.1 μg/ml Nucleoside-diphosphate kinase, 1.3 μg/ml Pyrophosphatase, 0.3 mM 20 amino acids each, 56 Abs_{260}/ml *E. coli* tRNA, 50 mM Hepes-KOH, pH 7.6, 100 mM potassium glutamate, 13 mM $Mg(OAc)_2$, 2 mM spermidine, 1 mM dithiothreitol, 2 mM adenosine-5¢-triphosphate (ATP), 2 mM guanosine-5′-triphosphate (GTP), 1 mM cytidine-5′-triphosphate (CTP), 1 mM uridine-5′-triphosphate (UTP), 20 mM creatine phosphate, 10 μg/ml 10-formyl-5, 6, 7, 8- tetrahydrofolic acid (FD).
2. WBT buffer: 50 mM Tris-acetate, pH 7.5, 150 mM NaCl, 50 mM $Mg(OAc)_2$, 0.5% (v/v) Tween20.
3. Blocking buffer: 50 mM Tris-acetate, pH 7.5, 150 mM NaCl, 50 mM $Mg(OAc)_2$, 0.5% (v/v) Tween20, 5% (w/v) Bovine Serum Albumin (BSA), 4% (w/v) BlockAce (Dainippon Sumitomo Pharma, Japan).

2.3. Preparation of Biotinylated Lysozyme

1. Lysozyme (Egg White, Seikagaku Corporation).
2. Ez-Link Sulfo-NHS-LC-LC-Biotin (Pierce).
3. Phosphate buffered saline (PBS) pH 7.4.
4. Protein Assay (Bio-Rad).
5. 10 mg/ml BSA; BSA powder is dissolved in H_2O.

2.4. In Vitro Selection

1. WBT buffer: 50 mM Tris-acetate, pH 7.5, 150 mM NaCl, 50 mM $Mg(OAc)_2$, 0.5% (v/v) Tween20.
2. Dynabeads M-270 Streptavidin (Invitrogen).
3. Elution buffer: 50 mM Tris-acetate, pH 7.5, 150 mM NaCl, 50 mM ethylenediamine tetraacetic acid (EDTA), 10 μg/ml total RNA from *Saccharomyces cerevisiae* (Sigma).

2.5. Reverse Transcription and PCR

1. SuperScript III One-Step Reverse Transcription and PCR (RT-PCR) System with Platinum *Taq* DNA Polymerase (Invitrogen).
2. MinElute PCR purification kit (Qiagen).

3. Methods

3.1. Preparation of mRNA for In Vitro Translation

1. Synthesize mRNA from 0.1 to 2 pmol of the template DNA by an in vitro transcription kit according to manufacture's manual.
2. Purify the transcribed mRNA using an RNA purification kit according to supplier's procedure.
3. Measure the absorbance of the purified mRNA solution at 260 nm by a spectrophotometer.
4. Calculate the concentration of mRNA using the following equation and dilute the sample to 1 pmol/μl with H_2O;

$$[\text{mRNA}]\text{pmol}/\mu\text{l} = (\text{Abs}_{260} \times 40)/(\text{N(length)} \times 330)$$

5. Freeze the mRNA solution by liquid nitrogen and store at −80°C.

3.2. In Vitro Translation

1. Prepare the PURE system reagent without three release factors and ribosome recycling factor in a 1.5-ml tube and add H_2O to 29 μl (see Note 2).
2. Add 1 μl of 1 pmol/μl mRNA solution to the PURE system reagent mixture (see Note 2).
3. Incubate the reaction mixture at 37°C for 20 min to form the mRNA–ribosome–polypeptide ternary complex (see Note 3).
4. To stop translation reaction, place the tube on ice and add 120 μl of ice-cold WBT buffer.
5. Centrifuge the reaction tube at 14,000 × *g* for 20 min to remove insoluble components.
6. Transfer the supernatant to a new 1.5 ml-tube and add 50 μl of blocking buffer (see Note 4).

3.3. Preparation of Biotinylated Lysozyme

1. Dissolve 10 mg of lysozyme powder in 1 ml of PBS.
2. Prepare 10 mM Ez-Link Sulfo-NHS-LC-LC-Biotin by dissolving 6 mg of reagent powder in 900 μl of H_2O (see Note 5).
3. Mix Lysozyme solution with 840 μl of 10 mM Ez-Link Sulfo-NHS-LC-LC-Biotin and incubate at room temperature for 30 min.
4. Remove excess nonreacted and hydrolyzed biotin reagent by dialysis against PBS.
5. Determine the concentration of biotinylated lysozyme by Protein Assay (Bio-Rad) using BSA as a standard.

3.4. In Vitro Selection

1. Add 3 pmol of biotinylated lysozyme to the diluted supernatant (200 μl) of translation mixture from **step 6** in Subheading 3.2 (*see* Note 6).
2. Incubate the mixture at room temperature for 1 h (*see* Note 7).
3. To isolate biotinylated lysozyme, add 30 μl of streptavidin-coated magnetic beads (Dynabeads M-270 Streptavidin) and the mixture is incubated at room temperature for further 30 min with gentle shaking.
4. Recover the beads and wash 10 times with 1 ml of WBT buffer (see Note 6).
5. Isolate the mRNA from the beads by incubation in 60 μl of Elution buffer at room temperature for 30 min.
6. Purify the eluted mRNA using an RNA purification kit according to supplier's procedure.

3.5. Reverse Transcription and PCR

1. Using purified mRNA as a template RNA, perform RT-PCR by SuperScript III One-Step RT-PCR System with Platinum *Taq* DNA Polymerase (Invitrogen) (see Note 8). Primers used in this step are shown in Fig. 18.1 (7).
2. Mix the reaction mixture as follows:

2x Reaction mix buffer	25.0 μl
Purified mRNA solution	1.5 μl
50 μM forward primer	1.0 μl
50 μM reverse primer	1.0 μl
SuperScript III RT/Platinum *Taq* mix	4.0 μl
RNase free water	to 50.0 μl

3. Perform RT-PCR as follows:

1 cycle,	55°C for 30 min (cDNA synthesis)
1 cycle,	94°C for 2 min (denaturation)
20–30 cycles,	94°C for 15 sec (denature), 68°C for 90 sec (extension)

4. Subject the PCR reaction mixture to agarose gel electrophoresis to analyze the amplified products. An example of results is shown in Fig. 18.2.
5. Purify the amplified DNA by MinElute PCR purification kit (see Note 9) and subject to an additional selection round, if necessary.

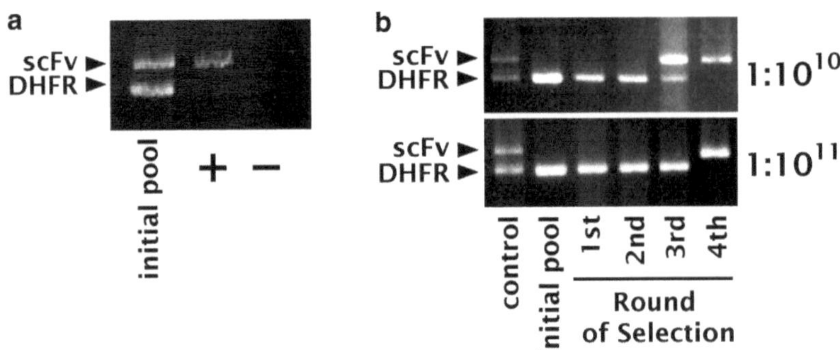

Fig. 18.2. Specific enrichment following single (**a**) or multiple (**b**) rounds of in vitro selection in the PURE ribosome display system was examined using mRNA of scFv against lysozyme and dihydrofolate reductase (DHFR) as a selection target and a competitor, respectively. (**a**) The equal amount of scFv and DHFR mRNA were translated and selected with lysozyme. The selection was performed in the presence (+) or absence (–) of lysozyme. (**b**) The initial mRNA pools were $1{:}10^{10}$ (*upper panel*) and $1{:}10^{11}$ (*lower panel*) mixtures of scFv and DHFR mRNAs. After each round of affinity selection, mRNA was isolated, subjected to RT-PCR and analyzed by agarose gel electrophoresis. The isolated mRNA was used for each subsequent selection step. (Reproduced from ref. (8) with permission from Elsevier Science).

4. Notes

1. The template DNA requires a T7 promoter sequence and ribosome binding site (Shine-Dalgano sequence) upstream of the open reading frame (ORF) and ORF must initiate with an initiation codon (ATG). To prevent steric hindrance between the displayed polypeptide and the ribosome, a spacer sequence such as gene III of a filamentous phage is introduced downstream of the polypeptide of interest. Furthermore, the SecM elongation arrest sequence is positioned downstream of the spacer sequence, to stabilize the mRNA–ribosome–polypeptide ternary complex. This sequence has shown to interact with the ribosomal polypeptide tunnel (9). To form the mRNA–ribosome–polypeptide ternary complex using the PURE system reagent without release factors, the ORF requires a termination codon (TAG, TGA or TAA) at 3′-terminus.
2. PURE system reagent containing 1–3 µM ribosome is usually used for a maximum yield of protein synthesis. In contrast, we have demonstrated that maximum mRNA recovery was achieved with 1 pmol of mRNA and 10 pmol of ribosome (0.33 µM) in 30 µl of reaction (8). Release factors are omitted from the reagent for formation and stabilization of the mRNA–ribosome–polypeptide ternary complex. The reagent composition can be easily modified according to the displayed polypeptide. For example, addition of molecular chaperones such as hsp70 and hsp60 can increase solubility of the displayed polypeptide. When the polypeptide with disulfide bonds is displayed and selected, addition of protein disulfide isomerase

with 1 mM oxidized glutathione and 0.1 mM reduced glutathione in place of dithiothreitol can improve correct folding of displayed polypeptide.

3. For formation of the mRNA–ribosome–polypeptide complex, 20-min incubation is sufficient. Longer incubation may destabilize the mRNA–ribosome–polypeptide ternary complex.
4. Unrelated RNA (e.g., *S. cerevisiae* RNA) is also added to blocking buffer to suppress nonspecific binding of mRNA to the target and magnet beads and to inhibit degradation of mRNA.
5. Because the NHS-ester moiety readily hydrolyzes, Ez-Link Sulfo-NHS-LC-LC-Biotin is dissolved in H_2O just before use.
6. Selection efficiency is also dependent on the ratio of the target to the displayed polypeptide and the stringency of washing step.
7. Because the mRNA–ribosome–polypeptide ternary complex is unstable in the cell-extract system, the reaction mixture must be placed at 4°C after translation reaction. In contrast, the mRNA–ribosome–polypeptide complex is highly stable in the PURE system, so it is possible to perform selection step at room temperature.
8. Two-step RT-PCR can be also performed.
9. If the diffuse band or by-products are identified on an agarose gel electrophoresis, the product with expected length is purified from a gel by MinElute Gel purification kit (Qiagen).

References

1. Hanes, J., Pluckthun, A. (1997) In vitro selection and evolution of functional proteins by using ribosome display. *Proc. Natl. Acad. Sci. USA* **94**, 4937–4942.
2. Hanes, J., Jermutus, L. and Pluckthun, A. (2000) Selecting and evolving functional proteins in vitro by ribosome display. *Methods Enzymol.* **328**, 404–430.
3. Hanes, J., Jermutus, L., Weber-Bornhauser, S., Bosshard, H.R. and Pluckthun, A. (1998) Ribosome display efficiently selects and evolves high-affinity antibodies in vitro from immune libraries. *Proc. Natl. Acad. Sci. USA* **95**, 14130–14135.
4. Binz, H. K., Amstutz, P., Kohl, A., Stumpp, M. T., Briand, C., Forrer, P., Grutter, M. G. and Pluckthun, A. (2004) High-affinity binders selected from designed ankyrin repeat protein libraries. *Nat. Biotechnol.* **22**, 575–582.
5. Zahnd, C., Amstutz, P., Pluckthun, A. (2007) Ribosome display: selecting and evolving proteins in vitro that specifically bind to a target. *Nat. Methods* **4**, 269–279.
6. Shimizu, Y., Inoue, A., Tomari, Y., Suzuki, T., Yokogawa, T., Nishikawa, K. and Ueda, T. (2001) Cell-free translation reconstituted with purified components. *Nat. Biotechnol.* **19**, 751–755.
7. Shimizu, Y., Kanamori, T. and Ueda, T. (2005) Protein synthesis by pure translation systems. *Methods* **36**, 299–304.
8. Ohashi, H., Shimizu, Y., Ying B. W. and Ueda T. (2007) Efficient protein selection based on ribosome display system with purified components. *Biochem. Biophys. Res. Commun.* **352**, 270–276.
9. Nakatogawa, H., Ito, K. (2002) The ribosomal exit tunnel functions as a discriminating gate. *Cell* **108**, 629–636.

Chapter 19

Incorporation of 3-Azidotyrosine into Proteins Through Engineering Yeast Tyrosyl-tRNA Synthetase and Its Application to Site-Selective Protein Modification

Takashi Yokogawa, Satoshi Ohno, and Kazuya Nishikawa

Abstract

An efficient method for site-selective introduction of 3-azidotyrosine into proteins has been developed. This method utilizes the yeast amber suppressor $tRNA^{Tyr}$/mutant tyrosyl-tRNA synthetase (Y43G) pair as the carrier of 3-azidotyrosine in an *Escherichia coli* cell-free translation system. Using rat calmodulin (CaM) as a model protein, we prepared an unnatural CaM molecule carrying a 3-azidotyrosine residue at the predetermined position 80. The synthesized CaM containing 3-azidotyrosine was site-specifically modified via azido group with a fluorescent alkyne derivative by click chemistry. This method will be useful to prepare not only a cross-linkable protein containing 3-azidotyrosine but also a fluorescent protein with a single fluorophore to facilitate the elucidation of molecular mechanisms of protein functions and protein-to-protein networks.

Key words: 3-Azidotyrosine, In vitro translation system, Suppression, Tyrosyl-tRNA synthetase, Site-selective chemical modification, Click chemistry

1. Introduction

In recent years, a general strategy for site-specific introduction of noncanonical amino acids into proteins in a cell-free translation system has been developed (1–5). This method utilizes chemically aminoacylated suppressor tRNAs, which incorporate noncanonical amino acids in response to the stop codons intentionally placed at predetermined positions of the template mRNA. However, the usefulness of this method seems to be limited because the chemically synthesized aminoacyl-suppressor tRNA can work only once in the translation process, and thus the final yields of the aimed

Y. Endo et al. (eds.), *Cell-Free Protein Production: Methods and Protocols*, Methods in Molecular Biology, vol. 607
DOI 10.1007/978-1-60327-331-2_19, © Humana Press, a part of Springer Science+Business Media, LLC 2010

proteins containing noncanonical amino acids are usually not enough for sample-intensive purposes.

If a mutant of an aminoacyl-tRNA synthetase which can recognize both a suppressor tRNA and a noncanonical amino acid as its substrates could be engineered, the resultant aminoacyl-tRNA would be used over and over again in a translation system and the yields of proteins should be much increased. For this reason, we attempted to convert an yeast $tRNA^{Tyr}$ to an amber suppressor tRNA and engineer yeast tyrosyl-tRNA synthetase (TyrRS) as the carrier of an extra (noncanonical) amino acid in the *Escherichia coli* translation system since we had observed that the pair of yeast amber suppressor $tRNA^{Tyr}$ and yeast TyrRS could function orthogonally in *E. coli* cells (6). Moreover, we have obtained a mutant of yeast TyrRS by genetic substitution of tyrosine at position 43 with glycine (Y43G), which can accept several 3-substituted tyrosine derivatives as substrates. Among these tyrosine analogs, 3-azidotyrosine would be the most promising substrate because the aryl-azido group can function as a cross-linker with UV-light and the azido group is known to have many specific "named reactions" in the field of organic chemistry such as the Staudinger-Bertozzi ligation, the Huisgen 1,3-dipolar cycloaddition and so on. Therefore, once such a protein carrying 3-azidotyrosine at a predetermined position could be produced, it could be directly used as a cross-linkable protein or be used as a site-selectively labeled protein after chemical modification with organic probes by applying one of these "named reactions." This modification reaction is highly specific because no other azido group is present in naturally occurring proteins.

Here, we describe an efficient method to prepare potentially useful proteins containing 3-azidotyrosine at an intended site using yeast amber suppressor $tRNA^{Tyr}$, a mutant TyrRS and an *E. coli* cell-free translation system.

2. Materials

2.1. Preparation of Amber Suppressor $tRNA^{Tyr}$ from Yeast

1. A plasmid designated as pGEMEX-supTyr (7). This plasmid carries a synthetic yeast amber suppressor $tRNA^{Tyr}$ gene between a T7 promoter and terminator.
2. *E. coli* strain having T7 RNA polymerase gene, for example, BL21(DE3).
3. LB-amp medium: 1% Bacto™ Tryptone, 0.5% Bacto™ yeast extract, 1% NaCl, 50 μg/ml ampicillin (and 1.7% agar for plate) (15 L in a culture vessel, 2 mL in a test tube and an LB-amp plate).

4. 0.3 M potassium acteate (KOAc) buffer (pH 4.75) (500 mL).
5. Water-saturated phenol (200 mL).
6. Q-Sepharose High Performance (GE Healthcare) column (ø2.5 cm × 5 cm; 25 mL resin).
7. QT0.4 buffer: 20 mM Tris–HCl (pH 7.6), 10 mM $MgCl_2$, 0.4 M NaCl (1 L).
8. QT0.6 buffer: 20 mM Tris–HCl (pH 7.6), 10 mM $MgCl_2$, 0.6 M NaCl (500 mL).
9. QA0.4 buffer: 20 mM KOAc (pH 4.75), 10 mM $MgCl_2$, 0.4 M NaCl (1 L).
10. QA0.6 buffer: 20 mM KOAc (pH 4.75), 10 mM $MgCl_2$, 0.6 M NaCl (500 mL).
11. 10× TBE.
12. Two sets of 12% native PAGE gel (8 cm width × 10 cm height × 1 mm thick): 12% polyacrylamide (acrylamide: BIS = 19:1), 1× TBE, 5% glycerol.
13. 6× glycerol loading solution: 0.25% bromophenol blue (BPB), 0.25% xylene cyanol (XC), 30% glycerol (1 mL).
14. Methylene blue: 0.1% methylene blue, 6% acetic acid (1 L). This solution can be used repeatedly.
15. Biotinylated probe (if needed): 5′-TCT CCC GGG GGC GAG TCG AAC GCC CGA TCT-Biotin-3′, which is complementary to the 3′ region of yeast $tRNA^{Tyr}$.

2.2. Preparation of Yeast Tyrosyl-tRNA Synthetase Mutant, Y43G

1. A plasmid designated as pETY43G (8). This plasmid has a synthetic yeast (*Saccharomyces cerevisiae* IFO1234) TyrRS gene with a tyrosine43 to glycine43 replacement and without any affinity tag between a T7 promoter and T7 terminator.
2. *E. coli* strain having T7 RNA polymerase gene, for example, BL21(DE3).
3. LB-amp medium: 1% Bacto™ Tryptone, 0.5%Bacto™ yeast extract, 1%NaCl, 50 µg/ml ampicillin (and 1.7% agar for plate) (1 L in a culture flask and an LB-amp plate).
4. HG0.05 buffer: 20 mM Hepes-KOH (pH 7.0), 1 mM $MgCl_2$, 50 mM KCl, 5% glycerol, 6 mM 2-mercaptoethanol (500 mL).
5. HG0.5 buffer: 20 mM Hepes-KOH (pH 7.0), 1 mM $MgCl_2$, 0.5 M KCl, 5% glycerol, 6 mM 2-mercaptoethanol (500 mL).
6. TG0.04 buffer: 20 mM Tris–HCl (pH 7.6), 1 mM $MgCl_2$, 40 mM KCl, 5% glycerol, 6 mM 2-mercaptoethanol (500 mL).

7. TG0.3 buffer: 20 mM Tris–HCl (pH 7.6), 1 mM $MgCl_2$, 0.3 M KCl, 5% glycerol, 6 mM 2-mercaptoethanol (500 mL).
8. HiTrap™ SP-HP (GE Healthcare) column (5 mL resin).
9. HiTrap™ Q-HP (GE Healthcare) column (5 mL resin).
10. Enzyme storage buffer: 20 mM Tris–HCl (pH 7.6), 1 mM $MgCl_2$, 40 mM KCl, 50% glycerol, 6 mM 2-mercaptoethanol (200 mL).
11. All the buffers and the columns should be stored at 4°C.

2.3. Acid-Urea Polyacrylamide Gel Electrophoresis (Acid-Urea PAGE) to Detect 3-Azidotyrosyl Suppressor $tRNA^{Tyr}$

1. 10× ATE buffer (pH 4.75): 900 mM acetic acid titrated with Tris base up to pH 4.75, 30 mM EDTA (500 mL). Store this buffer at 4°C.
2. 10% Acid-urea gel solution: 10% acrylamide (acrylamide: BIS = 19:1), 1× ATE, 7 M urea (500 mL). Store this solution at 4°C.
3. 5 mM 3-Azidotyrosine (1 mL): 3-azidotyrosine is a kind gift from Drs. Takamitsu Hosoya, Toshiyuki Hiramatsu and Masaaki Suzuki (Gifu university). Alternatively, it can be custom-ordered from Watanabe Chemical Industries (Hiroshima, Japan). Store this solution at –30°C in the dark (see Note 1).
4. 5× AAM: 500 mM Tris–HCl (pH 7.6), 50 mM $MgCl_2$, 200 mM KCl, 20 mM ATP (1 mL). Store this solution at –30°C.
5. 2× Acid-urea loading solution: 62.5 mM acetic acid titrated with Tris base up to pH 3.1, 0.05% BPB, 0.05% XC, 9 M urea (1 mL). Store this solution at –30°C.
6. EP buffer: 50 mM glutamic acid titrated with Tris base up to pH 5.1 (2 L). Store this buffer at 4°C.

2.4. Preparation of an E. coli Cell-Free Extract

1. *E. coli* Q13 (Hfr: met pnp rna tyr).
2. 2× YT medium: 1.6% Bacto™ Tryptone, 1% Bacto™ yeast extract, 0.5% NaCl (50 mL in a flask and 4 flasks, each containing 1 L of this medium).
3. Extraction buffer: 20 mM Hepes-KOH (pH 7.5), 10 mM $Mg(OAc)_2$, 20 mM NH_4OAc, 10 mM 2-mercapto-ethanol (500 mL).

2.5. Synthesis of Azidotyrosine Containing Protein Using In Vitro Translation System

1. 10× Protein synthesis buffer: 500 mM Hepes-KOH (pH 7.5), 77 mM $Mg(OAc)_2$, 275 mM NH_4OAc, 2 M KOAc, 17 mM DTT, 0.5 mM folinic acid, 2 mM each 19 amino acids excepting tyrosine (1 mL).
2. NTP Mix: 33.3 mM ATP, 22.2 mM GTP, 22.2 mM CTP, 22.2 mM UTP (1 mL).

3. 20% PEG (1 mL): dilute 40% polyethylene glycol solution (average molecular weight 8,000) (Sigma) with dH_2O (see Note 2) (1 mL).
4. 1 M creatine phosphate (Roche) (1 mL).
5. 170 A_{260}units/mL *E. coli* $tRNA^{Mix}$: This was prepared from Q13 cells by the method according to Zubay (9). *E. coli* $tRNA^{Mix}$ from Sigma is also usable.
6. 15 mg/mL creatine kinase (Roche) (1 mL): This enzyme was dissolved in Enzyme storage buffer (see Subheading 2.2).
7. 11 mg/mL T7 RNA polymerase: T7 RNA polymerase was prepared by the method according to Grodberg and Dunn (10). This enzyme was dissolved in 20 mM sodium phosphate buffer (pH 7.7), 1 mM EDTA, 1 mM DTT, 0.1 M NaCl, 50% glycerol. T7 RNA polymerase commercially obtained from any manufacturer is also usable.
8. 2.27 mM tyrosine (1 mL): accurate molar concentration of tyrosine can be calculated from absorbance at 275 nm (a molar extinction coefficient at 275 nm, $E_{275}^{1M} = 1,340$).
9. 1 mg/mL plasmid DNA, $pETCaM80_{am}$: This plasmid constructed as described previously (11) encodes a rat calmodulin (CaM) gene having an amber codon in frame at the 80th codon. The plasmid DNA was usually purified with GenoPure Plasmid Maxi Kit (Roche).
10. 10 A_{260}units/mL amber suppressor $tRNA^{Tyr}$: the amber suppressor $tRNA^{Tyr}$ (*see* Subheading 3.1) was diluted with dH_2O to a final concentration of 10 A_{260}units/mL.
11. 10 mg/mL mutant TyrRS (Y43G): This enzyme (*see* Subheading 3.2) was diluted with the Enzyme storage buffer to a final concentration of 10 mg/mL.
12. *E. coli* cell-free extract (*see* Subheading 3.4).
13. All buffers and solutions are stored at −30°C.

2.6. Purification of Azidotyrosine Containing CaM by Affinity Chromatography

1. Phenyl-Spharose CL-4B (GE Healthcare) suspension: the resin is suspended in dH_2O in the ratio of 1 to 1 (10 mL).
2. 1 M $CaCl_2$ (1 mL).
3. PolyPrep column (Bio-Rad).
4. Eq buffer: 50 mM Tris–HCl (pH 7.6), 5 mM $CaCl_2$, 0.1 M NaCl (50 mL).
5. Wash1 buffer: 50 mM Tris–HCl (pH 7.6), 0.1 mM $CaCl_2$ (50 mL).
6. Wash2 buffer: 50 mM Tris–HCl (pH 7.6), 0.1 mM $CaCl_2$, 0.5 M NaCl (50 mL).
7. Elution buffer: 50 mM Tris–HCl (pH 7.6), 1 mM EGTA (50 mL).

2.7. Site-specific Modification Via Azido Group Incorporated into the Protein

1. 50 mM Tris–HCl (pH8.0) (1 mL).
2. 0.2 mM Tris-(benzyltriazolylmethyl)amine (TBTA): dissolve 1 mg TBTA in 1.88 mL of dimethyl sulfoxide (DMSO) to make 1 mM solution, divide it into 10 μL aliquots and dry DMSO up (store each tube at −30°C). Just prior to use, dissolve a dried TBTA pellet again in 50 μL of DMSO to make 0.2 mM solution. After use, dry up DMSO in the remaining solution again and store it at −30°C.
3. 10 mM $CuSO_4$ (1 mL).
4. 20 mM L-Ascorbic acid (1 mL).
5. TAMRA alkyne (70 μL in DMSO): this reagent is included in the Click-iT™ Tetramethylrhodamine (TAMRA) Glycoprotein Detection Kit (Invitrogen) as component A.

3. Methods

3.1. Preparation of Amber Suppressor $tRNA^{Tyr}$ from Yeast

An amber suppressor $tRNA^{Tyr}$ from yeast is used as a "carrier" of 3-azidotyrosine in an *E. coli* translation system. The suppressor $tRNA^{Tyr}$ can be overexpressed in *E. coli* cells. Since this is used in an *E.coli* in vitro translation system, a slight contamination of *E. coli* tRNAs would not be a severe obstacle. The purification protocol of the suppressor $tRNA^{Tyr}$ overexpressed in *E. coli* cells is described below.

1. Transform an *E. coli* BL21(DE3) with pGEMEX-supTyr on a LB-amp plate.
2. Pick up a single colony with a toothpick and inoculate in 2 mL of LB-amp media. Do NOT keep the plate at 4°C. This inoculation should be done on the next day of transformation. Cultivate at 37°C until turbidity of media is observed (it will take about 5 h). Do NOT cultivate overnight (see Note 3).
3. Transfer 50 μL of the media to 15 L of LB-amp media and cultivate with aeration at 37°C until its optical density at 600 nm (OD_{600}) reaches between 0.5 and 1.0 (It will take about 10 h).
4. Add IPTG to a final concentration of 0.5 mM in order to induce the expression of the suppressor $tRNA^{Tyr}$. Keep cultivation at 37°C for 3 h.
5. Suspend the cells harvested by centrifugation (see Note 4) with 200 mL of 0.3 M KOAc (pH 4.75).
6. Add 200 mL of water-saturated phenol to the cell suspension and shake it vigorously for more than 1 h.

7. Recover 180 mL of the aqueous layer after centrifugation.
8. Add 200 mL of 0.3 M KOAc (pH 4.75) to the phenol layer, shake it and recover 200 mL of the aqueous layer again.
9. Recover total tRNAs from 380 mL of the combined aqueous layer by ethanol precipitation.
10. Dissolve tRNAs in 10 mL of dH_2O and loaded on the Q-Sepharose High Performance column (ø2.5 cm × 5 cm; 25 mL resin) pre-equilibrated with QT0.4 buffer.
11. Elute tRNAs with a gradient from QT0.4 buffer (500 mL) to QT0.6 buffer (500 mL) at a flow rate of 2.0 mL/min. Collect 100 eluted fractions (10 mL each of eluate per tube).
12. Withdraw a 10 μL aliquot from each fraction, add 2 μL each of 6× glycerol loading solution and apply to a native PAGE. Do NOT prerun (see Note 5) and run the gel at 200 V in 1× TBE buffer until XC goes to 3/4 and stain the gel with methylene blue.
13. Pool the fractions containing suppressor $tRNA^{Tyr}$ by comparing the mobility with that of the authentic sample and recover tRNAs by ethanol precipitation. If the expression of suppressor $tRNA^{Tyr}$ is insufficient and its localization is difficult, check each fraction by using northern hybridization with the biotinylated probe.
14. Dissolve tRNAs in 10 mL of dH_2O and load it on the Q-Sepharose High Performance column (ø2.5 cm × 5 cm; 25 mL resin) pre-equilibrated with QA0.4 buffer (see Note 6).
15. Elute tRNAs with a gradient from QA0.4 buffer (500 mL) to QA0.6 buffer (500 mL) at a flow rate of 2.0 mL/min. Collect 100 eluted fractions (10 mL each of eluate per tube).
16. Subject each fraction to a native PAGE to locate the fractions containing suppressor $tRNA^{Tyr}$.
17. Pool the fractions and recover the suppressor $tRNA^{Tyr}$ by ethanol precipitation.
18. Dissolve the suppressor $tRNA^{Tyr}$ in appropriate volume (less than 1 mL) of dH_2O and determine its concentration by measuring UV absorption at 260 nm.

3.2. Preparation of Yeast Tyrosyl-tRNA Synthetase Mutant (Y43G)

A wild type of yeast tyrosyl-tRNA synthetase (TyrRS) cannot aminoacylate the amber suppressor $tRNA^{Tyr}$ with 3-azidotyrosine, but a mutant of TyrRS we have developed can do so. This mutant TyrRS has a substitution of glycine for tyrosine at position 43 and possibly has also a mutation (asparagine to aspartic acid) at position 343 when compared with the reported amino acid sequence of yeast TyrRS (8). The mutant enzyme, designated as Y43G, can be overexpressed in *E. coli* cells. The purification protocol of mutant TyrRS (Y43G) overexpressed in *E. coli* cells is described below.

1. Transform an *E. coli* BL21 (DE3) with pETY43G on a LB-amp plate.
2. Pick up a single colony with a toothpick and inoculate directly in 1 L of LB-amp medium. Do NOT keep the plate at 4°C. This inoculation should be done on the next day of transformation (see Note 3).
3. Cultivate until its OD_{600} reaches between 0.5 and 1.0 (It will take about 6 h.) at 37°C.
4. Add IPTG to a final concentration of 0.5 mM in order to induce the expression of the mutant (Y43G). Keep cultivation at 37°C for 4 h.
5. Suspend the cells harvested by centrifugation (see Note 4) with 10 mL of HG0.05 buffer (see Note 7).
6. Disrupt cells by ultrasonic keeping the suspension cool and remove cell debris and ribosomal fractions by ultracentrifugation (100,000 × *g*, 4 h, at 4°C).
7. Load the supernatant on HiTrap™ SP-HP column (5 mL resin) pre-equilibrated with HG0.05 buffer.
8. Elute proteins with a gradient from HG0.05 buffer (50 mL) to HG0.5 buffer (50 mL) at a flow rate of 1.0 mL/min. Collect 50 eluted fractions (2 mL of eluate per tube).
9. Subject a 5 μL aliquot of each fraction to 10% SDS PAGE and stain the gel with Coomassie brilliant blue.
10. Pool the fractions containing Y43G by comparing the migration of the authentic TyrRS from yeast.
11. Dialyze the fractions (concentrated by ultrafiltration, if needed) against TG0.04 and load it on a HiTrap™ Q-HP column (5 mL resin) pre-equilibrated with TG0.04 buffer.
12. Elute proteins with a gradient from TG0.04 buffer (50 mL) to TG0.3 buffer (50 mL) at a flow rate of 1.0 mL/min. Collect 50 eluted fractions (2 mL of eluate per tube).
13. Subject each fraction to a 10% SDS PAGE to locate the fractions containing Y43G.
14. Pool the fractions containing Y43G by comparing the mobility with that of the authentic TyrRS.
15. Concentrate fractions by ultrafiltration till its volume is about 2 mL.
16. Dialyze the concentrated sample against Enzyme storage buffer and determine its concentration by measuring UV absorption at 280 nm. The purified Y43G has a molar extinction coefficient at 280 nm, E_{280}^{1M}, of 2.34×10^4 (Absorbance at 280 nm of 1 mg/mL enzyme solution equals 0.532 A_{280}).

3.3. Acid-Urea Polyacrylamide Gel Electrophoresis (Acid-Urea PAGE) to Detect 3-Azidotyrosyl Suppressor $tRNA^{Tyr}$

Amino acid-accepting activity of the amber suppressor $tRNA^{Tyr}$ and 3-azidotyrosine- activating activity of the mutant TyrRS (Y43G) can be monitored by using Acid-urea PAGE (8, 12). An aminoacyl-tRNA is more stable and its amino group of amino acid is positively charged under acidic conditions. Consequently, the electrophoretic mobility of an aminoacyl-tRNA in an acidic gel is slightly smaller than that of a corresponding deacylated tRNA. Here, we describe an improved method in which the samples loaded on the gel are effectively stacked.

1. Make a 10% Acid-urea polyacrylamide slab gel (8 cm width × 10 cm height × 1 mm thick). Use 3–4 times more amounts of ammonium persulfate and TEMED than usual to polymerize acrylamide under acidic condition.
2. Set the gel to an appropriate equipment and fill with EP buffer. Do NOT pre-run, otherwise glutamic acid intruding into the gel disturbs the electrophoretic pattern.
3. Set up the following reaction (total volume, 10 μl) and incubate it at 30°C for 30 min in the dark (see Note 1).

5× AAM	2 μL
Suppressor $tRNA^{Tyr}$	0.2 A_{260}units
5 mM 3-azidotyrosine	2 μL
Y43G	20 μg
dH_2O	up to 10 μL

4. Add 10 μL of 2× Acid urea loading solution and load 5 μL of sample on the acid urea polyacrylamide gel. Do NOT forget to load 0.05 A_{260}units of deacylated suppressor $tRNA^{Tyr}$ at BOTH sides of the sample slot in order to help comparison.
5. Run the gel at 150 V in EP buffer at 4°C in the dark until XC migrates to the bottom of the gel (It will take about 6 h). Stain the gel with methylene blue. A typical result of such an electrophoretic analysis is shown in Fig. 19.1.

3.4. Preparation of an *E. coli* Cell-Free Extract

To prepare an *E. coli* cell-free extract, we usually use an RNase I defective strain, Q13 (Hfr: met pnp rna tyr). When added in its extract, the suppressor $tRNA^{Tyr}$ appears to be more stable than in the extracts from other strains. Other RNase defective strains, for example A19 are also usable. The extracts from OmpT protease defective strains, for example BL21, are also used because it has the merit that the synthesized protein appears to be more stable. It is needless to say that the strains having amber suppressor tRNA, for example JM109, cannot be used in this reaction. Described below is the protocol to obtain the cell-free extract from *E. coli* Q13 cells mildly disrupted with Parr Cell Disruption Bomb.

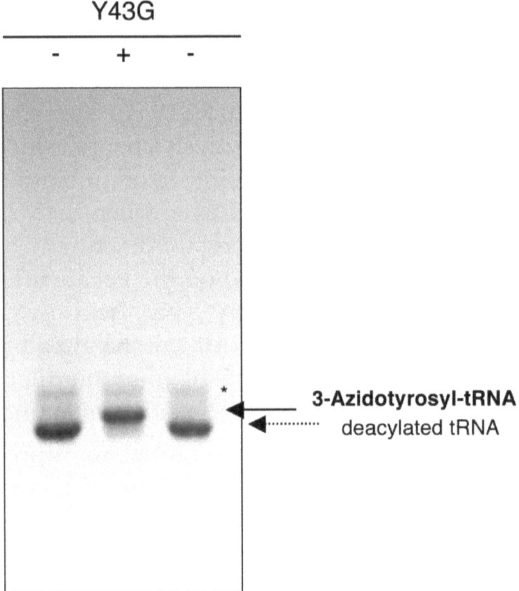

Fig. 19.1. Acid-urea PAGE to check the activities of an amber suppressor $tRNA^{Tyr}$ and a mutant TyrRS (Y43G). A main band observed in the middle lane (Y43G +) is slightly delayed, but fully shifted when compared with the bands in other lanes (Y43G –). The *solid* and *dotted arrows* indicate 3-azidotyrosyl-amber suppressor $tRNA^{Tyr}$ and deacylated amber suppressor $tRNA^{Tyr}$, respectively. This result means that the amber suppressor $tRNA^{Tyr}$ expressed in *E. coli* cells is fully active and the mutant TyrRS (Y43G) has the 3-azidotyrosine activating activity. The asterisk indicates a contaminated tRNA from *E. coli* endogenous tRNAs. Note that the band of contaminant tRNA is not shifted.

1. Pick up a single colony of strain Q13 with a toothpick and inoculate in 50 mL of 2× YT medium. Cultivate at 37°C overnight.
2. Transfer 10 mL of the medium to 1 L each of 2× YT medium and cultivate at 37°C until its OD_{600} reaches between 0.36 and 0.38 (It will take about 2 h.) Inoculate 4 × 1 L medium at the same time.
3. Cool down the medium immediately on ice and harvest cells by centrifugation.
4. Measure the wet weight of cells, add the same volume of Extraction buffer per cell weight (1 ml of Extraction buffer per 1 g of cells) and suspend cells well.
5. Pour the suspension into Parr Cell Disruption Bomb, apply pressure with compressed nitrogen gas till 1,500 psi, keep 10 min at 4°C and disrupt the cells by suddenly releasing the pressure.
6. Remove the cell debris by centrifugation (30,000 × *g*, 45 min, at 4°C).

7. Dialyze the supernatant against 400 mL of Extraction buffer overnight at 4°C to decrease the concentration of endogenous amino acids.
8. Divide the dialyzed cell-free extract into 300 μL aliquots, freeze with liquid nitrogen and store them at –80°C.

3.5. Synthesis of Azidotyrosine Containing Protein Using an In Vitro Translation System

An aimed protein containing 3-azidotyrosine at a desired site can be synthesized by adding the following materials to an *E. coli* in vitro translation system. They are 3-azidotyrosine, the amber suppressor $tRNA^{Tyr}$, the mutant TyrRS (Y43G) and a template DNA encoding the aimed protein with an in frame amber codon at a desired position. As a template DNA, we usually use an expression vector having T7 promoter as pET vectors (Novagen). A plasmid DNA has also a merit that it is not digested by endogenous exonucleases in a cell-free extract. Here, we describe a model case of protein synthesis using a rat CaM containing 3-azidotyrosine at the 80th codon as the target protein. The protocol for synthesis of CaM containing 3-azidotyrosine at the 80th site is described below. In principle, any proteins containing 3-azidotyrosine at any desired positions can be synthesized according to this protocol. Reaction mixture is basically based on the protocol of Kigawa et al. (13).

1. Mix contents one after another from the top of the list (total volume, 1 mL) under the light through UV-cut off filter or preventing samples from exposure to UV-light during mixing contents (see Note 1).

dH_2O	47 μL
20% PEG	200 μL
10× Protein synthesis buffer	100 μL
NTP Mix	38 μL
1 M creatine phosphate	80 μL
5 mM 3-azidotyrosine	100 μL
10 mg/mL Y43G	17 μL
10 A_{260}units/mL amber suppressor $tRNA^{Tyr}$	20 μL
170 A_{260}units/mL *E.coli* $tRNA^{Mix}$	20 μL
15 mg/mL creatine kinase	14 μL
11 mg/mL T7 RNA polymerase	9 μL
2.27 mM tyrosine	35 μL
E. coli cell-free extract	300 μL
1 mg/mL $pETCaM80_{am}$	20 μL

2. Incubate the reaction mixture at 30°C for 3 h.
3. Apply the reaction mixture to a column of affinity chromatography suitable to purify the translation product (*see* Subheadings 2.6 and 3.6).

3.6. Purification of Azidotyrosine Containing CaM by Affinity Chromatography

An in vitro synthesized protein is often designed to fuse with an adequate tag, for example hexa-histidine tag, in order to facilitate the purification step by affinity chromatography. Although the target protein in this case, CaM in itself has no such a tag, one-step purification procedure has been already established (14, 15). The Ca^{2+}-bound CaM strongly adsorbs to Phenyl-Sepharose CL-4B resin because of the exposed hydrophobic residues. On the other hand, the Ca^{2+}-deprived CaM changes its 3D-structure and the affinity to the resin is decreased. This property facilitates the easy separation of CaM, the aimed product in this synthesis. Described below is the protocol to purify the azidotyrosine containing CaM. It is desirable to manipulate samples under the light through UV-cut off filter or preventing samples from UV-light exposure.

1. Pour 400 μL of Phenyl-Sepharose CL-4B suspension (see Note 8) to a PolyPrep column (200 μL of bed volume) and equilibrate the resin with Eq buffer.
2. Add 25 μL of 1 M $CaCl_2$ to the 1 mL reaction mixture (*see* Subheading 3.5) and incubate at room temperature for 10 min.
3. Remove insolubles by centrifugation of the sample at 15,000 rpm (17,000 × g) for 10 min (room temperature).
4. Add 2 mL of Eq buffer to the supernatant and load it on the Phenyl-Sepharose CL-4B column. Perform this column chromatography at room temperature (see Note 9).
5. Wash the resin by adding 2 mL of Wash1 buffer after the sample has penetrated into the resin.
6. Wash the resin again by adding 2 mL of Wash2 buffer after Wash1 buffer has completely passed through the resin.
7. Elute CaM by adding 800 μL of Elution buffer.
8. Concentrate the eluate by ultrafiltration and apply some aliquot of the concentrated sample to 15% SDS–PAGE. Do NOT forget to load known amount of authentic CaM on another lane of the gel.
9. Run the gel till BPB goes to the bottom, stain the gel with Coomassie brilliant blue and roughly estimate the amount of synthesized CaM by visually comparing to the authentic sample.
10. A typical result of the purification of CaM containing 3-azidotyrosine is shown in Fig. 19.2.

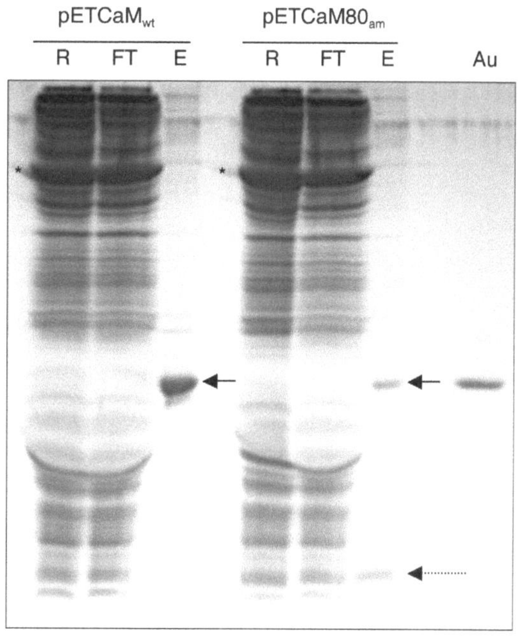

Fig. 19.2. Synthesis of azidotyrosine-containing protein using in vitro translation system. CaMs were synthesized from the template DNAs, pETCaM$_{wt}$ (without amber codon; as a control) and pETCaM80$_{am}$ (with amber codon at position 80) and purified according to Subheadings 3.5 and 3.6 and were subjected to 15% SDS PAGE. R, FT and E mean reaction mixture (loaded 1/100 of sample), flow through fraction from column (loaded 1/100 of sample) and elution fraction from column (loaded 1/10 of sample), respectively. *Solid arrows* indicate CaMs synthesized. A *dotted arrow* indicates possibly a truncated CaM terminated at the amber codon. Au means 1 μg of authentic CaM. The asterisks indicate Y43Gs added to the reaction mixtures.

3.7. Site-Specific Modification Via Azido Group Introduced into the Protein

Here, we introduce a convenient method to label the target protein containing 3-azidotyrosine with a commercial fluorescent alkyne derivative. There is no naturally occurring azido group in normal proteins. Since 3-azidotyrosine is site-specifically introduced into protein by this method, the protein has only one azido group. Although a protein containing 3-azidotyrosine itself is useful as a cross-linkable protein, the applicability of this protein will be much more expanded if the azido group is modified with more useful chemical compounds like fluorescent probes. An azido group is known to be engaged in a useful organic chemistry, for example the Staudinger-Bertozzi ligation, the Huisgen 1,3-dipolar cycloaddition and so on. The Huisgen 1,3-dipolar cycloaddition recently became very well-known as “click chemistry,” in which an azido and an alkyne groups make a 1,2,3-triazole ring in the presence of Cu(I). We have already reported that the CaM containing 3-azidotyrosine at a specific site was site-selectively biotinylated by the Staudinger-Bertozzi ligation (11). Therefore, we describe here the protocol to prepare site-selectively fluorescence-labeled proteins by click chemistry. Since this method is simple and easy to obtain a protein having a single fluorophore at a predetermined site in the protein molecule, it will prove to be very useful to many applications.

1. Mix contents one after another from the top of the list (total volume, 20 μL) under the light through UV-cut off filter or preventing samples from exposure to UV-light during mixing contents.

50 mM Tris–HCl (pH 8.0)	5 μL
dH_2O and protein solution (about 0.5 μg)	7.5 μL
TAMRA alkyne (in DMSO)	0.5 μL
0.2 mM TBTA	1 μL
10 mM $CuSO_4$	4 μL
20 mM L-Ascorbic acid	2 μL

2. Incubate the reaction mixture at 23°C for 30 min.
3. Subject the reaction mixture to 15% SDS–PAGE and run the gel till BPB goes to the bottom.
4. Detect fluorescence of TAMRA with an appropriate fluorescence imager or by taking a picture.
5. Stain the gel with Coomassie brilliant blue.
6. A typical result of labeling 3-azidotyrosine containing protein is shown in Fig. 19.3.

4. Notes

1. 3-Azidotyrosine is sensitive to UV-light. Be careful to avoid a long exposure to light of the samples containing 3-azidotyrosine.
2. Unless and otherwise stated, all solutions should be prepared with Milli-Q water (Millipore) that has a resistivity of 18.3 MΩ cm. We describe such water as "dH_2O" in this text.
3. In our experience, once an *E. coli* strain having a T7 RNA polymerase gene reaches a stationary phase, the expression level of a target gene is drastically decreased.
4. The harvested cells can be stored at –80°C for at least a month.
5. The polyacrylamide gel containing glycerol has a good effect diminishing the disturbance of run by salts contained in samples. However, prerunning sometimes disturbs the migration of samples.

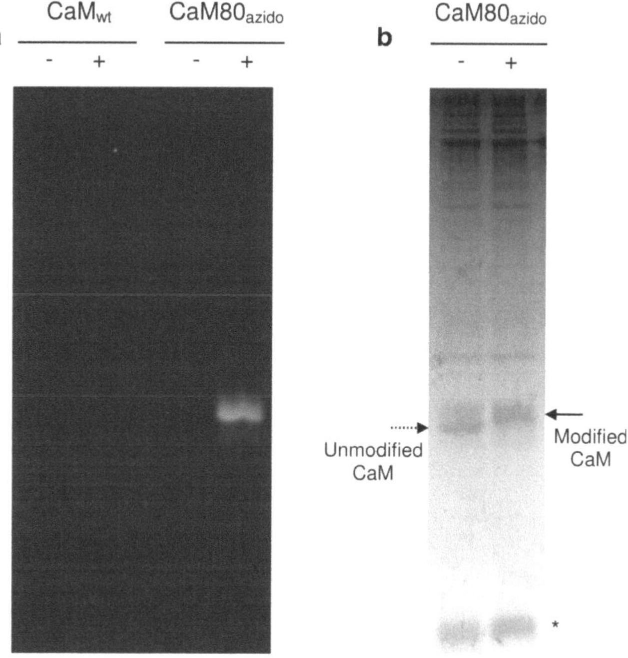

Fig. 19.3. Site-selective modification of CaM via azido group introduced into CaM. The fluorescence modification by click chemistry was performed as described in Subheading 3.7. (**a**) The fluorescent image under UV-light (366 nm) was obtained by taking a picture. Symbols, + and –, mean with and without modification, respectively. Although wild type CaM was synthesized in the reaction mixture containing 3-azidotyrosine, it was not labeled with tetramethylrhodamine (TAMRA) alkyne (CaM_{wt} +). However, CaM synthesized with $pETCaM80_{am}$ as a template was modified with TAMRA ($CaM80_{azido}$ +). The result indicates that only the CaM introduced with 3-azidotyrosine at the 80th position was labeled with TAMRA alkyene. (**b**) The gel was stained with Coomassie brilliant blue. *Solid* and *dotted arrows* indicate modified and unmodified CaMs, respectively. The asterisk shows the trancated CaM. The band of CaM modified with TAMRA alkyne ($CaM80_{azido}$ +) is slightly delayed, but fully shifted when compared with the band of unmodified CaM ($CaM80_{azido}$ –). This result suggests that all the labeled CaM molecules possess single TAMRA fluorophore at position 80.

6. RNAs are separated from each other mainly by the net negative charge when Q-Sepharose is used. Even if some tRNAs are eluted in a same fraction under neutral condition, these could possibly be separated under acidic condition due to the protonation of adenosines and cytosines.
7. Yeast TyrRS precipitates in a low-salt buffer. Keep the molar concentration of KCl higher than 40 mM.
8. Use Phenyl-Sepharose CL-4B resins because, for some reason, CaM cannot be purified very well with Phenyl-Sepharose High Performance resins (GE healthcare).
9. In our experience, CaM adsorbs to Phenyl-Sepharose CL-4B resins at room temperature better than at 4°C.

Acknowledgments

The authors would like to thank Akiyoshi Boku of our laboratory for technical assistance. This work was supported by an Industrial Technology Research Grant Program in 2003 from the New Energy and Industrial Technology Development Organization (NEDO) of Japan (to S.O.), and by Grants-in-Aid for Scientific Research from Ministry of Education, Culture, Sports, Science and Technology (MEXT) (to S.O. and T.Y.), and from Japan Society for the Promotion of Science (JSPS) (to T.Y and K.N.).

References

1. Noren, C.J., Anthony-Cahill, S.J., Griffith, M.C., and Schultz, P.G. (1989) A general method for site-specific incorporation of unnatural amino acids into protein. *Science* **244**, 182–188.
2. Mendel, D., Cornish, V.W., and Schultz, P.G. (1995) Site-directed mutagenesis with an expanded genetic code. *Annu. Rev. Biophys. Biomol. Struct.* **24**, 435–462.
3. Kanamori, T., Nishikawa, S., Shin, I., Schultz, P.G., and Endo, T. (1997) Probing the environment along the protein import pathways in yeast mitochondria by site-specific photocrosslinking. *Proc. Natl. Acad. Sci. USA* **94**, 485–490.
4. Hohsaka, T., Ashizuka, Y., Murakami, H., and Sisido, M. (1996) Incorporation of nonnatural amino acids into streptavidin through in vitro frame-shift suppression. *J. Am. Chem. Soc.* **118**, 9778–9779.
5. Sisido, M., Hohsaka, T. (2001) Introduction of specialty functions by the position-specific incorporation of nonnatural amino acids into proteins through four-base codon/anticodon pairs. *Appl. Microbiol. Biotechnol.* **57**, 274–281.
6. Ohno, S., Yokogawa, T., Fujii, I., Asahara, H., Inokuchi, H., and Nishikawa, K. (1998) Co-expression of yeast amber suppressor $tRNA^{Tyr}$ and tyrosyl-tRNA synthetase in *Escherichia coli*: possibility to expand the genetic code. *J. Biochem.* **124**, 1065–1068.
7. Fukunaga, J., Gouda, M., Umeda, K., Ohno, S., Yokogawa, T., and Nishikawa, K. (2006) Use of RNase P for efficient preparation of yeast $tRNA^{Tyr}$ transcript and its mutants. *J. Biochem.* **139**, 123–127.
8. Ohno, S., Yokogawa, T., and Nishikawa, K. (2001) Changing the amino acid specificity of yeast tyrosyl-tRNA synthetase by genetic engineering. *J. Biochem.* **130**, 417–423.
9. Zubay, G. (1962) The isolation and fractionation of soluble ribonucleic acid. *J. Mol. Biol.* **4**, 347–356.
10. Grodberg, J., Dunn, J. J. (1988) ompT encodes the *Escherichia coli* outer membrane protease that cleaves T7 RNA polymerase during purification. *J. Bacteriol.* **170**, 1245–1253.
11. Ohno, S., Matsui, M., Yokogawa, T., Nakamura, M., Hosoya, T., Hiramatsu, T., Suzuki, M., Hayashi, N., and Nishikawa, K. (2007) Site-selective post-translational modification of proteins using an unnatural amino acid, 3-azidotyrosine. *J. Biochem.* **141**, 335–343.
12. Köhrer, C., Rajbhandary, U. L. (2008) The many applications of acid urea polyacrylamide gel electrophoresis to studies of tRNAs and aminoacyl-tRNA synthetases. *Methods* **44**, 129–138.
13. Kigawa, T., Yabuki, T., Matsuda, N., Matsuda, T., Nakajima, R., Tanaka, A., and Yokoyama, S. (2004) Preparation of *Escherichia coli* cell extract for highly productive cell-free protein expression. *J. Struct. Funct. Genomics* **5**, 63–68.
14. Gopalakrishna, R., Anderson, W. B. (1982) Ca^{2+}-induced hydrophobic site on calmodulin: application for purification of calmodulin by phenyl-Sepharose affinity chromatography. *Biochem. Biophys. Res. Commun.* **104**, 830–836.
15. Hayashi, N., Matsubara, M., Takasaki, A., Titani, K., and Taniguchi, H. (1998) An expression system of rat calmodulinusing T7 phage promoter in *Escherichia coli*. *Protein Expr. Purif.* **12**, 25–28.

Chapter 20

Synthesis of Functional Proteins Within Liposomes

Takeshi Sunami, Tomoaki Matsuura, Hiroaki Suzuki, and Tetsuya Yomo

Abstract

In living cells, biochemical reaction systems are enclosed in the small lipidic compartments. To experimentally simulate various biochemical reactions occurring in extant cells, cell-sized lipid vesicles (liposomes) are used to reconstruct an artificial model cell. We present methods for encapsulation of the protein synthesis system inside liposomes and for measurement of the in liposome synthesis reaction using a fluorescence-activated cell sorter. These techniques would enable us to perform detailed analysis of the biochemical reactions occurring in the microcompartments and have the potential to reveal the role of compartmentalization in cellular systems.

Key words: Liposome, Flow cytometry, Fluorescence-activated cell sorter, Cell-free protein synthesis, Green fluorescent protein, β-Galactosidase, β-Glucuronidase

1. Introduction

Liposomes are polymolecular aggregates formed in aqueous solution on dispersion of bilayer-forming amphiphilic lipid molecules. Liposomes are mostly spherical in shape and contain one or more lamellae of the lipid bilayer. The bilayer membrane is formed by self-assembly of lipids, where the nonpolar acyl chain forms the hydrophobic interior of the bilayer and the polar head group is in contact with the aqueous phase. Depending on the method of preparation, the diameter of the liposomes may range from tens of nanometers to a few hundred micrometers (1). As the interior of liposomes is aqueous solution, various water-soluble chemical and biochemical materials can be encapsulated. Liposomes have been used extensively as drug carriers for administration of pharmaceutical compounds to achieve therapeutic effects (2) and as a lipid-based transfection vehicle for injection of genetic materials into cells (3). Our field of interest is reconstruction of artificial

Y. Endo et al. (eds.), *Cell-Free Protein Production: Methods and Protocols,* Methods in Molecular Biology, vol. 607
DOI 10.1007/978-1-60327-331-2_20, © Humana Press, a part of Springer Science+Business Media, LLC 2010

cell systems to model the characteristics of extant cells. For this purpose, giant liposomes, with diameter larger than 1 μm, are tailored to encapsulate biochemical reaction systems (4).

Protein expression is an essential reaction for all living organisms. Thus, as the first step in reconstructing an artificial cell, synthesis of functional proteins within liposomes has been examined in pioneering works including studies by our group (e.g., Poly(Phe) (5), green fluorescent protein (GFP) (6–8), T7 RNA polymerase (8), or α-hemolysin (9) have been synthesized using cell-free synthesis systems). In addition, we demonstrated that liposomes can be used as a tool for directed evolution (a method used in protein engineering) (10), similar to water/oil/water (w/o/w) emulsions (11–13). We also constructed a simplified self-encoding system within liposomes (14). With all these technologies we have developed, it will be possible to quantify gene evolution under conditions in which the lipid membrane interacts with the encapsulated biochemical materials and synthesized proteins.

In this chapter, we introduce a method for synthesizing proteins within liposomes prepared by the freeze-dried empty liposomes (FDEL) method (15) and measuring the reaction using a fluorescence-activated cell sorter (FACS). Since Bangham first discovered liposomes (16), many types of liposome preparation method have been proposed, such as the simple hydration method, reversed-phase evaporation method (17), electroformation method (18), and water-in-oil (w/o) emulsion method (19). We have been using the FDEL method to encapsulate a cell-free protein synthesis system in liposomes because of its high encapsulation efficiency even for large biological polymers. This feature is essential as the protein synthesis system contains numbers of components with various properties and concentrations. Furthermore, this method is applicable for almost any lipid composition. Thus, liposomes composed of arbitrary lipid mixtures can be applied not only for protein synthesis but also for other biochemical reactions. In the following, we describe various points where care should be taken in the preparation of liposomes encapsulating a cell-free protein synthesis system.

For evaluation of the characteristics of liposomes, several analytical methods have been often applied. Dynamic light scattering (DLS) is a major technique that can determine the size distribution of liposomes in nanometer scale. Light microscopy is used to directly observe the size, shape, and fluorescence intensity of each liposome. However, these conventional methods are not suitable to simultaneously determine the multiple parameters of large amounts of liposome containing the reaction system, which is necessary to analyze the volume dependency of the reaction. Thus, we employed FACS to simultaneously analyze the multiple properties of individual liposomes in a high-throughput fashion

(10, 20). This single-vesicle-based assay would enable us to perform detailed kinetic analysis of biochemical reactions occurring in liposomes. With the aid of this technology, it becomes possible to elucidate the role of compartmentalization in cell systems, including the effects of the membrane, which is expected to be one of the major differences from bulk experiments.

2. Materials

2.1. Plasmid Preparation

1. OligoDNA_1 5′-TTGGATCCATGCCTTCTGAACAATG GAA.
2. OligoDNA_2 5′-TAAATTAAGCTTTTATTATTATTTTTG ACACC.
3. OligoDNA_3 5′-AATTCGAGGCCCTGAGGGCCAGGAG GCCTCCTGGCCTATGCGGCCGCA.
4. OligoDNA_4 5′-AGCTTGCGGCCGCATAGGCCAGGA GGCCTCCTGGCCCTCAGGGCCTCG.
5. pSV-β-Galactosidase Control Vector (pSV-bGal) (Promega Corporation, Madison, WI).
6. pET-21a (Novagen, San Diego, CA).
7. pCA24NuidA (ASKA library (21) JW1609, kindly provided by the National Institute of Genetics, Japan).
8. QIAGEN Plasmid Midi Kit (Qiagen, Hilden, Germany).

2.2. Liposome Preparation

1. 1-Palmitoyl-2-Oleoyl-*sn*-Glycero-3-Phosphocholine (POPC) (Avanti Polar Lipids, Inc., Alabaster, AL).
2. Cholesterol (Nacalai Tesque, Inc., Kyoto, Japan).
3. Distearoyl phosphatidyl ethanolamine-poly (ethylene glycol) 5000 (DSPE-PEG5000) (kindly provided by NOF Corporation, Tokyo, Japan).
4. Diethyl ether and Dichloromethane (Wako Pure Chemical Industries Ltd, Osaka, Japan).
5. Ultrasonic disrupter (Tomy Seiko Co., Ltd, Tokyo, Japan).
6. Freeze dryer (Labconco Corporation, Kansas, MO).
7. Polycarbonate filters (Whatman, Brentford, UK).

2.3. Protein Synthesis in Liposomes

1. BD FACSAria (Becton Dickinson, San Jose, CA).
2. *E. coli* T7 S30 Extract System for Circular DNA (Promega Corporation).
3. PURESYSTEM classic II (Post Genome Institute Co., Ltd, Tokyo, Japan).

4. RTS 100 *E. coli* HY Kit (Roche, Basel, Switzerland).
5. Allophycocyanin (APC) (Invitrogen Corporation, Carlsbad, CA).
6. Plasmids encoding proteins; pETG5tag (*see* Subheading 3.1.1), pET-bGal (see Subheading 3.1.2), and pET-uidA (*see* Subheading 3.1.3).
7. 5-(Pentafluorobenzoylamino)fluorescein di-β-D-galactopyranoside (PFB-FDG) (Invitrogen Corporation).
8. Fluorescein di-β-D-glucuronide (FDGlcU) (Invitrogen Corporation).
9. 5-(pentafluorobenzoylamino)fluorescein di-β-D-glucuronide (PFB-FDGlcU) (Invitrogen Corporation).
10. Protease from *Streptomyces griseus* (Sigma-Aldrich Corporation, St. Louis, MO).
11. BD FACSFlow (Becton Dickinson).

3. Methods

3.1. Plasmid Preparation

In this section, we describe the large-scale preparation method of three different DNA templates used for *in liposome* protein synthesis reaction (see Note 1). pETG5tag is a plasmid encoding the GFP gene under the control of the T7 promoter (Fig. 20.1a). GFP is composed of 238 amino acids, which is frequently used as a fluorescent marker for protein expression (22). pET-bGal is a plasmid encoding the β-galactosidase gene (*lacZ*) under the control of the T7 promoter (Fig. 20.1b). β-Galactosidase is a hydrolytic enzyme that forms a tetramer of four noncovalently linked subunits (23) composed of 1,021 amino acids (24), which catalyzes the hydrolysis of β-galactosides into monosaccharides. As it also hydrolyzes other substrates producing colored or fluorescent

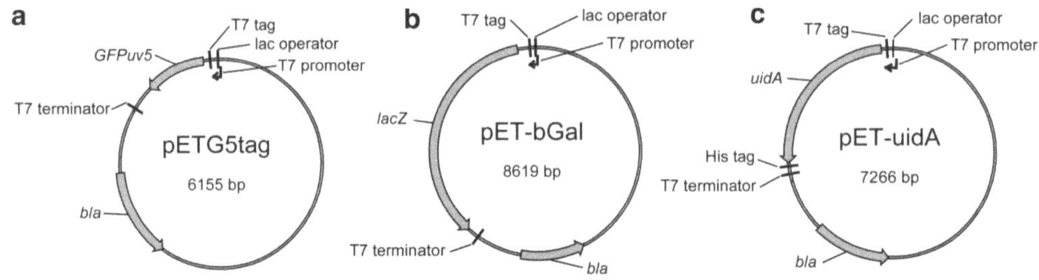

Fig. 20.1. Schematic diagrams of three plasmids used for protein synthesis. (**a**) pETG5tag: DNA template for GFP synthesis. (**b**) pET-bGal: DNA template for β-galactosidase synthesis. (**c**) pET-uidA: DNA template for β-glucuronidase synthesis.

compounds, this enzyme has been used to monitor a wide range of biological processes (25). pET-uidA is a plasmid encoding the β-glucuronidase gene (*uidA*) under the control of the T7 promoter (Fig. 20.1c). β-Glucuronidase is also a hydrolytic enzyme, composed of 603 amino acids, which catalyzes the cleavage of a wide variety of β-glucuronides (26). In molecular biology, β-glucuronidase has been used as a reporter gene to monitor gene expression, because it transforms certain substrates into colored or fluorescent products, which can then be detected (27).

3.1.1. pETG5tag Construction

1. The plasmid pETG5tag (10) carries the T7 promoter, Shine–Dalgarno sequence, and T7-tag sequence (MTGGQQMGR) followed by GFPuv5 (28). GFPuv5 is an improved variant of GFPuv (Clontech, Palo Alto, CA) with F64L/S65T/S208L/I167T mutations (numbering based on wild-type GFP).

3.1.2. pET-bGal Construction

1. PCR was performed with two primers (oligoDNA_1, oligoDNA_2) and the template (pSV-bGal) to amplify the *lacZ* gene.
2. A plasmid encoding β-galactosidase (pET-bGal) was prepared by inserting the *lacZ* gene obtained by digesting the PCR product with *Bam*HI and *Hin*dIII and ligating the insert into the plasmid pET-21a also digested with the same restriction enzymes.

3.1.3. pET-uidA Construction

1. The plasmid pET-21a-SfiI was prepared by ligating pET-21a digested with *Eco*RI and *Hin*dIII, with hybridized oligoDNA (hybrid of oligoDNA_3 and oligoDNA_4) with an *Sfi*I restriction site in the middle.
2. A plasmid encoding β-glucuronidase (pET-uidA) was prepared by inserting the *uidA* gene obtained by digesting pCA24NuidA with *Sfi*I and ligating the insert into the plasmid pET-21a-SfiI also digested with the same restriction enzyme. Note that the his-tag is located at the C terminus of β-glucuronidase.

3.1.4. Large-Scale Plasmid Preparation

1. *E. coli* DH5α was transformed with plasmid (pETG5tag, pET-bGal, or pET-uidA), spread on Luria-Bertani (LB) agar plates containing ampicillin (50 μg/ml), and then incubated overnight at 37°C.
2. A colony was picked and grown in 5-ml overnight culture in LB medium with ampicillin (50 μg/ml) at 37°C.
3. 200 μl of culture was added to 200 ml of LB medium with ampicillin (50 μg/ml), incubated overnight at 37°C, and cells were harvested.
4. Plasmids were prepared using a QIAGEN Plasmid Midi Kit according to the manufacturer's instructions (see Note 2).

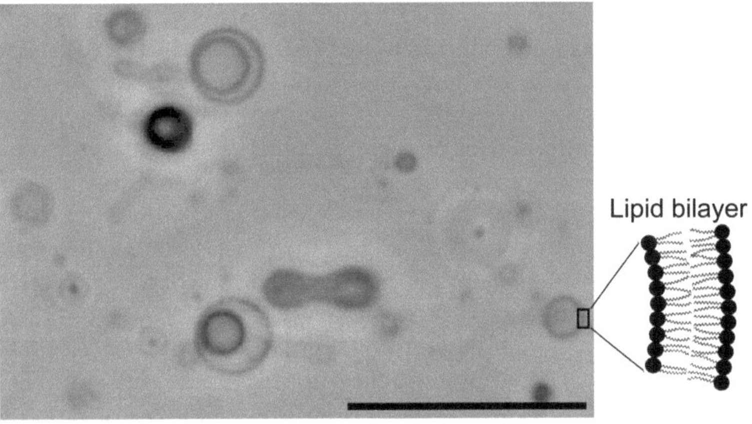

Fig. 20.2. Phase-contrast micrograph of liposomes prepared by the FDEL method. These liposomes were composed of POPC, cholesterol, and DSPE-PEG5000 (58:39:3, molar ratio).

3.2. Preparation of Freeze-Dried Empty Liposomes

In this section, we describe how to prepare liposomes in the FDEL method (15) (Fig. 20.2). This method has three main advantages. First, FDELs can be prepared on a large scale, and the lyophilized liposomes are stable under an argon atmosphere for several months. Therefore, we can perform experiments using lipids from the same lot many times. Second, this method enables rapid and easy encapsulation of the reaction mixture into liposomes; all that is required is to add the solution to the lyophilized liposomes and then mix gently. Third, the liposomes prepared by this method are stable enough to withstand mechanical stress as they are mostly multilamellar; in measurement and sorting processes by FACS, liposomes are subjected to the shear flow through a narrow orifice and the osmotic stress owing to the sheath flow. Therefore, liposomes prepared by the FDEL method with particular properties (i.e., those with strong green fluorescence) can be sorted from the initial collection of liposomes. As explained above, the FDEL method is suitable for effective encapsulation of a protein synthesis system in liposomes and quantitative evaluation of reaction product within liposomes by FACS.

1. The lipid mixture (12 μmol, molar ratio of POPC:cholesterol:DSPE–PEG5000 = 58:39:3) dissolved in organic solvent (diethyl ether:dichloromethane = 1:1 v/v) was subjected to rotary evaporation in a pear-shaped flask under vacuum to yield a thin lipid film.
2. Deionized water (1 ml) was added to the lipid film under argon gas.
3. After 15 min, the lipid film was vortexed to disperse the liposomes.

4. The liposome dispersion was homogenized on ice by sonication with an ultrasonic disruptor and extruded through a polycarbonate filter with a pore size of 0.4 μm.
5. Aliquots of 40 μl of the solution was transferred to Eppendorf tubes and lyophilized.
6. The lyophilized empty liposomes were stored at −20°C under argon gas atmosphere.

3.3. Protein Synthesis in Liposomes

In this section, we describe the method for synthesizing proteins within liposomes. Cell-free protein synthesis reaction was carried out using PURESYSTEM classic II according to the manufacturer's instructions. The most outstanding property of this system is the high yield of protein when using a low concentration of DNA template. We tested various cell-free protein synthesis systems, including those prepared in our laboratory (29, 30), which usually required 500 ng of DNA in a 50-μl reaction mixture for sufficient protein production. However, with PURESYSTEM the DNA concentration could be reduced to 1 ng/μl without affecting the yield of protein, presumably because of the low levels of DNase, RNase, and protease activity (Fig. 20.3). This feature is vital for synthesis of proteins from a single DNA template within liposomes and allows the one-to-one correspondence of genotype and phenotype in each liposome, which is necessary for directed evolution experiments.

A mixture containing the DNA template and cell-free protein synthesis reaction system was encapsulated in liposomes by adding the mixture to the freeze-dried liposome membrane. Synthesis

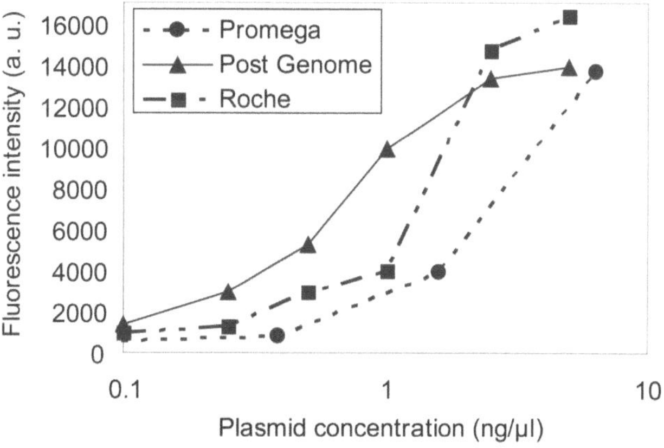

Fig. 20.3. Yield of GFP using the DNA template (pETG5tag) in a test tube with different cell-free protein synthesis systems. We used three cell-free protein synthesis kits: *E. coli* T7 S30 Extract System for Circular DNA (Promega, *filled circles*), PURESYSTEM classic II (Post Genome, *filled triangles*), and RTS 100 *E. coli* HY Kit (Roche, *filled squares*). Fluorescence intensity of synthesized GFP (*vertical axis*) after 1 h of incubation at 30°C (Roche) or 37°C (Promega and Post Genome) was plotted as a function of DNA template concentration (*horizontal axis*, logarithmic scale).

of GFP, β-galactosidase, or β-glucuronidase in liposomes was detected by measuring the final fluorescent product of the reaction using FACS. GFP was synthesized from the DNA template (pETG5tag) (Fig. 20.4a), and the green fluorescence of GFP was detected. In the cases of β-galactosidase and β-glucuronidase (Fig. 20.4b), these hydrolytic enzymes were synthesized from the DNA templates (pET-bGal and pET-uidA, respectively), which then hydrolyzed the substrate (PFB-FDG and PFB-FDGlcU, respectively). As the hydrolysis reaction of these enzymes proceeds rapidly, the green fluorescence of the hydrolysis product (both substrates turns into PFB-fluorescein) acts as a reporter of the protein synthesis reaction. When we performed these *in liposome* biochemical reactions, APC (red fluorescent protein) was added to the reaction mixtures as a marker for estimation of internal aqueous volume of each liposome (10). Therefore, we can evaluate the amount of the reaction product from the green fluorescence intensity and internal aqueous volume of each liposome from the red fluorescence intensity simultaneously.

3.3.1. GFP Synthesis in Liposomes

1. Aliquots of 10 μl of the reaction mixture, consisting of PURESYSTEM, 500 nM APC, and 1 ng/μl plasmid DNA (pETG5tag), were added to the lyophilized liposomes on ice and vortexed for 10 s. Then, an aliquot of 1.5 μl of the liposome suspension was transferred into a new Eppendorf tube.
2. An aliquot of 28.5 μl of the dilution mixture, consisting of PURESYSTEM and 1 mg/ml protease (see Note 3), was added to the liposome suspension and mixed gently on ice (see Note 4).

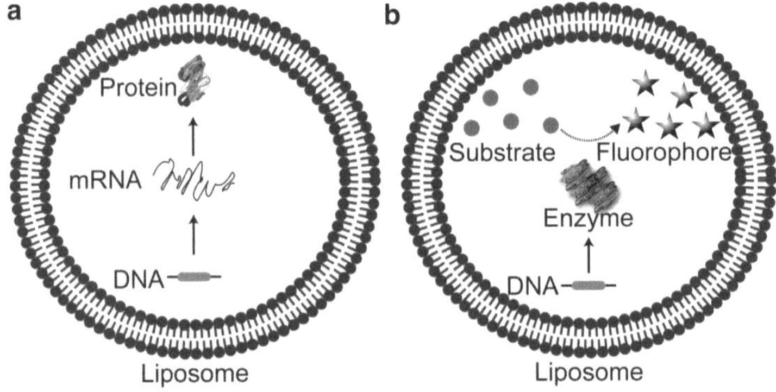

Fig. 20.4. Two detection schemes of protein synthesis within liposomes. (**a**) GFP synthesis. The mRNA is transcribed from the DNA template, and GFP is translated from the mRNA using the biochemical materials included in the cell-free protein synthesis system within the liposomes. (**b**) Enzyme synthesis and subsequent substrate hydrolysis. First, the enzyme (β-galactosidase, β-glucuronidase) is synthesized from the DNA template (pET-bGal, pET-uidA) by the cell-free protein synthesis system within the liposomes. Second, the synthesized enzyme hydrolyzes the substrate (PFB-FDG, PFB-FDGlcU) to green fluorophore (PFB-fluorescein).

3. To initiate the translation reaction, the liposome solution was incubated at 37°C for 3 h.
4. The liposome suspension was diluted by approximately 20-fold with FACSFlow and analyzed by FACS (see Note 5).

3.3.2. β-Galactosidase Synthesis in Liposomes

1. Aliquots of 10 μl of the reaction mixture, consisting of PURESYSTEM, 500 nM APC, 50 μM PFB-FDG (*see* Notes 6 and 7), and 1 ng/μl plasmid DNA (pET-bGal), were added to the lyophilized liposomes on ice and vortexed for 10 s. An aliquot of 1.5 μl of the liposome suspension was then transferred to a new tube. The subsequent procedures were identical to those described for GFP synthesis.

3.3.3. β-Glucuronidase Synthesis in Liposomes

1. Aliquots of 10 μl of the reaction mixture, consisting of PURESYSTEM, 500 nM APC, 50 μM PFB-FDGlcU (*see* Notes 6 and 7), and 1 ng/μl plasmid DNA (pET-uidA), were added to the lyophilized liposomes on ice and vortexed for 10 s. An aliquot of 1.5 μl of the liposome suspension was then transferred to a new tube. The subsequent procedures were identical to those described for GFP synthesis.

3.4. FACS Measurement of Liposomes

In this section, we describe the quantitative measurement and analysis method of the final reaction product produced in liposomes as the result of protein synthesis reaction. In our experimental scheme, the intensities of red and green fluorescent signals are proportional to the whole volume of reaction mixture encapsulated in a liposome and the amount of product produced (GFP or PFB-fluorescein), respectively. These fluorescence signals in two different colors emitted from individual liposomes were measured simultaneously with FACS. In detail, APC was excited with a HeNe laser (633 nm) and the emission was detected through a 660 ± 10 nm bandpass filter. GFP and PFB-fluorescein were excited with a 488-nm semiconductor laser and the emission was detected through a 530 ± 15 nm bandpass filter.

Examples of the data obtained by FACS are shown in Fig. 20.5. Here, the intensities of green and red fluorescent signals are plotted on the *x* and *y* axes, respectively. Each dot in the figure represents measured data of individual liposomes. In FACS analysis, data points are usually plotted on a logarithmic scale to show the wide ranges of measured quantities over multiple orders of magnitude. The plot of liposomes containing the protein synthesis system without the DNA template showed almost no green fluorescence signal, while the distribution of red signal had a wide distribution (Fig. 20.5a). When 0.5 nM of pETG5tag plasmid was encapsulated together with the protein synthesis system, a large number of green fluorescent liposomes were detected (Fig. 20.5b). This result indicated that large amounts of functional

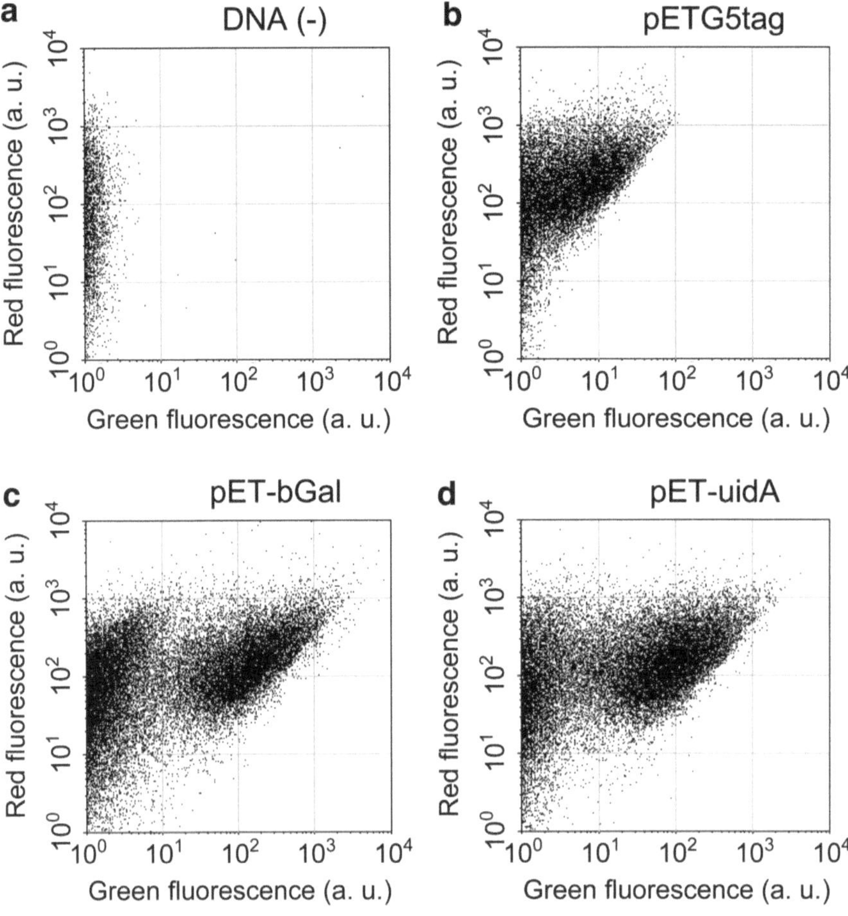

Fig. 20.5. Two-dimensional dot plots of the intensity of green (*horizontal axis*) and red fluorescence (*vertical axis*, internal aqueous volume marker) signals from liposomes containing cell-free protein synthesis system. Each *dot* represents a single liposome detected by FACS. These four figures represent the measurements after 3 h incubation at 37°C. A total of 100,000 data points were used in each plot. (**a**) Liposomes in which no protein synthesis occurred without any DNA template. (**b**) Liposomes in which GFP was synthesized from the DNA template (2 ng/μl pETG5tag). (**c**) Liposomes in which β-galactosidase was synthesized from the DNA template (1 ng/μl pET-bGal), and the substrate (50 μM PFB-FDG) was hydrolyzed to green fluorophore by the synthesized β-galactosidase. (**d**) Liposomes in which β-glucuronidase was synthesized from the DNA template (1 ng/μl pET-uidA), and the substrate (50 μM PFB-FDGlcU) was hydrolyzed to green fluorophore by the synthesized β-glucuronidase.

GFP, which emits green fluorescence, were synthesized within liposomes. Similarly, with pET-bGal and pET-uidA, the population of liposomes emitting green fluorescence clearly appeared as a result of hydrolysis of fluorogenic substrate (Fig. 20.5c, d, respectively). These results indicated that enzymes (β-galactosidase or β-glucuronidase) were synthesized in their functional form and subsequent substrate hydrolysis occurred within liposomes.

In addition to the examples presented above, we have demonstrated the reconstitution of a cascading genetic network in liposomes in which the protein product of the first stage (T7 RNA polymerase) is required to drive the protein synthesis of the second stage (GFP) (8). Moreover, we have shown that a self-encoding system can be reconstituted within liposomes (14). In the self-encoding system, the β-subunit of Qβ replicase is first synthesized from the RNA template. Second, the antisense strand of the RNA template, which encodes the β-galactosidase gene, is replicated by Qβ replicase, a heterotetramer of synthesized β-subunit and three host proteins included in the cell-free protein synthesis system. Third, β-galactosidase is synthesized from the antisense strand and the fluorogenic substrate is hydrolyzed by the synthesized β-galactosidase. As explained above, not only protein synthesis but also more complex biochemical reaction systems have been reconstituted within liposomes prepared using the FDEL method. Although these systems are still potentially available for the realization of a minimal cell and many challenges remain before, we can gain an in-depth understanding of the origin of life, this *in liposome* protein synthesis method and respective liposome evaluation method by FACS can be used as powerful tools to access systems akin to living cells in nature.

3.4.1. Measurement Parameters and Threshold Parameters

1. Forward scatter (FSC) signal (Ex.: 488 nm, Type: Area, Voltage: 200).
2. Side scatter (SSC) signal (Ex.: 488 nm, Em.: 488 ± 5 nm, Type: Area, Voltage: 400).
3. Green channel (FITC channel) signal (Ex.: 488 nm, Em.: 530 ± 15 nm, Type: Area, Voltage: 560).
4. Far red channel (APC channel) signal (Ex.: 633 nm, Em.: 660 ± 10 nm, Type: Area, Voltage: 680).
5. Threshold parameters (FSC: 200, SSC: 200, Threshold operator: And).

4. Notes

1. Both the linear DNA amplified from the plasmid by PCR and RNA synthesized from the plasmid by T7 RNA polymerase can be used as a template for protein synthesis.
2. The degree of purification largely affects the amount of protein synthesis from the plasmid. We recommend not using the QIAprep Spin Miniprep Kit (QIAGEN, Hilden, Germany) but the QIAGEN Plasmid Midi Kit.

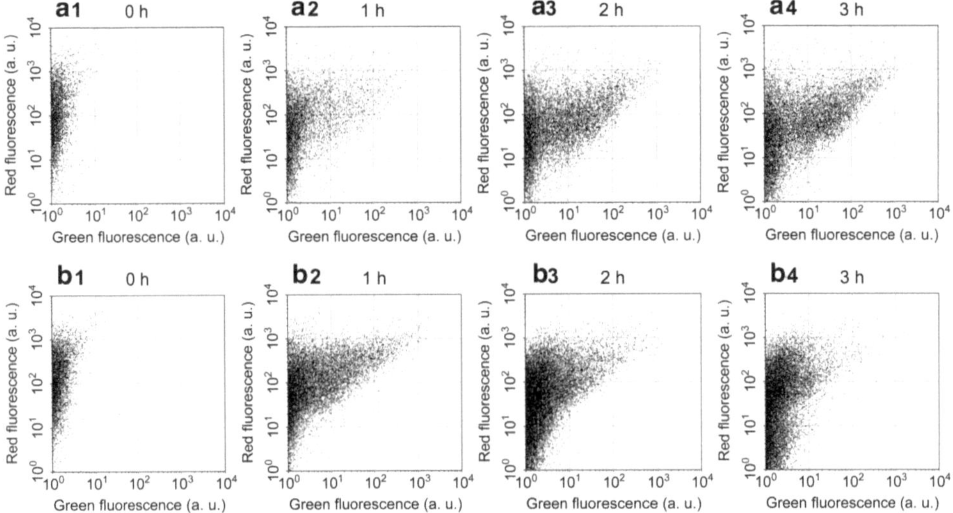

Fig. 20.6. The reaction time course of β-glucuronidase synthesis and subsequent substrate hydrolysis within liposomes. These figures represent the measurements after 0, 1, 2, and 3 h, from *left* to *right*, of incubation at 37°C. Intensity of green fluorescence (*horizontal axis*) from liposomes containing cell-free protein synthesis system was plotted with their red fluorescence (*vertical axis*, internal aqueous volume marker). Each dot represents single liposome detected by FACS. A total of 100,000 data points were used in each plot. (**a**) β-Glucuronidase was synthesized from the DNA template (5 ng/μl pET-uidA) and the substrate (50 μM PFB-FDGlcU) was hydrolyzed to green fluorophore by the synthesized β-glucuronidase within liposomes. (**b**) The same reaction with the different fluorogenic substrate (50 μM FDGlcU). The green fluorescence intensity decreased after 3 h incubation due to the leakage of the hydrolyzed substrate.

3. Protease (final conc. 1 mg/ml) was included in the dilution mixture to completely suppress protein synthesis that may occur outside liposomes. Not only protease but also RNase and EDTA can be used for suppression of protein synthesis, as these inhibitors hardly penetrate into liposomes from the external solution.
4. We confirmed that highly concentrated lipids (final conc. over 20 mM) in the reaction mixture strongly inhibited protein synthesis in a test tube and within liposomes. Therefore, the liposome suspension must be diluted with the solution that also contains essential materials for protein synthesis because of leakage of the requisite materials from liposomes.
5. Before measurement of liposomes by FACS, liposome suspension was diluted with FACSFlow, a sheath solution for use with BD FACSAria, to reduce the liposome concentration (less than 20,000 events/s). The liposomes prepared by the FDEL method can be diluted with FACSFlow without collapse of liposomes, although isotonic buffer is desirable.
6. A number of fluorogenic substrates are available for β-galactosidase, such as FDG (Invitrogen), CMFDG (Invitrogen),

PFB-FDG (Invitrogen), etc., or β-glucuronidase, such as FDGlcU (Invitrogen), PFB-FDGlcU (Invitrogen), etc. While these substrates can be used in a test tube, FDG and FDGlcU cannot be used for the reaction within liposomes. It seems that the hydrolysis product from these substrates leaks from liposomes (Fig. 20.6b), while the hydrolysis products from PFB-FDG, CMFDG, and PFB-FDGlcU do not (Fig. 20.6a).

7. The stock solution (final conc. 10 mM) of fluorogenic substrate (PFB-FDG, PFB-FDGlcU) was prepared in dimethylsulfoxide (DMSO). DMSO markedly affects the amount of protein synthesis, so the amount of DMSO in the reaction mixture should be less than 1% of the total volume.

Acknowledgments

This research was supported in part by "Special Coordination Funds for Promoting Science and Technology: Yuragi Project" and "Global COE (Centers of Excellence) Program" of the Ministry of Education, Culture, Sports, Science, and Technology, Japan.

References

1. Walde, P. and Ichikawa, S. (2001) Enzymes inside lipid vesicles: preparation, reactivity and applications. *Biomol. Eng.* **18,** 143–177.
2. Lian, T. and Ho, R. J. (2001) Trends and developments in liposome drug delivery systems. *J. Pharm. Sci.* **90,** 667–680.
3. Felgner, P. L., Gadek, T. R., Holm, M., Roman, R., Chan, H. W., Wenz, M., Northrop, J. P., Ringold, G. M., and Danielsen, M. (1987) Lipofection: a highly efficient, lipid-mediated DNA-transfection procedure. *Proc. Natl. Acad. Sci. U.S.A.* **84,** 7413–7417.
4. Luisi, P. L., Ferri, F., and Stano, P. (2006) Approaches to semi-synthetic minimal cells: a review. *Naturwissenschaften* **93,** 1–13.
5. Oberholzer, T., Nierhaus, K. H., and Luisi, P. L. (1999) Protein expression in liposomes. *Biochem. Biophys. Res. Commun.* **261,** 238–241.
6. Yu, W., Sato, K., Wakabayashi, M., Nakaishi, T., Ko-Mitamura, E. P., Shima, Y., Urabe, I., and Yomo, T. (2001) Synthesis of functional protein in liposome. *J. Biosci. Bioeng.* **92,** 590–593.
7. Nomura, S. M., Tsumoto, K., Hamada, T., Akiyoshi, K., Nakatani, Y., and Yoshikawa, K. (2003) Gene expression within cell-sized lipid vesicles. *Chembiochem* **4,** 1172–1175.
8. Ishikawa, K., Sato, K., Shima, Y., Urabe, I., and Yomo, T. (2004) Expression of a cascading genetic network within liposomes. *FEBS Lett.* **576,** 387–390.
9. Noireaux, V. and Libchaber, A. (2004) A vesicle bioreactor as a step toward an artificial cell assembly. *Proc. Natl. Acad. Sci. U.S.A.* **101,** 17669–17674.
10. Sunami, T., Sato, K., Matsuura, T., Tsukada, K., Urabe, I., and Yomo, T. (2006) Femtoliter compartment in liposomes for in vitro selection of proteins. *Anal. Biochem.* **357,** 128–136.
11. Bernath, K., Hai, M., Mastrobattista, E., Griffiths, A. D., Magdassi, S., and Tawfik, D. S. (2004) In vitro compartmentalization by

double emulsions: sorting and gene enrichment by fluorescence activated cell sorting. *Anal. Biochem.* **325,** 151–157.

12. Aharoni, A., Amitai, G., Bernath, K., Magdassi, S., and Tawfik, D. S. (2005) High-throughput screening of enzyme libraries: thiolactonases evolved by fluorescence-activated sorting of single cells in emulsion compartments. *Chem. Biol.* **12,** 1281–1289.
13. Mastrobattista, E., Taly, V., Chanudet, E., Treacy, P., Kelly, B. T., and Griffiths, A. D. (2005) High-throughput screening of enzyme libraries: in vitro evolution of a beta-galactosidase by fluorescence-activated sorting of double emulsions. *Chem. Biol.* **12,** 1291–1300.
14. Kita, H., Matsuura, T., Sunami, T., Hosoda, K., Ichihashi, N., Tsukada, K., Urabe, I., and Yomo, T. (2008) Replication of genetic information with self-encoded replicase in liposomes. *Chembiochem* **9,** 2403–2410.
15. Kikuchi, H., Suzuki, N., Ebihara, K., Morita, H., Ishii, Y., Kikuchi, A., Sugaya, S., Serikawa, T., and Tanaka, K. (1999) Gene delivery using liposome technology. *J. Control. Release* **62,** 269–277.
16. Bangham, A. D. and Horne, R. W. (1964) Negative staining of phospholipids and their structural modification by surface-active agents as observed in the electron microscope. *J. Mol. Biol.* **8,** 660–668.
17. Szoka, F., Jr. and Papahadjopoulos, D. (1978) Procedure for preparation of liposomes with large internal aqueous space and high capture by reverse-phase evaporation. *Proc. Natl. Acad. Sci. U.S.A.* **75,** 4194–4198.
18. Angelova, M. I. and Dimitrov, D. S. (1986) Liposome electroformation. *Faraday Discuss. Chem. Soc.* **81,** 303–311.
19. Pautot, S., Frisken, B. J., and Weitz, D. A. (2003) Production of unilamellar vesicles using an inverted emulsion. *Langmuir* **19,** 2870–2879.
20. Sato, K., Obinata, K., Sugawara, T., Urabe, I., and Yomo, T. (2006) Quantification of structural properties of cell-sized individual liposomes by flow cytometry. *J. Biosci. Bioeng.* **102,** 171–178.
21. Kitagawa, M., Ara, T., Arifuzzaman, M., Ioka-Nakamichi, T., Inamoto, E., Toyonaga, H., and Mori, H. (2005) Complete set of ORF clones of Escherichia coli ASKA library (a complete set of E. coli K-12 ORF archive): unique resources for biological research. *DNA Res.* **12,** 291–299.
22. Chalfie, M., Tu, Y., Euskirchen, G., Ward, W. W., and Prasher, D. C. (1994) Green fluorescent protein as a marker for gene expression. *Science* **263,** 802–805.
23. Benito, A., Feliu, J. X., and Villaverde, A. (1996) Beta-galactosidase enzymatic activity as a molecular probe to detect specific antibodies. *J. Biol. Chem.* **271,** 21251–21256.
24. Fowler, A. V. and Zabin, I. (1977) The amino acid sequence of beta-galactosidase of Escherichia coli. *Proc. Natl. Acad. Sci. U.S.A.* **74,** 1507–1510.
25. Nolan, G. P., Fiering, S., Nicolas, J. F., and Herzenberg, L. A. (1988) Fluorescence-activated cell analysis and sorting of viable mammalian cells based on beta-D-galactosidase activity after transduction of Escherichia coli lacZ. *Proc. Natl. Acad. Sci. U.S.A.* **85,** 2603–2607.
26. Jefferson, R. A., Burgess, S. M., and Hirsh, D. (1986) beta-Glucuronidase from Escherichia coli as a gene-fusion marker. *Proc. Natl. Acad. Sci. U.S.A.* **83,** 8447–8451.
27. Jefferson, R. A., Kavanagh, T. A., and Bevan, M. W. (1987) GUS fusions: beta-glucuronidase as a sensitive and versatile gene fusion marker in higher plants. *EMBO J.* **6,** 3901–3907.
28. Ito, Y., Kawama, T., Urabe, I., and Yomo, T. (2004) Evolution of an arbitrary sequence in solubility. *J. Mol. Evol.* **58,** 196–202.
29. Matsuura, T., Yamaguchi, M., Ko-Mitamura, E. P., Shima, Y., Urabe, I., and Yomo, T. (2002) Importance of compartment formation for a self-encoding system. *Proc. Natl. Acad. Sci. U.S.A.* **99,** 7514–7517.
30. Matsuura, T. and Pluckthun, A. (2003) Selection based on the folding properties of proteins with ribosome display. *FEBS Lett.* **539,** 24–28.

INDEX

Q

R

S

T

U

V

W

X

MIX
Papier aus verantwortungsvollen Quellen
Paper from responsible sources
FSC® C105338

If you have any concerns about our products,
you can contact us on
ProductSafety@springernature.com

In case Publisher is established outside the EU,
the EU authorized representative is:
Springer Nature Customer Service Center GmbH
Europaplatz 3, 69115 Heidelberg, Germany

Printed by Libri Plureos GmbH
in Hamburg, Germany